Prevention, Diagnosis and Treatment of Skin Cancer

Prevention, Diagnosis and Treatment of Skin Cancer

Editor

Aimilios Lallas

MDPI • Basel • Beijing • Wuhan • Barcelona • Belgrade • Manchester • Tokyo • Cluj • Tianjin

Editor
Aimilios Lallas
First Department of
Dermatology
Aristotle University
Thessaloniki
Greece

Editorial Office
MDPI
St. Alban-Anlage 66
4052 Basel, Switzerland

This is a reprint of articles from the Special Issue published online in the open access journal *Cancers* (ISSN 2072-6694) (available at: www.mdpi.com/journal/cancers/special_issues/Skin_Cancers).

For citation purposes, cite each article independently as indicated on the article page online and as indicated below:

LastName, A.A.; LastName, B.B.; LastName, C.C. Article Title. *Journal Name* **Year**, *Volume Number*, Page Range.

ISBN 978-3-0365-6681-8 (Hbk)
ISBN 978-3-0365-6680-1 (PDF)

Contents

About the Editor

Aimilios Lallas

Aimilios Lallas, MD, MSc, PhD, is an Associate Professor of Dermatology at the First Department of Dermatology of Aristotle University in Thessaloniki, Greece. He specializes in skin cancer diagnosis with non-invasive techniques, as well as in the management of skin cancer patients. Dr. Lallas is currently the President of the International Dermoscopy Society.

His main research interests are dermoscopy of skin tumors, applications of the method in general dermatology and improvements in the management of oncologic patients. He is an author of 373 scientific papers published on PubMed Central, most of them on dermoscopy and skin cancer, and several books and book chapters on dermoscopy. The total number of citations of his papers exceeds 11.400, and his h-index is 52. He is a co-investigator in several phase III clinical trials on skin cancer treatment. He has been awarded several scholarships and scientific awards.

Over recent years, Dr. Lallas has established scientific collaboration with numerous colleagues from several countries and supervised several fellows on skin cancer diagnosis and management. He is an invited speaker in several domestic and international congresses and meetings, mainly on dermoscopy and on skin cancer diagnosis and management. He is particularly involved in teaching activities on dermoscopy, having organized and participated in numerous domestic and international courses.

Article

Results of a Primary Skin-Cancer-Prevention Campaign in Early Childhood on Sun-Related Knowledge and Attitudes in Southern Hungary

Zsuzsanna Horváth, Csernus A. Evelin, Péter Oláh, Rolland Gyulai and Zsuzsanna Lengyel *

Department of Dermatology, Venerology and Oncodermatology, Medical School, University of Pécs, 7632 Pécs, Hungary; horvath.zsuzsanna2@pte.hu (Z.H.); csernus.evelin@pte.hu (C.A.E.); olah.peter@pte.hu (P.O.); gyulai.rolland@pte.hu (R.G.)
* Correspondence: lengyel.zsuzsanna@pte.hu

Simple Summary: Primary skin-cancer-prevention campaigns among young children are important as this is the age when individuals are developing behaviors. Our aim was to evaluate sun-protection knowledge and behavior among caregivers in daycare centers and kindergartens and to determine if educational lectures are positively influential. In daycare centers, we discovered that measures of sun protection (e.g., hat, sunscreen, and shaded areas) are more likely to be available when compared to kindergartens. Knowledge regarding sun safety has improved following our initial presentation, however, not significantly. Sun safety policies did not exist in any of the facilities, presenting an urgent need for their implementation.

Abstract: Avoidance of ultraviolet (UV) exposure in early childhood is important for reducing the lifetime risk of developing skin cancer. The goal of the present prospective, multicenter pilot study was to assess the sun-protection practices in kindergartens and daycare centers and to evaluate sun protection knowledge and behavior among caregivers employed in the surveyed facilities. The study consisted of two parts. A baseline questionnaire was completed by the caregivers in relation to knowledge regarding basic sun protection and sun protection practices of the participating facilities. Afterward, a thirty-minute presentation was hosted in reference to this topic. Six months following the presentation, a follow-up questionnaire was distributed among the caregivers, evaluating the attitude-related and behavioral changes towards children. A total of 153 caregivers from five daycare centers (children between 6 months and 3 years of age) and sixteen kindergartens (children between 3 and 7 years of age) willfully participated in our study. According to our results, the main source of information regarding sun protection originated from different types of media. We found that staying in shaded areas and the use of protective clothing were not frequent in the facilities. Following our presentation regarding skin types and sunscreen use, protective measures improved, but not significantly ($p = 0.222$). The majority (92.31%) of caregivers distributed the information throughout their environment and also to parents. Sun protection knowledge is necessary; however, motivation among caregivers and parents and involvement of children is also relevant. Hence, a continuous, repetitive educational program regarding sun-smart behavior is deemed essential.

Keywords: primary prevention; children; daycare centers; kindergarten; ultraviolet radiation

Citation: Horváth, Z.; Evelin, C.A.; Oláh, P.; Gyulai, R.; Lengyel, Z. Results of a Primary Skin-Cancer-Prevention Campaign in Early Childhood on Sun-Related Knowledge and Attitudes in Southern Hungary. *Cancers* **2021**, *13*, 3873. https://doi.org/10.3390/cancers13153873

Academic Editors: Georg T. Wondrak and Aimilios Lallas

Received: 16 June 2021
Accepted: 27 July 2021
Published: 31 July 2021

Publisher's Note: MDPI stays neutral with regard to jurisdictional claims in published maps and institutional affiliations.

1. Introduction

The incidence of melanoma and non-melanoma cancers has shown a persistent increase worldwide in recent decades [1,2], placing an increasing burden on the health care system. Recently, incidence rates of melanoma were documented as high as 35 cases in 100,000 in Australia and around 14 cases in 100,000 in both the USA and UK [1]. In 2019, 24 new melanoma cases per 100,000 inhabitants were diagnosed in Hungary according to the National Health Insurance Fund of Hungary (NHIF) [3]. A recent study found significant changes in melanoma incidence trends in Hungary. Between 2011 and 2015, a significant mean annual increase was detected in the overall melanoma incidence, while between 2015 and 2019, a relevant mean annual decrease was seen. The authors assumed that joining the Euromelanoma campaign supported existing Hungarian seasonal screening programs to reach a wider population leading to the decline [3].

Solar and artificial UV exposure is the main risk factor for the development of melanoma and non-melanoma skin cancers. UV exposure in childhood and adolescence elevates the individual's lifetime risk of developing skin cancer more than exposure in adulthood [4]. The majority of lifetime sun exposure occurs from early childhood to adolescent ages, and 50% to 80% of lifetime cumulative sun exposure occurs prior to 18 years of age [5]. In particular, early-childhood sun exposure and blistering sunburn prior to the age of 20 have been shown to be a determinant risk factor regarding malignant melanoma [6–8], and it may increase the risk by nearly two-fold [9]. Children spend more time outdoors, toddlers are harder to restrict to shaded areas during outdoor activities, and the skin of small children is more vulnerable to the effects of UV radiation (biological vulnerability) [10–13]. Other risk factors are also involved in skin tumor formation such as genetic predisposition, fair skin, the presence of atypical melanocytic naevi, and high numbers of melanocytic naevi (latter two in melanoma) [14–16]. Among all the risk factors, UV radiation is the only exogenous factor and therefore the one targeted for modification.

Since the 1980s, several health-education campaigns and skin-cancer-prevention programs have evolved worldwide [17]; however, few studies are available referencing the very young, pre-school-aged children [18,19]. As of this writing, no organized primary skin-cancer-prevention program exists in Hungary for pre-school-aged children.

Health education campaigns can be effective in terms of improving knowledge, attitude, and behavior among young children [20,21]. Adult health behavior is often established during childhood. According to recent publications, proper sun-protection habits of the mother led to less sun exposure among children [7,9,22,23].

The effectiveness of melanoma prevention is dependent upon how it is accepted by the population. Fun and at the same time deterrent examples are often used in campaigns [21,24–26]. Educational purposes can be achieved through environmental intervention (brochures, handouts, and books), presentations, and behavioral intervention (role play, and games), the latter of which has proven effective regarding children [15,26].

In many EU countries (including Hungary), the use of tanning booths is prohibited to minors [27]; however, there is a lingering misconception according to which acquiring a suntan indoors offers protection from skin cancer and the harmful effects of UV light [28]. We find it immensely important to educate children, parents, and caregivers, to set a positive example by shunning the use of artificial UV light.

All three stakeholders (caregivers, parents, and children) should be targeted to achieve effective implementation of sun-safe behavior [29]. A study in German childcare centers demonstrated a significant improvement in knowledge of sun-related issues among staff members and parents following a relatively brief training session. Children were also taught sun safety messages through a playful approach. The effectiveness regarding children's learning was not assessed in the study, yet 82% of caregivers believed that children learned something beneficial regarding sun protection [30,31].

No data were available in reference to sun-protection policies and habits applied in Hungarian daycare centers and kindergartens when the study was designed. Therefore, our primary aim was to survey any existing sun-safety measures among the two types of institutions. Our secondary goal was to evaluate the knowledge of caregivers regarding sun-safety and whether educational presentations can improve such knowledge, since they spend eight to ten hours among our youngest ones five days out of a week.

2. Materials and Methods

2.1. Participating Facilities

The study was conducted in the year 2017 and included five daycare centers (children between 6 months and 3 years) and sixteen kindergartens (children between 3 and 7 years). All childcare institutions were located in the city of Pécs, in the southern part of Hungary, with a Mediterranean climate. The sunshine duration in Pécs between May and August varies between 245 and 289 h. The yearly average sunshine duration is 2080 h [32].

The kindergartens were all "green kindergartens", implying that they follow an environmentally conscious education program. All facilities are financed by the city government and located in neighborhoods with average-income households.

Throughout our investigation, two self-administered questionnaires and a multimedia presentation were used, which were developed by the authors of this article. All institutions were personally visited by the authors. Participation of the staff members was voluntary and anonymous.

2.2. Questionnaires and Educational Intervention

The baseline questionnaire consisted of twenty-eight questions. The first questions (Q3-13) of this questionnaire evaluated the caregivers' knowledge regarding basic sun protection information. The second part assessed the sun protection practices of the kindergarten or daycare center (e.g., the time frame, when children are taken outside during the summer, availability of shaded areas, and the use of sunscreen, sunhat, and sunglasses) and the knowledge of correct sunscreen use by caregivers (Figure S1).

Following our assessment of the first questionnaire, an approximately thirty-minute-long presentation was hosted by a dermatologist (one of the authors). The lecture highlighted information regarding the basics of the physical properties (ambient UV exposure via reflectance of UV radiation, glass, and water penetration), and the human biological effects of UV radiation including its cancerogenic and photoaging effects (highlighted with clinical pictures). The different skin types, UV index, knowledge regarding ozone depletion and sun protection skills were presented, along with a special emphasis in reference to shade seeking, wearing proper clothing (broad-brimmed hats, sunglasses, and long-sleeved clothing with ultraviolet protection factor—UPF) and the proper use of sunscreens. The importance of avoiding sunbeds, avoiding midday hours spent in direct sunlight, and successful primary prevention campaigns were also mentioned (SunSmart and SunWise programs).

Six months (including summer months) following our presentation, a follow-up questionnaire was distributed to the caregivers of the same institutions in early autumn, who participated in our presentations. This questionnaire evaluated the attitudinal and behavioral change in relation to children and of themselves concerning sun safety and the acquired knowledge regarding sun protection skills. Positive environmental changes in the proportion of shaded areas in surveyed institutions were also evaluated (Figure S2).

The tests were scored to determine knowledge regarding two topic areas prior to and following the presentation. Knowledge regarding skin types was measured by a match-test question in which correct answers were awarded with one point out of a maximum six points. Proper sunscreen application was scored with single choice test questions, and correct answers were also awarded one point. Questions regarding UV light properties were measured using single-choice test questions (Q3-6-7-8-9) in the first test and multiple-choice test in the second questionnaire (Q2). Correct answers were awarded with one point each, in which the percentage of correct answers was compared with various knowledge traits. The pre-test was filled out by 153 participants (57 daycare workers and 96 kindergarten employees). The post test was completed by staff members who attended the intervention, a total of 143 participants (56 daycare workers and 87 kindergarten employees). The total number of answerers depicted in the tables vary, since not all questions were answered by all participants.

2.3. Statistical Analyses

The responses were analyzed, and odds ratios were calculated. Samples were tested for normality using the Kolmogorov–Smirnov test, and hypotheses were tested with independent *t*-test, Mann–Whitney U test, and chi-square tests, as appropriate. A 95% confidence interval was used ($p = 0.05$) to determine statistical significance (*p*-values are also provided).

3. Results

3.1. Baseline Questionnaire

One hundred and fifty-three female caregivers (daycare center $n = 57$, kindergarten $n = 96$) completed our first questionnaire. The mean age of the respondents of the first questionnaire was 46.5 years (21–62 years), (daycare centers, 43.5 years (21–60 years); kindergartens, 48.4 years (25–62 years)).

3.1.1. Source of Information Regarding Sun Protection

Analysis of the results of the first questionnaire on basic sun protection knowledge revealed that virtually none of the participants received any form of education on sun protection during their training—not even in the "green kindergartens".

The most common (68.37%) source to gather information regarding sun protection was the different means of media (TV, radio, internet, and the newspaper). The caregivers employed in the nurseries were more likely to acquire information from a healthcare professional as compared to kindergartens (from general practitioners (GPs) 28.07% vs. 9.38%, from dermatologists 45.61% vs. 35.42%, respectively) (Figure 1).

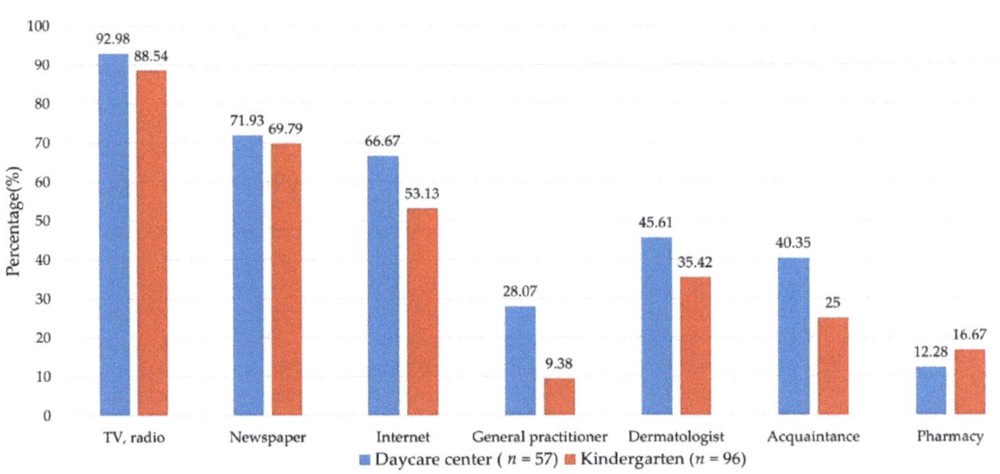

Figure 1. Information sources among caregivers concerning sun safety.

3.1.2. Knowledge Regarding UV Light Properties and Sun Protection

Correct answers were given by 75.8% of the caregivers to the questions on the effects of UV radiation on the skin, UV index, the time interval that should be avoided while exposed to direct sunlight, and sun protection practices. However, caregivers experienced difficulties determining the different skin types: only 32.0% ($n = 49$) of all caregivers were able to identify all of them. Caregivers employed in daycare centers scored better results: 42.11% reached maximum points, while in kindergartens it was 26%. In ascertaining the knowledge regarding sunscreen use, only 11.9% of all caregivers reached the maximum points allocated. A larger proportion of caregivers in the daycare centers (15.69%) achieved maximum points in this section of the questionnaire than caregivers in kindergartens (9.64%) (Table 1). The average knowledge regarding sun safety of caregivers in the daycare centers was slightly increased (mean of ranks: 72.3 in daycare centers vs. 64.5 in kindergartens), although the difference was not significant ($p = 0.228$) (not shown in table).

Table 1. Improvement in different knowledge traits, from baseline to follow-up questionnaire.

	Daycare Center		Kindergarten		Total		*p*-Value *
Knowledge of sunscreen use	Baseline ($n = 51$)	Follow-up ($n = 56$)	Baseline ($n = 83$)	Follow-up ($n = 87$)	Baseline ($n = 134$)	Follow-up ($n = 143$)	
mean performance (%)	67.2	62.5	61.5	65.1	64.0	49.2	0.222
performance 100% *n* (%)	8 (15.7)	5 (8.9)	8 (9.6)	16 (18.4)	16 (11.9)	21 (14.7)	
Knowledge of skin type	Baseline ($n = 57$)	Follow-up ($n = 56$)	Baseline ($n = 96$)	Follow-up ($n = 87$)	Baseline ($n = 153$)	Follow-up ($n = 143$)	
mean performance (%)	61.0	73.0	39.0	58.0	47.0	64.0	0.307
performance 100% *n* (%)	24 (42.1)	31 (55.4)	25 (26.0)	33 (37.9)	49 (32.03)	64 (44.8)	

* Mann–Whitney test.

3.1.3. Sun-Protection Practices in the Facilities

According to the directors of all institutions, no written sun protection policy existed in the facilities. Analysis of the sun-protection practices in the institutions revealed that children stay outdoors between 09:00 and 11:00 and between 15:00 and 17:00 h. Additionally, if the outdoor temperature rises above 30 °C, children stay inside. This practice is verbally communicated to the staff.

All centers named trees as the main source of shading within their own institutions. Almost all (98%) caregivers claimed that the institution yard is properly shaded but staff members only rated shade over play structures rather than the entire outdoor area. In contrast, the authors noted that most of these institutions lack properly shaded areas. On average, half of the entire outdoor area in the daycare centers was shaded, while in the kindergartens it was only one-third. However, daycare centers were smaller in size, and on average, the number of attending children is 50 or less, while 100–150 children are cared for in the surveyed kindergartens. In evaluating the availability of sun-protection measures in the facilities, we found that the prevalent measures used in sun protection among children were distributed as follows: in most of the centers only a few of the children wear sunglasses (daycare centers: 1.75% and kindergartens: 15.63%), a larger proportion of children have sunhats in the daycare centers (daycare centers: 89.74%, kindergartens 73.96%), and the availability of sunscreens are more common in the daycare centers (daycare centers: 68.42%, kindergartens 36.5%) (Table 2). The general availability of sunhat and sunscreen differed significantly between daycare centers and kindergartens ($p < 0.001$, chi-square test), and a similar difference is indicated in sunglasses availability as well ($p = 0.007$). Sunglasses are often considered hazardous to young children and their use is not encouraged; the latter was communicated by several of the staff members as a side note (Table 2).

Nearly all caregivers help those in their care in putting on hats (100%) and aid in applying sunscreen (98.7%) when going outdoors, if indeed, and generally, the child comes equipped with his/her own personal hat and sunscreen. The majority of caregivers encourage parents to bring a hat and sunscreen to the child centers (sunhat 86.3% and sunscreen 88.9%).

The authors were interested in knowing if children having skin types I and II are more likely to have access to sun-protection measures provided by their parents. Less than half of the staff members believed there is a correlation (daycare centers: 45.61%, kindergartens: 45.83%) (Figure 2).

Table 2. Availability of sunscreen, sunhat, and sunglasses based on the judgment of caregivers.

	Daycare Center (*n* = 57)	Kindergarten (*n* = 96)	Total (*n* = 153)
Availability of sunhat to children, *n* (%)			
Almost every child	36 (63.2)	15 (15.6)	51 (33.3)
Approx. half of the children	15 (26.3)	56 (58.3)	71 (46.4)
A few children	3 (5.3)	23 (24.0)	26 (17.0)
Availability of sunscreen to children, *n* (%)			
Almost every child	25 (43.9)	2 (2.1)	27 (17.6)
Approx. half of the children	14 (24.6)	33 (34.4)	47 (30.7)
A few children	18 (31.6)	57 (59.4)	75 (49.0)
Availability of sunglasses to children, *n* (%)			
Almost every child	0 (0.0)	2 (2.1)	2 (1.3)
Approx. half of the children	1 (1.8)	13 (13.5)	14 (9.2)
A few children	54 (94.7)	79 (82.3)	133 (86.9)

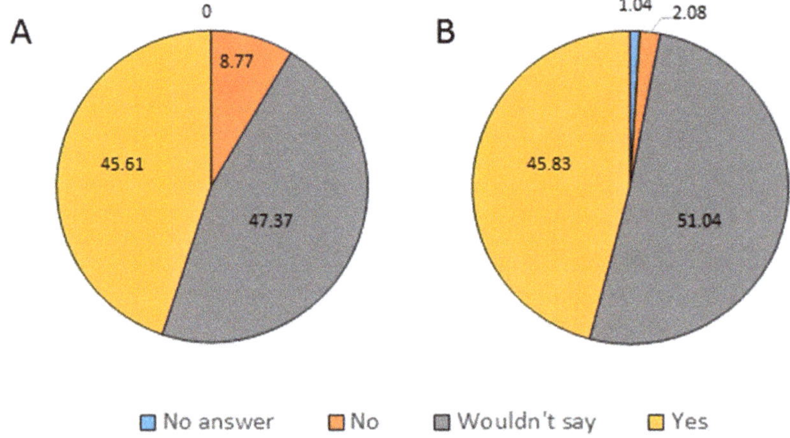

Figure 2. Availability of sunhat and sunscreen to fair-skinned children based on caregiver's judgment in daycare centers (**A**) and kindergartens (**B**).

Ninety-eight percent of all caregivers believe sun safety is important; however, the willingness to learn more regarding sun safety was only 65.4%.

3.2. Follow-Up Questionnaire

One hundred and forty-three caregivers (daycare centers *n* = 56, kindergartens *n* = 87) participated in our presentation and filled out the second questionnaire. The mean age of the responders of the second questionnaire was 43.7 years (22–60 years) (daycare centers, 39.8 years (22–57 years): kindergartens, 46.6 years (24–60 years)). Distinctly, 91.6% of all caregivers found the presentation educative and useful in the second questionnaire.

3.2.1. Knowledge Regarding UV Light Properties, Sun Protection, and Changes in Behavior

A total of 80.4% of all caregivers achieved at least 60% correct answers in the second questionnaire regarding the properties and biological effects associated with UV light. The knowledge of the meaning of UV index was shown to be accurate, as 91.5% correctly responded (daycare center: 82.7% and kindergarten: 93.7%).

In general, caregivers scored higher levels of accuracy regarding questions about sunscreen use (14.7%), yet the improvement was not significant (baseline test 11.9%, $p = 0.222$). Additionally, an insignificant improvement was detected regarding the identification of the different skin types: 44.7% of all caregivers reached maximum points in the second questionnaire (baseline test 32%, $p = 0.307$) (Table 1).

A total of 40.6% ($n = 58$) of caregivers claimed to have changed their sun-protection habits, including giving up sunbathing, confining themselves to shady areas, using the appropriate amount of sunscreen, donning sunhats and sunglasses and/or protective clothing. A total of 43.1% ($n = 25$) of these caregivers changed one, two (27.6%, $n = 16$), three (22.4%, $n = 13$), and four (3.4%, $n = 2$) of their sun protection habits, respectively (Table 3).

Table 3. Changes in sun-protection habits among caregivers in child centers.

	Daycare Center ($n = 56$)	Kindergarten ($n = 87$)	Total $n = 143$, n (%)
positive change in behavior	21	37	58 (40.6)
no sunbathing	16	25	41 (28.67)
stay in shade	6	10	16 (11.19)
used hat and sunglasses	7	14	21 (14.68)
used sunscreen	5	3	8 (5.59)
appropriate sunscreen use	6	11	17 (11.89)
against efforts was sunburnt	1	0	1 (0.70)
no change in behavior	35	49	84 (58.7)
never tanned before	11	19	30 (20.98)
tans anyway	16	23	39 (27.27)
no answer	0	1	1 (0.70)

Eighty-four (58.5%) caregivers did not alter their sun-protection habits, among those resolute in keeping to their former habits. Thirty (20.9%) reportedly had never tanned, and 27% ($n = 39$) continued sunbathing despite the newly acquired information. Fifteen (17.86%) caregivers did not specify the reason for their unaltered sun-protection habit; however, respondents claimed they kept doing it since they enjoy outdoors, including exposure to the sun, they have no history of sunburn, and make adequate use of sunscreens. Almost all caregivers stated that they shared the acquired information on sun safety with family, friends, and colleagues (92.3%). The majority of caregivers raised awareness in regard to parents highlighting the major factors of sun safety (68.5%).

3.2.2. Sun-Protection Practices in the Facilities

The majority (95.8%) of caregivers in daycare centers and kindergartens both agreed that they consciously paid more attention to the sun safety of children, with an emphasis on restricting play to shaded areas of the yard and using sunhats and sunscreens.

All directors stated they were not able to develop the proportion of shaded area as large as they wanted. The most common cause was the lack of financial support.

4. Discussion

There are a substantial number of studies evaluating the effect of sun-protection programs targeting preschool-aged children and caregivers and showing favorable results in education and awareness [19,25,26,30,31], although there is a paucity of such programs according to literature data in Eastern Europe; therefore, the authors find it important to survey and help to develop conscious and sun-safe behavior in caregivers and small children.

This manuscript provides preliminary results of our primary skin cancer prevention program. Our primary prevention program sets the goal of educating the proper form of sun protection first among caregivers and later the parents of children aged between 6 months and to the end of the kindergarten years, as well as the goal of involving children in the learning process in a playful way. It is important that caregivers (including parents) in charge of the youngest children are thoroughly educated regarding sun protection since their actions and behaviors set the foundation for lifelong correct sun-protection practices for children within their care [11,17,23,33,34].

Additionally, recent studies underline the necessity of primary prevention awareness campaigns targeting children, since even adults in risk groups (sailors and agriculture employees) show a lack of information and disinterest in sun protection measures [35,36].

In our study, we found that no education program in reference to sun protection currently exists for training caregivers of preschool children in the surveyed facilities. However, similarly to previous studies [22,28], our results demonstrated that caregivers possessed relatively accurate knowledge regarding the biological effects of UV light. However, this premise does not necessarily translate into appropriate sun-safe behavior [37]. The knowledge level regarding different skin types improved following our presentation, but the difference was not significant. These results are in agreement with the German "Sun-Pass" project, on which the percentage of staff members naming the skin types correctly increased only slightly [31]. Daycare center caregivers achieved higher scores on the queries regarding sunscreen use and skin type. This insignificant difference may be explained by data that show that caregivers in daycare centers are more likely to acquire their information from doctors compared to employees in kindergartens. On the other hand, post-intervention scores regarding sunscreen use decreased in the daycare centers, which may be biased by the number of responders (51 pre-test vs. 56 post-test). Overall, the knowledge of sunscreen use did not significantly improve, suggesting the correct form of sunscreen use has yet to be taught, as has been previously emphasized in another study [38]. In our study, the average knowledge regarding sun safety of caregivers in the daycare centers showed a slight increase; however, the difference was not significant. Not being able to demonstrate a significant change after intervention, as shown by other studies [30,31], might have been influenced by the fact that the duration of time between the two tests was too long (6 months), and therefore recurring education is necessary within six months or less. As a part of an Italian primary prevention program among primary school children, parents were asked to complete pre- and post-intervention tests regarding the use of sun-protection measures. No significant difference was stated concerning the investigated factors, with the exception of a slight increase in the use of sunhats and sunglasses (elapsed time between the two tests was approx. 6 months). However, the authors were able to document a slight improvement regarding behavior and reduced sunburn rates over almost two decades [39].

Additionally, in our study, we found that less than half of the staff members believed that there is a correlation between sunscreen availability for fair-skinned children provided by their parents (daycare centers: 45.61%, kindergartens: 45.83%). Our results can be compared with a survey conducted in the USA involving parents of preschool-aged children; in that study, 72.6% of parents' children who sunburn easily used sunscreen, compared to 42.1% of those who tan more easily, and only 12% of parents of African-American children used sunscreen [38]. A 2001 Florida study found that parents of light-skinned children scored higher on UV-related knowledge than parents of children with darker skin types (74% vs. 26%, $p < 0.05$) [40]. Therefore, we recommend that some educational materials should directly target fair-skinned individuals to achieve a maximal sun-protection benefit. The identification of the Fitzpatrick skin type might be an easily applicable useful tool by caregivers and parents alike to identify higher-risk children [14,16,41].

The most common source to gather information regarding sun protection was the different means of media in our surveyed—all female participants—population at 68.4%. This is in accordance with previous findings, where it was found that the common sources of information among the surveyed individuals were television (79%), magazines (52%), newspapers (49%), health professionals (35%), and family and friends (31%). In addition, female participants showed heightened awareness of skin protection information [17]. This premise underscores the necessity and responsibility of professionals to use different media sources as means of sharing information. Today, as social media and apps are widely used, their application in primary prevention campaigns may increase campaign efficacy, although this has yet to be investigated. These channels can be used to deliver and promote the most important messages (e.g., self-skin examination method) population-wide and may be prioritized in the near future [42–44].

None of the participating childcare facilities had a sun safety policy, which would be much needed in order to standardize sun safety measures in centers caring for our young ones, as mentioned in former studies [19,45]. Interestingly, sun protection practices were superior in institutions with written policies [19,30,45].

Although knowledge regarding UV index was shown to be high among caregivers in both daycare centers and kindergartens, no UV index policies regulated playtime outdoors in any of the surveyed facilities. Similar findings were recently published [46]. However, caregivers were attentive in keeping children indoors during midday hours.

Most of the institutes lack a proper shaded area, and access to good-quality sunscreen is considerably limited, as stated in a previous study [22]. We found among the surveyed kindergartens that outdoor playing areas are shaded to a lesser extent compared to the daycare centers. This phenomenon may be affected by the generally larger size of kindergartens and increased protective attitude towards younger age groups in daycare centers. While most caregivers at first thought yards belonging to child centers were properly shaded, many facilities erected sun sails and planted trees following our presentation. Shade is an important UV minimizing method and is needed in all outdoor areas where children gather (not only over "play areas" such as sandboxes). To create an effective shaded area, shade audits are promoted to assess the proper size and the location of the area for maximum protection and taking factors into account that influence shade UV protection (e.g., UV reflectivity). Such audits and documents regarding shade planning have been developed for child care centers and schools, e.g., in the United States and Australia [47–49]. Emphasizing the importance of this component of sun protection in educational materials and in primary prevention campaigns is important to enhance sun safety. Lack of funding to deploy shade can be a hurdle as stated in previous studies [45]. Similarly, in several locations in our study, caregivers cited that necessary changes cannot be executed due to the lack of financial support.

We also ascertained that the availability of sunscreens, sunglasses, and protective clothing was limited in the facilities. Wearing sunhats (79.7%) and applying sunscreens (48.4%) were more common in daycare centers and kindergartens than the use of sunglasses (10.5%), although the general availability of these items was significantly higher in daycare centers based on the caregivers' judgment ($p < 0.001$). In Germany, a recently conducted study found that caregivers in nearly all institutions use sunscreen (98.8%) and sunhats (98.4%), and the usage of sunglasses was 2.4%. The results were based upon the referral of the directors of the nurseries [45]. The differences in the results between the two institutions we examined in our study might be explained by the more intense preventive attitude of caregivers (including parents) towards children under the age of three.

None of the child centers surveyed in this study supply sunscreen to children on a regular basis, due to its high cost. It falls mostly to families to provide sunscreen or, on occasion, sponsors who supply sunscreen products to facilities. According to our results, the knowledge regarding sunscreen use (re-applying and application of sufficient amount) is not appropriate among caregivers. Building sun shelters, remaining in shaded areas, and wearing hats and long-sleeved clothes are the pillars regarding sun protection, since sunscreen use alone is deemed insufficient and plants a false sense of security [22,38].

It is important to develop a general sun-safety policy that should be effectively implemented in the facilities promptly. Ideally, such a policy will not only issue guidance and set the rules regarding how and when children are taken outdoors, yet also focus on the availability of sufficient shaded areas proportionately sized to the number of children enrolled at the facility. It can be one of the first steps towards increasing sun protection awareness among governmentally funded Hungarian child care centers. Based on our results, we observed, among facilities taking care of younger children (daycare centers) the availability of sun-protection measures (e.g., hat, and sunscreen) are higher, and typically provided by the parents, the proportion of shaded areas is generally higher also compared to kindergartens.

Our study experienced several limitations, such as the reliance on self-reporting and the representativeness of the data. Even though the study was multicentric and took place in the fifth-largest city in Hungary, we cannot conclude that the sample was representative for all daycare centers and kindergartens in the region. Furthermore, the child care centers were situated in areas with average incomes, but the caregivers were not asked to divulge their place of residence. Currently, Hungarian institutions caring for pre-school-aged children encounter significant staff fluctuations, which also shows in our sample size obtained in a half-year time interval; this may also explain the stagnation and limited improvement in our results. Furthermore, privately funded institutions did not participate, although access to sun-protective measures is not only determined by financial support [45].

In conclusion, sun protection knowledge is required; however, in itself, it is not sufficient for the successful implementation of proper sun-protection behavior in kindergartens and daycare centers. According to our results, over one fourth (27%) of the caregivers did not change their sun protection habits and professed to bask in the sun despite the information presented, which is in accordance with a recent Danish study, where, after the first measures of the campaign, sunbed use was only slightly reduced (odds ratio (OR) = approx. 0.9 versus OR = approx. 0.3 after eight years of repetition) [42]. However, forty percent of the participants changed something in their habits, and the majority of caregivers passed along their information regarding sun protection to the parents. Furthermore, employees paid more attention to the sun safety of children after the presentation. These findings show a positive effect on sun-safety behavior, but most importantly, highlight the relevance of continuous education on this topic from reliable sources [39,42,50,51].

To improve and stabilize sun-protection behavior in child care centers in our region, in addition to the implementation of written institutional sun-safety policies (including shading), the availability of private and institutional funds and an incorporated repetitive educational program for caregivers is needed. In consideration of training, published literature, multi-media presentations, and online workshops with periodical mandatory testing can be used to promote and verify sufficient and lasting knowledge. The continuing education should be taught by a reliable individual, such as a well-trained nurse, repeated each spring and provided memos regarding sun protection sent to caregivers every 3–4 months. Furthermore, the motivation and involvement of parents is necessary as they play an important role in enhancing sun safety among children [7,9,22,40,50]. This study documented the limited knowledge and practices regarding sun safety, helped to increase awareness in caregivers, and allowed for recommendations for improving sun protection outcomes for children in our region.

Supplementary Materials: The following are available online at https://www.mdpi.com/article/10.3390/cancers13153873/s1, Figure S1: Baseline questionnaire, Figure S2: Follow-up questionnaire.

Author Contributions: Z.H., C.A.E. and Z.L. conceived the work; Z.H., C.A.E. and Z.L. drafted the original manuscript and generated the figures; Z.H., C.A.E., P.O., R.G. and Z.L. revised the manuscript; Z.H., C.A.E., P.O., R.G. and Z.L. revised the figures, table, and references; R.G. and Z.L. supervised the work. All authors have read and agreed to the published version of the manuscript.

Funding: This research received no external funding.

Institutional Review Board Statement: Ethical review and approval were waived for this study, due to the fact that study participation was anonymous and voluntary and did not contain personal medical data.

Informed Consent Statement: Patient consent was waived due to the fact that study participation was anonymous and voluntary.

Data Availability Statement: The data presented in this study are available on request from the corresponding author.

Conflicts of Interest: The authors declare no conflict of interest.

References

1. Ferlay, J.; Soerjomataram, I.; Dikshit, R.; Eser, S.; Mathers, C.; Rebelo, M.; Parkin, D.M.; Forman, D.; Bray, F. Cancer incidence and mortality worldwide: Sources, methods and major patterns in GLOBOCAN. 2012. *Int. J. Cancer* **2015**, *136*, E359–E386. [CrossRef] [PubMed]
2. Fartasch, M.; Diepgen, T.L.; Schmitt, J.; Drexler, H. The Relationship between Occupational Sun Exposure and Non-Melanoma Skin Cancer, Clinical Basics, Epidemiology, Occupational Disease Evaluation, and Prevention. *Dtsch. Arztebl. Int.* **2012**, *109*, 715–720. [CrossRef] [PubMed]
3. Liszkay, G.; Kiss, Z.; Gyulai, R.; Oláh, J.; Holló, P.; Emri, G.; Csejtei, A.; Kenessey, I.; Benedek, A.; Polányi, Z.; et al. Changing Trends in Melanoma Incidence and Decreasing Melanoma Mortality in Hungary between 2011 and 2019: A Nationwide Epidemiological Study. *Front. Oncol.* **2021**, *12*, 612459. [CrossRef] [PubMed]
4. Oliveria, S.A.; Saraiya, M.; Gerller, A.C.; Heneghan, M.K.; Jorgensen, C. Sun exposure and risk of melanoma. *Arch. Dis. Child.* **2006**, *91*, 131–138. [CrossRef] [PubMed]
5. Stern, R.S.; Weinstein, M.C.; Baker, S.G. Risk reduction for nonmelanoma skin cancer with childhood sunscreen use. *Arch. Dermatol.* **1986**, *122*, 537–545. [CrossRef] [PubMed]
6. Glanz, K.; Saraiya, M.; Wechsler, H. Guidelines for school programs to prevent skin cancer. *MMWR* **2002**, *51*, 1–16. [CrossRef] [PubMed]
7. Grob, J.J.; Guglielmina, C.; Gouvernet, J.; Zarour, H.; Noé, C.; Bonerandi, J.J. Study of sunbathing habits in children and adolescents: Application to the prevention of melanoma. *Dermatology* **1993**, *186*, 94–98. [CrossRef] [PubMed]
8. Zanetti, R.; Franceschi, S.; Rosso, S.; Colonna, S.; Bidoli, E. Cutaneous melanoma and sunburns in childhood in a southern European population. *Eur. J. Cancer* **1992**, *28A*, 1172–1176. [CrossRef]
9. Townsend, J.S.; Pinkerton, B.; McKenna, S.A.; Higgins, S.M.; Tai, E.; Steele, C.B.; Derrick, S.R.; Brown, C. Targeting children through school-based education and policy strategies: Comprehensive cancer control activities in melanoma prevention. *J. Am. Acad. Dermatol.* **2011**, *65*, S104–S113. [CrossRef] [PubMed]
10. Paller, A.S.; Hawk, J.L.; Honig, P.; Giam, Y.C.; Hoath, S.; Mack, M.C.; Stamatas, G.N. New insights about infant and toddler skin: Implications for sun protection. *Pediatrics* **2011**, *128*, 92–102. [CrossRef]
11. Lebbé, C.; Robert, C.; Ricard, S.; Sassolas, B.; Grange, F.; Saiag, P.; Lhomel, C.; Mortier, L. Evolution of sun-protection measures for children. *J. Eur. Acad. Dermatol. Venereol.* **2015**, *29*, 20–22. [CrossRef] [PubMed]
12. Balk, S.J. Ultraviolet radiation: A hazard to children and adolescents. *Pediatrics* **2011**, *127*, e791–e817. [CrossRef] [PubMed]
13. Marks, R. Campaigning for melanoma prevention: A model for a health education program. *J. Eur. Acad. Dermatol. Venereol.* **2004**, *18*, 44–47. [CrossRef] [PubMed]
14. Fehér, K.; Cercato, M.C.; Prantner, I.; Dombi, Z.; Burkali, B.; Paller, J.; Ramazzotti, V.; Sperduti, I.; Nádasi, E.; Parragi, K. Skin cancer risk factors among primary school children: Investigations in Western Hungary. *Prev. Med.* **2010**, *51*, 320–324. [CrossRef]
15. MacKie, R.M. Incidence, risk factors and prevention of melanoma. *Eur. J. Cancer* **1998**, *34*, 3–6. [CrossRef]
16. Gellén, E.; Janka, E.; Tamás, I.; Ádám, B.; Horkay, I.; Emri, G.; Remenyik, É. Pigmented naevi and sun protection behaviour among primary and secondary school students in an Eastern Hungarian city. *Photodermatol. Photoimmunol. Photomed.* **2016**, *32*, 98–106. [CrossRef]
17. Gavin, A.; Boyle, R.; Donnelly, D.; Donnelly, C.; Gordon, S.; McElwee, G.; O'Hagan, A. Trends in skin cancer knowledge, sun protection practices and behaviors in the Northern Ireland population. *Eur. J. Public Health* **2012**, *22*, 408–412. [CrossRef] [PubMed]
18. Saraiya, M.; Glanz, K.; Briss, P.A.; Nichols, P.; White, C.; Das, D.; Smith, S.J.; Tannor, B.; Hutchinson, A.B.; Wilson, K.M.; et al. Interventions to prevent skin cancer by reducing exposure to ultraviolet radiation: A systematic review. *Am. J. Prev. Med.* **2004**, *27*, 422–466. [CrossRef]
19. Ettridge, K.A.; Bowden, J.A.; Rayner, J.M.; Wilson, C.J. The relationship between sun protection policy and associated practices in a national sample of early childhood services in Australia. *Health Educ. Res.* **2011**, *26*, 53–62. [CrossRef]

20. Dadlani, C.; Orlow, S.J. Planning for a brighter future: A review of sun protection and barriers to behavioral change in children and adolescents. *Dermatol. Online J.* **2008**, *14*, 1. [CrossRef]

21. Bastuji-Garin, S.; Grob, J.J.; Grognard, C.; Grosjean, F.; Guillaume, J.C. Melanoma prevention: Evaluation of a health education campaign for primary schools. *Arch. Dermatol.* **1999**, *135*, 936–940. [CrossRef] [PubMed]

22. Stanton, W.R.; Janda, M.; Baade, P.D.; Anderson, P. Primary prevention of skin cancer: A review of sun protection in Australia and internationally. *Health Promot. Int.* **2004**, *19*, 369–378. [CrossRef] [PubMed]

23. Mortier, L.; Lepesant, P.; Saiag, P.; Robert, C.; Sassolas, B.; Grange, F.; Lhomel, C.; Lebbe, C. Comparison of sun protection modalities in parents and children. *J. Eur. Acad. Dermatol. Venereol.* **2015**, *29*, 16–19. [CrossRef]

24. Richard, M.A.; Martin, S.; Gouvernet, J.; Folchetti, G.; Bonerandi, J.J.; Grobb, J.J. Humour and alarmism in melanoma prevention: A randomized controlled study of three types of information leaflet. *Br. J. Dermatol.* **1999**, *140*, 909–914. [CrossRef] [PubMed]

25. Seidel, N.; Stoelzel, F.; Garzarolli, M.; Herrmann, S.; Breitbart, E.W.; Berth, H.; Baumann, M.; Ehninger, G. Sun protection training based on a theater play for preschoolers: An effective method for imparting knowledge on sun protection? *J. Cancer Educ.* **2013**, *28*, 435–438. [CrossRef] [PubMed]

26. Loescher, L.J.; Emerson, J.; Taylor, A.; Christensen, D.H.; McKinney, M. Educating preschoolers about sun safety. *Am. J. Public Health* **1995**, *85*, 939–943. [CrossRef] [PubMed]

27. Grange, F.; Mortier, L.; Crine, A.; Robert, C.; Sassolas, B.; Lebbe, C.; Lhomel, C.; Saiag, P. Prevalence of sunbed use, and characteristics and knowledge of sunbed users: Results from the French population-based Edifice Melanoma survey. *J. Eur. Acad. Dermatol. Venereol.* **2015**, *29*, 23–30. [CrossRef] [PubMed]

28. Saiag, P.; Sassolas, B.; Mortier, L.; Grange, F.; Robert, C.; Lhomel, C.; Lebbé, C. EDIFICE Melanoma survey: Knowledge and attitudes on melanoma prevention and diagnosis. *J. Eur. Acad. Dermatol. Venereol.* **2015**, *29*, 11–15. [CrossRef] [PubMed]

29. Walker, D.K. Skin Protection for (SPF) Kids Program. *J. Pediatr. Nurs.* **2012**, *27*, 233–342. [CrossRef] [PubMed]

30. Aulbert, W.; Parpart, C.; Schulz-Hornbostel, R.; Hinrichs, B.; Krüger-Corcoran, D.; Stockfleth, E. Certification of sun protection practices in a German child day-care centre improves children's sun protection–the 'SunPass' pilot study. *Br. J. Dermatol.* **2009**, *161*, 5–12. [CrossRef]

31. Stöver, L.A.; Hinrichs, B.; Petzold, U.; Kuhlmei, H.; Baumgart, J.; Parpart, C.; Rademacher, O.; Stockfleth, E. Getting in early: Primary skin cancer prevention at 55 German kindergartens. *Br. J. Dermatol.* **2012**, *167*, 63–69. [CrossRef]

32. Available online: http://www.met.hu/eghajlat/magyarorszag_eghajlata/varosok_jellemzoi/Pecs/ (accessed on 27 February 2017). (In Hungarian).

33. Berwick, M.; Erdei, E.; Hay, J. Melanoma epidemiology and public health. *Dermatol. Clin.* **2009**, *27*, 205–214. [CrossRef]

34. Eisinger, F.; Morère, J.F.; Pivot, X.; Grange, F.; Lhomel, C.; Mortier, L.; Robert, C.; Saiag, P.; Sassolas, B.; Viguier, J. Melanoma risk-takers: Fathers and sons. *J. Eur. Acad. Dermatol. Venereol.* **2015**, *29*, 35–38. [CrossRef] [PubMed]

35. Zalaudek, I.; Conforti, C.; Corneli, P.; Jurakic Toncic, R.; di Meo, N.; Pizzichetta, M.A.; Fadel, M.; Mitija, G.; Curiel-Lewandrowski, C. Sun-protection and sun-exposure habits among sailors: Results of the 2018 world's largest sailing race Barcolana' skin cancer prevention campaign. *J. Eur. Acad. Dermatol. Venereol.* **2020**, *34*, 412–418. [CrossRef]

36. Tizek, L.; Schielein, M.C.; Schuster, B.; Ziehfreund, S.; Biedermann, T.; Zink, A. Effects of an unconventional skin cancer prevention campaign: Impacts on the sun protection behavior of outdoor workers. *Hautarz* **2020**, *71*, 455–462. (In German) [CrossRef] [PubMed]

37. Robert, C.; Lebbe, C.; Ricard, S.; Saiag, P.; Grange, F.; Mortier, L.; Lhomel, C.; Sassolas, B. Personal vs. intrinsic melanoma risk awareness: Results of the EDIFICE Melanoma survey. *J. Eur. Acad. Dermatol. Venereol.* **2015**, *29*, 31–34. [CrossRef] [PubMed]

38. Hall, H.I.; Jorgensen, C.M.; McDavid, K.; Kraft, J.M.; Breslow, R. Protection from sun exposure in US white children ages 6 months to 11 years. *Public Health Rep.* **2001**, *116*, 353–361. [CrossRef]

39. Stanganelli, I.; Naldi, L.; Cazzaniga, S.; Gandini, S.; Magi, S.; Quaglino, P.; Ribero, S.; Simonacci, M.; Pizzichetta, M.A.; Spagnolo, F.; et al. Sunburn-related variables, secular trends of improved sun protection and short-term impact on sun attitude behavior in Italian primary schoolchildren: Analysis of the educational campaign "Il Sole Amico" ("The sun as a friend"). *Medicine* **2020**, *99*, e18078. [CrossRef]

40. Black, C.; Grise, K.; Heitmeyer, J.; Readdick, C.A. Sun Protection: Knowledge, Attitude, and Perceived Behavior of Parents and Observed Dress of Preschool Children. *Fam. Consum. Sci. Res. J.* **2001**, *30*, 93–109. [CrossRef]

41. Fitzpatrick, T.B. The validity and practicality of sun-reactive skin Type-I through Type-VI. *Arch. Dermatol.* **1988**, *124*, 869–871. [CrossRef]

42. Køster, B.; Meyer, M.K.H.; Andersson, T.M.-L.; Engholm, G.; Dalum, P. Sunbed use 2007-2015 and skin cancer projections of campaign results 2007-2040 in the Danish population: Repeated cross-sectional surveys. *BMJ Open* **2018**, *8*, e022094. [CrossRef] [PubMed]

43. Criado, P.R.; Ocampo-Garza, J.; Brasil, A.L.D.; Belda, W., Jr.; Di Chiacchio, N.; de Moraes, A.M.; Parada, B.M.; Rabay, F.O.; Moraes, O., Jr.; Rios, R.S.; et al. Skin cancer prevention campaign in childhood: Survey based on 3676 children in Brazil. *J. Eur. Acad Dermatol. Venereol.* **2018**, *32*, 1272–1277. [CrossRef] [PubMed]

44. Thornton, C.M.; Piacquadio, D.J. Promoting sun awareness: Evaluation of an educational children's book. *Pediatrics* **1996**, *98*, 52–55. [PubMed]

45. Fiessle, C.; Pfahlberg, A.B.; Uter, W.; Gefeller, O. Shedding light on the Shade: How Nurseries Protect their children from Ultraviolet radiation. *Int. J. Environ. Res. Public Health* **2018**, *15*, 1793. [CrossRef] [PubMed]

46. Perez, M.; Donaldson, M.; Jain, N.; Robinson, J.K. Sun Protection Behaviors in Head Start and Other Early Childhood Education Programs in Illinois. *JAMA Dermatol.* **2018**, *154*, 336–340. [CrossRef]
47. Parisi, A.V.; Turnbull, D.J. Shade provision for UV minimization: A review. *Photochem. Photobiol.* **2014**, *90*, 479–490. [CrossRef]
48. Available online: https://www.cancercouncil.com.au/wp-content/uploads/2011/04/Guidelines_to_shade_WEB2.pdf (accessed on 10 July 2021).
49. Available online: https://www.cdc.gov/cancer/skin/pdf/shade_planning.pdf (accessed on 9 July 2021).
50. O'Riordan, D.L.; Geller, A.C.; Brooks, D.R.; Zhang, Z.; Miller, D.R. Sunburn reduction through parental role modeling and sunscreen vigilance. *J. Pediatr.* **2003**, *142*, 67–72. [CrossRef] [PubMed]
51. Saraiya, M.; Glanz, K.; Briss, P.; Nichols, P.; White, C.; Das, D. Preventing skin cancer: Findings of the Task Force on Community Preventive Services On reducing Exposure to Ultraviolet Light. *MMWR Recomm. Rep.* **2003**, *17*, 1–12. [CrossRef]

MDPI

Review

Beyond Nicotinamide Metabolism: Potential Role of Nicotinamide *N*-Methyltransferase as a Biomarker in Skin Cancers

Roberto Campagna [1,†], Valentina Pozzi [1,†], Davide Sartini [1,*], Eleonora Salvolini [1], Valerio Brisigotti [2], Elisa Molinelli [2], Anna Campanati [2], Annamaria Offidani [2] and Monica Emanuelli [1,3]

[1] Department of Clinical Sciences, Polytechnic University of Marche, 60100 Ancona, Italy; r.campagna@univpm.it (R.C.); v.pozzi@staff.univpm.it (V.P.); e.salvolini@univpm.it (E.S.); m.emanuelli@univpm.it (M.E.)
[2] Department of Clinical and Molecular Sciences, Polytechnic University of Marche, 60100 Ancona, Italy; valeriobrisigotti@hotmail.it (V.B.); e.molinelli@pm.univpm.it (E.S.); a.campanati@univpm.it (A.C.); a.m.offidani@univpm.it (A.O.)
[3] New York-Marche Structural Biology Center (NY-MaSBiC), Polytechnic University of Marche, 60131 Ancona, Italy
[*] Correspondence: d.sartini@univpm.it; Tel.: +39-071-2204676
[†] These authors contributed equally to this work.

Citation: Campagna, R.; Pozzi, V.; Sartini, D.; Salvolini, E.; Brisigotti, V.; Molinelli, E.; Campanati, A.; Offidani, A.; Emanuelli, M. Beyond Nicotinamide Metabolism: Potential Role of Nicotinamide *N*-Methyltransferase as a Biomarker in Skin Cancers. *Cancers* **2021**, *13*, 4943. https://doi.org/10.3390/cancers13194943

Academic Editor: Francesca Ricci

Received: 23 August 2021
Accepted: 28 September 2021
Published: 30 September 2021

Publisher's Note: MDPI stays neutral with regard to jurisdictional claims in published maps and institutional affiliations.

Simple Summary: Skin cancers (SC) are a frequent type of malignancy in white populations and include malignant melanoma and non-melanoma skin cancer. Due to their increasing incidence rate worldwide, aggressive behavior, and usually late diagnosis, they represent an important challenge for health care systems. Therefore, identifying new biomarkers suitable for diagnosis, as well as for prognosis and targeted therapy is mandatory. Nicotinamide *N*-methyltransferase (NNMT) is an enzyme that plays a key role in the progression of several malignancies. There is increasing evidence that NNMT is also involved in the malignant behavior of SC. Therefore, this review aims to summarize the current state of the art regarding NNMT role in SC and to support future studies focused on exploring the diagnostic and prognostic potential of NNMT in skin malignancies, as well as its suitability for targeted therapy.

Abstract: Skin cancers (SC) collectively represent the most common type of malignancy in white populations. SC includes two main forms: malignant melanoma and non-melanoma skin cancer (NMSC). NMSC includes different subtypes, namely, basal cell carcinoma (BCC), squamous cell carcinoma (SCC), Merkel cell carcinoma (MCC), and keratoacanthoma (KA), together with the two pre-neoplastic conditions Bowen disease (BD) and actinic keratosis (AK). Both malignant melanoma and NMSC are showing an increasing incidence rate worldwide, thus representing an important challenge for health care systems, also because, with some exceptions, SC are generally characterized by an aggressive behavior and are often diagnosed late. Thus, identifying new biomarkers suitable for diagnosis, as well as for prognosis and targeted therapy is mandatory. Nicotinamide *N*-methyltransferase (NNMT) is an enzyme that is emerging as a crucial player in the progression of several malignancies, while its substrate, nicotinamide, is known to exert chemopreventive effects. Since there is increasing evidence regarding the involvement of this enzyme in the malignant behavior of SC, the current review aims to summarize the state of the art as concerns NNMT role in SC and to support future studies focused on exploring the diagnostic and prognostic potential of NNMT in skin malignancies and its suitability for targeted therapy.

Keywords: skin cancer; melanoma; non-melanoma skin cancer; nicotinamide *N*-methyltransferase; biomarker

1. Introduction

Skin cancers (SC) are the neoplasms with the highest incidence in white populations, and their incidence has gradually intensified in the last decade [1,2]. The term "skin cancer" identifies two main forms, namely, malignant melanoma and non-melanoma skin cancer (NMSC). NMSCs include several subtypes, such as basal cell carcinoma (BCC), squamous cell carcinoma (SCC), Merkel cell carcinoma (MCC), and the two pre-neoplastic conditions Bowen disease (BD) and actinic keratosis (AK) [1,3].

Among all SC, malignant melanoma, that arises from altered pigment cells i.e., the melanocytes, is considered to be the most aggressive type [4]. In fact, while malignant melanoma represents only 1% of all SC, it is responsible for the majority of SC-related deaths [5]. Although the predominant part of malignant melanomas involves the skin, in particular, 25% of cutaneous melanomas affects the head and neck, the neoplasm may also arise in mucosal surfaces, the meninges, and the uveal tract [6]. Due to its aggressiveness, it is extremely important to diagnose malignant melanoma when it is in early stages, since the 5-year survival rate is 99% if the disease is diagnosed when still localized, while it drops to 27% if the disease is already metastatic at the time of diagnosis [5,7]. Several risk factors have been associated with malignant melanoma development such as a family history of SC, male sex, fair skin, amount of moles, and age, while the main environmental risk factor is ultraviolet (UV) exposure [7–13]. Indeed, cutaneous melanoma develops primarily in Caucasian people as the consequence of chronic sun exposure [14,15].

A diagnosis of malignant melanoma is facilitated by the ABCDEF criteria, which include lesion Asymmetry, Border irregularity, Color variegation, Diameter > 6 mm, Evolution of a nevus, and a nevus characteristic of Looking Funny, describing a malignant nevus that does not match in appearance with the other nevi variants displayed by a patient [16]. Upon diagnosis, the stage of malignant melanoma is identified considering the rules created by the American Joint Committee on Cancer (AJCC) to manage patient treatment and prognosis. Following these rules, melanomas are classified into five distinct stages, from 0 (melanoma in situ) to IV (metastatic melanoma), distinguished by a worsening prognosis [17]. A variation of the classical TNM system is used by AJCC criteria to characterize melanoma (from early-stage to late-stage) by analyzing the tumor thickness with or without ulceration, nodal involvement, and presence of metastasis [17]. Once diagnosed, surgical resection represents the best opportunity for the definitive cure of a primary melanoma. Other therapeutical options include radiotherapy, chemotherapy, immunotherapy, and targeted therapies [7,18–23]. However, while surgery can achieve 99% of success if the diagnosed melanoma is in situ, melanomas at advanced stages are problematic to be treated [24]. This occurs due to the presence of metastases (lymph nodes, lungs, brain, liver, and bone are the most frequent metastatic sites), to an intrinsic resistance towards most of the therapies available nowadays, and to the high genomic heterogeneity that characterizes melanocytic tumors [25]. In this regard, the identification of novel biomarkers that could be used as prognostic or predictive markers and as objectives of targeted therapies is of utmost importance. Although, in the last years, several biomarkers have been proposed (e.g., microphthalmia-associated transcription factor, cyclooxygenase-2, chondroitin sulfate proteoglycan 4, human melanoma black-45), none of them has become of routine use in the clinical practice, with the exception of BRAF and MEK that are targets of specific inhibitors used with success in the clinical practice, but to only a small subset of patients responds [26].

BCCs arise from basal keratinocytes and account for approximately 75–80% of NMSCs but they are characterized by a more benign behavior, having a very limited metastatic potential [27]. However, it is the most common malignancy in humans, and several histological subtypes have been described, each of them characterized by different clinical features, outcomes, and prognosis [28,29]. The nodular subtype is the most common type, distinguished by the presence of large nodules of tumor cells within the dermis, and represents a low-risk type.

On the contrary, infiltrating BCC is a more aggressive variant, consisting of narrow tumor cords and nests of atypical basaloid cells with an infiltrative growth pattern. Since it displays a high risk of recurrence, this variant requires a more careful approach with an accurate evaluation of its surgical margins [3]. Regardless of the subtype, due to the fact that aging is one the main risk factors, and given that the global population age is increasing, an increase in its related morbidity and local recurrence rates is expected [30].

SCC develops from stratum spinosum keratinocytes and is classified as well, moderately and poorly differentiated. Analogously to BCC, the increase in the average population age and in the exposure to UV light is resulting in an increase in SCC diagnosis [31,32]. SCC occurs often in the head and neck region, where it emerges as an erythematous scaling nodule and plaque, eventually ulcerated. SCC can be classified in several subtypes that differ for their histological appearance and prognosis. The most common subtype displays atypical keratinocytes invading the dermis [33,34]. Depth of invasion (tumor thickness >2 mm), tumor size (diameter >2 cm), acantholysis, perineural and vascular involvement are considered negative prognostic factors. Furthermore, an association between the site of the primary tumor and prognosis was demonstrated, since head and neck SCCs are more prone to metastasize compared to tumors that arises on the extremities or trunk [35,36]. Indeed, despite the fact that surgical excision is curative for most of SCCs, a subset of patients will undergo relapse and eventually will develop metastasis. Hence, the identification of biomarkers with a prognostic value that could support SCC treatment and promote an adjuvant targeted therapy is a primary goal.

Keratoacanthoma (KA) is a malignancy displaying a bi-phasic growth pattern characterized by a fast growing phase generally followed by involution. As most of the other SCs, UV exposure is a risk factor, since KA mostly occurs on sun-damaged skin [37]. There is not a consensus among authors on whether KA is a variant of SCC or a separate entity; however, a certain differential diagnosis between these two malignancies is of primary need, since KA, unlike SCC, is characterized by a good prognosis due to its inclination to spontaneous involution [37,38].

As concerns BD, it is believed to be an in situ SCC, whereas AK can be considered a precancerous lesion which may develop in SCC. Although both diseases exhibit a close association with SCC, they are characterized by different histopathological characteristics [39].

MCC is usually present as erythematous nodule characterized by rapid growth. In the past, MCC was considered to derive from skin Merkel cells, and this explains its name. However, nowadays it is believed that MCC arises from skin precursors of epithelial, lymphoid, or fibroblastic type. Although it represents <1% of all NMSCs, it displays a very aggressive behavior reflected by the presence of clinical or pathological node disease in up to 48% of the patients at diagnosis, while 10% of them already display a metastatic stage at diagnosis [40]. The combination of surgery and radiotherapy is considered the first line of treatment; nonetheless, since recurrence rates are high, about 40% of patients will undergo recurrence within 2 years of diagnosis [41].

Since early and accurate diagnosis and prognosis have a crucial impact on the outcome of these diseases, clinical practice is constantly looking for new genetic and molecular markers that could facilitate an early diagnosis or an accurate setting of the prognosis for both cancerous and non-cancerous diseases, in order to reduce morbidity and improve patients' survival [30,42–46]. This is particularly relevant for SC, since the number of new cases is expected greatly increase in the next future due to increasing UV exposure and population age [1,47,48].

2. Nicotinamide (NAM) in SC

NAM is a form of vitamin B3 largely utilized for the management of several chronic dermatoses, which includes rosacea, acne, blistering immune disorders, atopic dermatitis, and cutaneous neoplasms [49]. NAM is the precursor of nicotinamide adenine dinucleotide (NAD$^+$), a co-enzyme of redox reactions crucial for the production of adenosine triphosphate (ATP). For this reason, it is a master influencer of cellular metabolism, regulating multiple pathways involved in both cellular survival and apoptosis [50].

UV exposure induces damage of cellular DNA, triggering strand breaks, crosslinks, and base modifications that are repaired by various repair systems, which in turn consume ATP [51,52]. Therefore, UV exposure leads to ATP consumption, which in turn induces a kind of cellular energy crisis, and, since repair systems need high levels of ATP to work properly, the consequence is an accumulation of molecular aberrations and genome instability [53]. Therefore, it was reported that NAM exerts UV protective effects due its involvement in cellular energy pathways, as a precursor of NAD^+ [54]. In detail, UV irradiation causes the block of glycolysis by activating poly-ADP-ribose-polymerase 1 (PARP-1), and this event ultimately inhibits NAD^+ production [55,56]. Furthermore, NAM inhibits skin carcinogenesis regulating the proteins p53 and sirtuins. Indeed, when DNA damage is too extensive, p53 is activated by NAD^+ and triggers cell cycle arrest and apoptosis. Since NAM is able to restore the intracellular levels of NAD^+, it was reported that NAM can modulate the p53 pathway [57]. Moreover, NAM was proposed to be a negative regulator of SIRT1, a NAD^+-dependent histone deacetylase able to inhibit p53, thus preventing apoptosis [58,59].

NAM was demonstrated to play a key role in preventing skin carcinogenesis by counteracting UV-induced immunosuppression. Indeed, UVB-induced DNA damage stimulates cutaneous antigen-presenting cells (APC) to produce interleukin (IL)-10, which downregulates the immune response [60]. According to this hypothesis, the topic use of NAM in UV-irradiated mice could counteract the cutaneous carcinogenesis process. NAM supplementation through the diet was able to decrease SC incidence in UV-exposed mice with a dose-dependent effect [61]. Another study demonstrated that a group of people taking oral nicotinamide had a minor diminution of delayed-type hypersensitivity when exposed to UV light for three consecutive days compared to the placebo group. The protective effect of NAM was ascribable to its activity of counteracting immunosuppression by restoring the sufficient energy levels demanded by cells for repairing DNA damage and preventing PARP overactivation [62].

In order to evaluate the chemopreventive effect of NAM in high-risk patients, the Phase III double-blind ONTRAC Study was designed [63]. In this study, patients were selected for having had two or more NMSC in the last 5 years. In the period of treatment, which lasted 12 months, the mean incidence of NMSC observed in the nicotinamide-treated group was 1.8, while that in the placebo group was 2.4. In particular, the mean number of BCCs was 1.3 in the nicotinamide-treated group and 1.9 in the placebo group, with a smaller rate of 20% after adjustment for medical center of treatment and 5-year BCC history. The mean number of SCCs was 0.5 in the nicotinamide-treated group and 0.7 in the placebo group, with a smaller rate of 30% after adjustment for center and 5-year SCC history [63].

As regards malignant melanoma, studies performed in vitro demonstrated that NAM is able to improve the repair rate of the nucleotide excision system and can increase the percentage of melanocytes undergoing DNA repair after UV exposure [64]. Subsequent in vitro studies on melanoma cell lines demonstrated that NAM supplementation, equal to the orally administered doses utilized in the ONTRAC study, did not boost cell viability, proliferation, or invasiveness. Nonetheless, NAM was able to induce an immune response directed to the existing melanomas in vivo [65]. Taken together, these findings led to the hypothesis that oral NAM does not worsen melanoma pathogenesis but, on the contrary, might be useful in melanoma chemoprevention.

3. Nicotinamide *N*-Methyltransferase

Nicotinamide *N*-methyltransferase (NNMT) is an enzyme that catalyzes the *N*-methylation of nicotinamide, using *S*-adenosyl-L-methionine (SAM) as a methyl donor, thus yielding *N*1-methylnicotinamide (MNA) as a product and releasing *S*-adenosyl-L-homocysteine (SAH) [66]. Since it can also methylate other pyridines and other structural analogs, it plays a pivotal role not only in nicotinamide homeostasis but also in the biotransformation and detoxification of several xenobiotic compounds [67,68]. Furthermore, it was demonstrated that NNMT takes part also in several crucial metabolic pathways.

Nicotinamide is the precursor of nicotinamide adenine dinucleotide (NAD$^+$), a co-enzyme of redox reactions required for ATP production. Therefore, the amount of nicotinamide inside the cell available for energy metabolism can be regulated by NNMT activity (Figure 1), and therefore, the catalytic activity of the enzyme can affect and modulate multiple pathways of cellular survival and apoptosis [50]. In addition, by influencing the SAM/SAH ratio inside the cell, it can indirectly impact gene expression [69]. In the last two decades, NNMT has been the focus of a number of studies that demonstrated the involvement of this enzyme in the progression of numerous malignancies including oral squamous cell carcinoma (OSCC), papillary thyroid cancer, lung cancer, gastric cancer, pancreatic cancer, colorectal cancer, clear cell renal cell carcinoma (ccRCC), breast cancer, bladder urothelial carcinoma (BUC), and ovarian clear cell carcinoma [70–85].

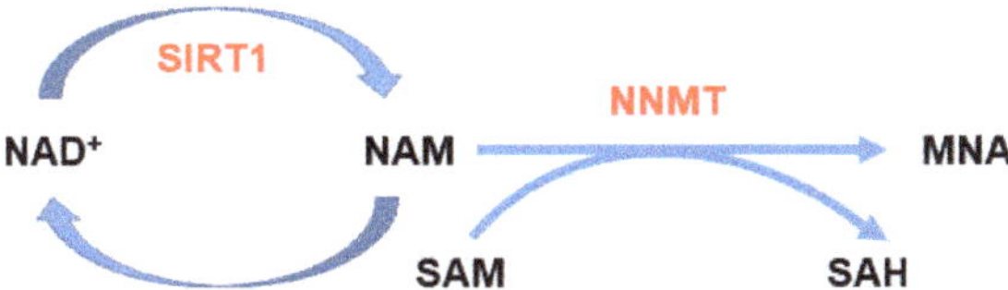

Figure 1. Nicotinamide (NAM) metabolism. NAM can be methylated by nicotinamide *N*-methyltransferase (NNMT) utilizing *S*-adenosyl-L-methionine (SAM) as a methyl donor, which in turn is converted to *S*-adenosyl-L-homocysteine (SAH). NNMT activity can affect NAD$^+$ biosynthesis and thus ATP production, since it can regulate the amount of NAM converted into NAD$^+$. Furthermore, by modulating the intracellular SAM/SAH ratio, it can indirectly impact gene expression.

The analysis of NNMT expression levels in ccRCC demonstrated that the amount of upregulated enzyme is inversely correlated with tumor size, suggesting that the enzyme could play a role in cancer progression [86]. Similar results were obtained in OSCC, for which NNMT upregulation was negatively correlated with the parameters pT, lymph node metastasis, pathological and histological grading; this evidence led to hypothesize its potential involvement in tumor growth and differentiation [87,88].

NNMT expression levels were also found to be notably upregulated in exfoliated cells isolated from the urine of BUC patients compared to that of controls, and an inverse correlation between enzyme expression and histological grade was demonstrated, an observation that suggested the remarkable diagnostic accuracy of a urine test based on the detection of NNMT levels [83]. Consistently with these findings, an increased level of NNMT was detected in saliva samples of OSCC patients compared to controls, a finding that suggested the use of NNMT as a salivary biomarker for the early and non-invasive diagnosis of oral cancer [89]. Taken together, these studies demonstrate that the NNMT enzyme has a remarkable potential as a diagnostic and prognostic biomarker in a wide spectrum of malignancies.

4. Involvement of NNMT in SC

Given the increasing evidence that NNMT relevantly contributes to cancer progression, several studies have been performed in order to explore its potential involvement also in SC. A summary of the results of all studies is reported in Table 1.

Table 1. Summary of studies exploring the role of NNMT in SC.

Type of Skin Cancer	Diagnostic Potential	Prognostic Potential	Therapeutical Target	Reference
Cutaneous malignant melanoma	Yes	Yes (inverse correlation)	N.A.	[89]
Cutaneous malignant melanoma	Yes	Yes (inverse correlation)	N.A.	[90]
Oral malignant melanoma	No	Yes (positive correlation)	N.A.	[90]
Human malignant melanoma cell lines	N.A.	N.A.	Yes	[91]
Basal cell carcinoma	Yes	Yes (inverse correlation)	N.A.	[92]
Squamous cell carcinoma	No	Yes (inverse correlation)	N.A.	[92]
Keratoacanthoma	Yes	Yes (inverse correlation)	N.A.	[93]
Human skin squamous carcinoma cell lines SCC12/13	N.A.	Yes	N.A.	[94]

Ganzetti et al. were the first authors to investigate the role of NNMT in malignant melanoma. In this retrospective study, a total of 34 primary melanomas and 34 melanocytic non-congenital non-atypical compound and dermal nevi, used as the control group, were analyzed by immunohistochemistry. In this work, a significantly higher NNMT expression level was found in cutaneous malignant melanoma samples compared to benign nevi [90]. An analysis of NNMT expression in melanoma samples from the Pan-Cancer Analysis of Whole Genomes (PCAWG) (https://www.ebi.ac.uk/gxa/home) (accessed on 14 September 2021) confirmed these findings, showing that the enzyme displayed an expression level of 26 transcript per million (TPM), while in other types of cancers, characterized by a marked overexpression of NNMT, such as bladder, lung, and breast cancer, an expression level of 21, 161, and 71 TPM, respectively, was detected. These results demonstrate that NNMT overexpression in melanoma is remarkable. Furthermore, Ganzetti et al. demonstrated that the NNMT levels measured in the melanoma samples resulted to be inversely correlated to Breslow thickness, Clark level, the presence and number of mitoses, and ulceration, thus suggesting that the enzyme has a good potential to be used as a prognostic biomarker [90]. A subsequent immunohistochemical study regarding NNMT expression levels in cutaneous melanoma confirmed these findings [91]. In the same study, the authors also analyzed the enzyme expression level in samples of patients with oral malignant melanoma, an exceptionally rare and aggressive variant of the neoplasm of the head and neck region, which notoriously displays a poor prognosis. The immunohistochemical analysis revealed that NNMT expression was significantly higher in cutaneous malignant melanoma samples, but oral malignant melanoma samples exhibited more strongly stained cells, thus suggesting a potential involvement of NNMT in oral malignant melanoma. Furthermore, the findings presented in this work indicated that NNMT levels, measured in both oral malignant melanoma and cutaneous melanoma samples, showed a potential association with the presence of ulcers, which was contrasting in the two neoplasms, since the staining intensity was higher in ulcerated oral malignant melanoma samples, while the ulcerated cases of cutaneous melanoma displayed a reduction of NNMT levels. Finally, statistical analysis revealed an inverse correlation between the percentage of NNMT-positive cells in the tumor samples and the disease-free survival time in oral malignant melanoma patients, indicating that NNMT could be an efficient prognostic factor for this malignancy [91].

In another study, the functional role of NNMT was investigated through shRNA-mediated silencing of the enzyme in the melanoma cell lines A375 and WM-115 [92]. Following NNMT knockdown, cell proliferation, migration, and chemosensitivity were evaluated. The data obtained revealed that enzyme silencing triggered a significant reduction of cell proliferation and migration in A375 melanoma cells. Furthermore, enzyme downregulation sensitized melanoma cells to the chemotherapeutic dacarbazine. In addition, similar effects on cell proliferation and chemosensitivity were obtained in WM-115 melanoma cells, upon enzyme silencing. These findings led to the hypothesis that NNMT might be involved in promoting mechanisms of chemoresistance. A subsequent study explored the role of NNMT also in NMSC. A total of 79 specimens (40 BCC and 39 SCC cases) were subjected to immunohistochemical analysis to evaluate the enzyme expression levels, with the healthy tissue margins used as a control [94]. In the BCC cohort, NNMT expression was significantly higher in tumor specimens than in normal tissue margins. Interestingly, immunopositivity was higher in nodular BCC compared to the infiltrative BCC subtype. This expression pattern was also exhibited by BCCs displaying both nodular and infiltrative features within a single tumor. Therefore, the obtained data suggest an inverse correlation between NNMT expression and tumor aggressiveness [94].

Regarding SCC, the study analyzed both tumor samples from the head and neck region and lesions affecting the rest of the body. Unexpectedly, the findings reported in this study showed a significant lower NNMT expression in cancer cells compared to healthy margin tissues. Interestingly, the fraction of immuno-positive cells was markedly higher in SCC specimens excised from extremities and trunk compared to specimens from the head and neck, thus reinforcing the hypothesis of an inverse correlation between enzyme expression and tumor aggressiveness [94]. Altogether, these findings suggest that NNMT may be a potential prognostic biomarker for these neoplasms.

A subsequent study analyzed differences in protein expression between the human skin squamous carcinoma cell lines SCC12 and SCC13, with the aim to identify which genes determine the high invasive potential displayed by the SCC12 cell line compared with the poorly invasive SCC13 cell line [93].

Using matrix-assisted laser desorption/ionization time-of-flight mass spectrometry (MALDI–TOF-MS), the authors identified NNMT as an upregulated protein in the SCC12 cell line. Therefore, shRNA silencing of the enzyme was performed in order to evaluate the impact of NNMT knockdown on cell proliferation, migration, and invasion. NNMT downregulation strongly inhibited the proliferation and density-dependent growth of SCC cells, as well as their migration and invasion. Furthermore, the impact of NNMT knockdown on epithelial–mesenchymal transition (EMT)-associated gene expression was investigated through the RT^2 Profiler PCR Array. The results showed that NNMT silencing was able to downregulate 10 of the 84 EMT-related genes analyzed, namely the genes coding for MMP9, SPP1, and versican core protein (VCAN), which play a role in the modulation of extracellular matrix (ECM) structure and function. Furthermore, the mRNA expression level of Slug, a key effector of EMT, was also repressed by NNMT silencing. In the light of the above-mentioned findings, NNMT was proposed as a novel prognostic biomarker and therapeutic target for patients with SCC [93].

Another study evaluated the differential expression of NNMT in cutaneous KA and SCC on 48 samples through immunohistochemistry [95]. The reported results demonstrated a significantly higher NNMT expression level in KA compared to SCC. In detail, the percentage of NNMT-positive cells was significantly lower in head and neck SCC compared to SCC samples from the rest of the body. It is noteworthy that, according to previous studies, tumors with a less favorable prognosis displayed reduced NNMT levels [86,87,90,94]. Since KA and well-differentiated SCC are difficult to be distinguished from a histopathological perspective, the observed differences in NNMT expression may be exploited to perform a prompt differential diagnosis between these two pathological conditions. Indeed, while KA is characterized by an excellent prognosis due to its natural tendency to involute, SCC is characterized by a very aggressive behavior. Therefore, these findings reinforce the idea that NNMT may be a novel biomarker suitable for both the early diagnosis and prognosis of these neoplasms, and for designing targeted therapeutic strategies.

5. Conclusions and Future Perspectives

Even though the contribution of NNMT to the cancer progression was demonstrated in a large number of malignancies, the effective role of this enzyme in cancer cells still needs to be fully elucidated.

The above-mentioned studies demonstrated that NAM exerts a positive role in counteracting carcinogenesis. On the other hand, the enzyme NNMT, the master regulator of intracellular NAM, seems to be involved in tumor progression. Notably, given the chemopreventive role of NAM, it is conceivable that NNMT may exert a primary role in the first step of carcinogenesis, irreversibly methylating NAM, thus generating MNA. Since overexpression of NNMT was reported in most SC, the enzyme activity may determine a drop in the intracellular levels of NAM, resulting in UV sensitization of cells and impairing the mechanisms involved in cycle cell arrest and DNA repair, as discussed above. All these events may be responsible for neoplastic cell transformation over time. In this regards, further studies are required in order to elucidate whether NNMT overexpression is responsible for the neoplastic transformation of cells or whether it is a consequence of the altered gene expression pattern of the neoplastic cell.

Nevertheless, it clearly appears that NNMT might be an excellent candidate as a diagnostic and prognostic marker in skin cancers. The studies performed to date are promising, but further analyses are required in order to widen the cohort of patients taken into consideration, thus confirming the suitability of the enzyme as a biomarker in the clinical practice.

A large number of studies were focused on exploring the impact of NNMT downregulation in several cancer models, leading to the discovery that the suppression of this enzyme prevents cancer cell proliferation, invasion, and metastasis, as well as chemoresistance [50,92]. Moreover, it was suggested that the enzyme may contribute to the radioresistance of cancer cells [96,97]. In the light of the above-mentioned considerations, NNMT can also be considered a promising molecule for targeted therapy.

One of the encouraging frontiers that has recently drawn much attention is the development of specific inhibitors of the enzyme, which are providing encouraging results [98–103]. It remains to be seen whether the strong preclinical evidence of small-molecule inhibitors against NNMT could still be translated in clinical practice for patients' treatment. Therefore, appropriate studies should be performed in this direction.

Author Contributions: Conceptualization, R.C., M.E., A.O.; software, V.B., E.M.; resources, E.S., A.O., M.E., D.S.; writing—original draft preparation, R.C., V.P.; writing—review and editing, R.C., D.S., A.C., A.O., M.E.; visualization, R.C., V.B., E.M.; supervision, A.C., A.O., M.E.; project administration, E.S.; funding acquisition, E.S., A.O., M.E. All authors have read and agreed to the published version of the manuscript.

Funding: This research received no external funding.

Conflicts of Interest: The authors declare no conflict of interest.

References

1. Leiter, U.; Keim, U.; Garbe, C. Epidemiology of Skin Cancer: Update 2019. *Adv. Exp. Med. Biol.* **2020**, *1268*, 123–139. [CrossRef]
2. Park, Y.J.; Kwon, G.H.; Kim, J.O.; Kim, N.K.; Ryu, W.S.; Lee, K.S. A retrospective study of changes in skin cancer characteristics over 11 years. *Arch. Craniofacial Surg.* **2020**, *21*, 87–91. [CrossRef]
3. Samarasinghe, V.; Madan, V. Nonmelanoma skin cancer. *J. Cutan. Aesthet. Surg.* **2012**, *5*, 3–10. [CrossRef]
4. Isola, A.L.; Eddy, K.; Chen, S. Biology, Therapy and Implications of Tumor Exosomes in the Progression of Melanoma. *Cancers* **2016**, *8*, 110. [CrossRef]
5. Siegel, R.L.; Miller, K.D.; Fuchs, H.E.; Jemal, A. Cancer Statistics, 2021. *CA Cancer J. Clin.* **2021**, *71*, 7–33. [CrossRef] [PubMed]
6. Rigel, D.S.; Carucci, J.A. Malignant melanoma: Prevention, early detection, and treatment in the 21st century. *CA Cancer J. Clin.* **2000**, *50*, 215–236, quiz 237–240. [CrossRef]

7.   Matthews, N.H.; Li, W.Q.; Qureshi, A.A.; Weinstock, M.A.; Cho, E. Epidemiology of Melanoma. In *Cutaneous Melanoma: Etiology and Therapy*; Ward, W.H., Farma, J.M., Eds.; Codon Publications: Brisbane, Australia, 2017.

8.   White, L.P. Studies on melanoma. II. Sex and survival in human melanoma. *Plast. Reconstr. Surg.* **1959**, *24*, 224. [CrossRef]

9.   Scoggins, C.R.; Ross, M.I.; Reintgen, D.S.; Noyes, R.D.; Goydos, J.; Beitsch, P.D.; Urist, M.M.; Ariyan, S.; Sussman, J.J.; Edwards, M.J.; et al. Gender-Related Differences in Outcome for Melanoma Patients. *Ann. Surg.* **2006**, *243*, 693–700. [CrossRef] [PubMed]

10. Shain, A.H.; Bastian, B.C. From melanocytes to melanomas. *Nat. Rev. Cancer* **2016**, *16*, 345–358. [CrossRef] [PubMed]

11. Holly, E.A.; Kelly, J.W.; Shpall, S.N.; Chiu, S.H. Number of melanocytic nevi as a major risk factor for ma-lignant melanoma. *J. Am. Acad. Dermatol.* **1987**, *17*, 459–468. [CrossRef]

12. Alexandrov, L.B.; Nik-Zainal, S.; Wedge, D.C.; Aparicio, S.A.; Behjati, S.; Biankin, A.V.; Bignell, G.R.; Bolli, N.; Borg, A.; Borresen-Dale, A.L.; et al. Signatures of mutational processes in human cancer. *Nature* **2013**, *500*, 415–421. [CrossRef]

13. Ford, D.; Bliss, J.M.; Swerdlow, A.J.; Armstrong, B.K.; Franceschi, S.; Green, A.; Holly, E.A.; Mack, T.; Mackie, R.M.; Østerlind, A.; et al. Risk of cutaneous melanoma associated with a family history of the disease. *Int. J. Cancer* **1995**, *62*, 377–381. [CrossRef] [PubMed]

14. Crombie, I.K. Variation of melanoma incidence with latitude in North America and Europe. *Br. J. Cancer* **1979**, *40*, 774–781. [CrossRef] [PubMed]

15. Diepgen, T.L.; Drexler, H.; Schmitt, J. Epidemiology of occupational skin cancer due to UV-irradiation. *Hautarzt* **2012**, *63*, 769–777. [CrossRef] [PubMed]

16. Daniel Jensen, J.; Elewski, B.E. The ABCDEF Rule: Combining the "ABCDE Rule" and the "Ugly Duckling Sign" in an Effort to Improve Patient Self-Screening Examinations. *J. Clin. Aesthet. Dermatol.* **2015**, *8*, 15.

17. Gershenwald, J.E.; Scolyer, R.A.; Hess, K.R.; Sondak, V.K.; Long, G.; Rossi, C.R.; Lazar, A.J.; Faries, M.B.; Kirkwood, J.M.; McArthur, G.; et al. Melanoma staging: Evidence-based changes in the American Joint Committee on Cancer eighth edition cancer staging manual. *CA A Cancer J. Clin.* **2017**, *67*, 472–492. [CrossRef]

18. Mishra, H.; Mishra, P.K.; Ekielski, A.; Jaggi, M.; Iqbal, Z.; Talegaonkar, S. Melanoma treatment: From con-ventional to nanotechnology. *J. Cancer Res. Clin. Oncol.* **2018**, *144*, 2283–2302. [CrossRef]

19. Dimitriou, F.; Long, G.; Menzies, A. Novel adjuvant options for cutaneous melanoma. *Ann. Oncol.* **2021**, *32*, 854–865. [CrossRef]

20. Gogas, H.; Bafaloukos, D.; Bedikian, A.Y. The role of taxanes in the treatment of metastatic melanoma. *Melanoma Res.* **2004**, *14*, 415–420. [CrossRef]

21. Rosenberg, S.A.; Lotze, M.T.; Yang, J.C.; Topalian, S.L.; Chang, A.E.; Schwartzentruber, D.J.; Aebersold, P.; Leitman, S.; Linehan, W.M.; Seipp, C.A.; et al. Prospective Randomized Trial of High-Dose Interleukin-2 Alone or in Conjunction With Lymphokine-Activated Killer Cells for the Treatment of Patients With Advanced Cancer. *J. Natl. Cancer Inst.* **1993**, *85*, 622–632. [CrossRef]

22. Eggermont, A.M.; Sileni, V.C.; Grob, J.-J.; Dummer, R.; Wolchok, J.D.; Schmidt, H.; Hamid, O.; Robert, C.; Ascierto, P.A.; Richards, J.M.; et al. Prolonged Survival in Stage III Melanoma with Ipilimumab Adjuvant Therapy. *N. Engl. J. Med.* **2016**, *375*, 1845–1855. [CrossRef] [PubMed]

23. Hodi, F.S.; O'Day, S.J.; McDermott, D.F.; Weber, R.W.; Sosman, J.A.; Haanen, J.B.; Gonzalez, R.; Robert, C.; Schadendorf, D.; Hassel, J.C.; et al. Improved survival with ipilimumab in patients with metastatic mela-noma. *N. Engl. J. Med.* **2010**, *363*, 711–723. [CrossRef] [PubMed]

24. Campagna, R.; Bacchetti, T.; Salvolini, E.; Pozzi, V.; Molinelli, E.; Brisigotti, V.; Sartini, D.; Campanati, A.; Ferretti, G.; Offidani, A.; et al. Paraoxonase-2 Silencing Enhances Sensitivity of A375 Melanoma Cells to Treatment with Cisplatin. *Antioxidants* **2020**, *9*, 1238. [CrossRef] [PubMed]

25. Sabel, M.S.; Liu, Y.; Lubman, D.M. Proteomics in Melanoma Biomarker Discovery: Great Potential, Many Obstacles. *Int. J. Proteom.* **2011**, *2011*, 1–8. [CrossRef]

26. Eddy, K.; Shah, R.; Chen, S. Decoding Melanoma Development and Progression: Identification of Therapeutic Vulnerabilities. *Front. Oncol.* **2020**, *10*, 626129. [CrossRef]

27. Diepgen, T.L.; Mahler, V. The epidemiology of skin cancer. *Br. J. Dermatol.* **2002**, *146*, 1–6. [CrossRef]

28. Apalla, Z.; Nashan, D.; Weller, R.B.; Castellsague, X. Skin Cancer: Epidemiology, Disease Burden, Patho-physiology, Diagnosis, and Therapeutic Approaches. *Dermatol. Ther.* **2017**, *7*, 5–19. [CrossRef]

29. Bacchetti, T.; Salvolini, E.; Pompei, V.; Campagna, R.; Molinelli, E.; Brisigotti, V.; Togni, L.; Lucarini, G.; Sartini, D.; Campanati, A.; et al. Paraoxonase-2: A potential biomarker for skin cancer aggressiveness. *Eur. J. Clin. Investig.* **2020**, *51*, e13452. [CrossRef]

30. Nikolouzakis, T.K.; Falzone, L.; Lasithiotakis, K.; Kruger-Krasagakis, S.; Kalogeraki, A.; Sifaki, M.; Spandidos, D.A.; Chrysos, E.; Tsatsakis, A.; Tsiaoussis, J. Current and Future Trends in Molecular Biomarkers for Diag-nostic, Prognostic, and Predictive Purposes in Non-Melanoma Skin Cancer. *J. Clin. Med.* **2020**, *9*, 2868. [CrossRef]

31. Woo, W.A.; M-Falero, B.S.; Keohane, S.G. Summary of the updated 2020 guidelines for cutaneous squamous cell carcinoma. *Clin. Exp. Dermatol.* **2021**, *46*, 1174–1177. [CrossRef]

32. Parekh, V.; Seykora, J.T. Cutaneous Squamous Cell Carcinoma. *Clin. Lab. Med.* **2017**, *37*, 503–525. [CrossRef]

33. Gurudutt, V.V.; Genden, E.M. Cutaneous squamous cell carcinoma of the head and neck. *J. Skin Cancer* **2011**, *2011*, 502723. [CrossRef]

34. Sartini, D.; Campagna, R.; Lucarini, G.; Pompei, V.; Salvolini, E.; Mattioli-Belmonte, M.; Molinelli, E.; Brisigotti, V.; Campanati, A.; Bacchetti, T.; et al. Differential immunohistochemical expression of paraoxonase-2 in actinic keratosis and squamous cell carcinoma. *Hum. Cell* **2021**, 1–3. [CrossRef]

35. Paolino, G.; Donati, M.; Didona, D.; Mercuri, S.R.; Cantisani, C. Histology of Non-Melanoma Skin Cancers: An Update. *Biomedicines* **2017**, *5*, 71. [CrossRef]

36. Claveau, J.; Archambault, J.; Ernst, D.; Giacomantonio, C.; Limacher, J.; Murray, C.; Parent, F.; Zloty, D. Multidisciplinary Management of Locally Advanced and Metastatic Cutaneous Squamous Cell Carcinoma. *Curr. Oncol.* **2020**, *27*, 399–407. [CrossRef]

37. Nirenberg, A.; Steinman, H.; Dixon, A. Keratoacanthoma: Update on the Debate. *Am. J. Dermatopathol.* **2020**, *43*, 305–307. [CrossRef]

38. Mandrell, J.C.; Santa Cruz, D. Keratoacanthoma: Hyperplasia, benign neoplasm, or a type of squamous cell carcinoma? *Semin. Diagn. Pathol.* **2009**, *26*, 150–163. [CrossRef] [PubMed]

39. Yanofsky, V.R.; Mercer, S.E.; Phelps, R.G. Histopathological Variants of Cutaneous Squamous Cell Carcinoma: A Review. *J. Ski. Cancer* **2010**, *2011*, 1–13. [CrossRef] [PubMed]

40. Angeles, C.V.; Sabel, M.S. Immunotherapy for Merkel cell carcinoma. *J. Surg. Oncol.* **2021**, *123*, 775–781. [CrossRef]

41. Van Veenendaal, L.M.; Van Akkooi, A.C.J.; Verhoef, C.; Grünhagen, D.J.; Klop, W.M.C.; Valk, G.D.; Tesselaar, M.E.T. Merkel cell carcinoma: Clinical outcome and prognostic factors in 351 patients. *J. Surg. Oncol.* **2018**, *117*, 1768–1775. [CrossRef] [PubMed]

42. Altobelli, E.; Latella, G.; Morroni, M.; Licini, C.; Tossetta, G.; Mazzucchelli, R.; Profeta, V.F.; Coletti, G.; Leocata, P.; Castellucci, M.; et al. Low HtrA1 expression in patients with long-standing ulcerative colitis and colorectal cancer. *Oncol. Rep.* **2017**, *38*, 418–426. [CrossRef]

43. Trager, M.H.; Geskin, L.J.; Samie, F.H.; Liu, L. Biomarkers in melanoma and non-melanoma skin cancer prevention and risk stratification. *Exp. Dermatol.* **2020**. [CrossRef] [PubMed]

44. Avellini, C.; Licini, C.; Lazzarini, R.; Gesuita, R.; Guerra, E.; Tossetta, G.; Castellucci, C.; Giannubilo, S.R.; Procopio, A.; Alberti, S.; et al. The trophoblast cell surface antigen 2 and miR-125b axis in urothelial bladder cancer. *Oncotarget* **2017**, *8*, 58642–58653. [CrossRef]

45. Galley, J.D.; Mar, P.; Wang, Y.; Han, R.; Rajab, A.; Besner, G.E. Urine-derived extracellular vesicle miRNAs as possible biomarkers for and mediators of necrotizing enterocolitis: A proof of concept study. *J. Pediatr. Surg.* **2021**. [CrossRef]

46. Gesuita, R.; Licini, C.; Picchiassi, E.; Tarquini, F.; Coata, G.; Fantone, S.; Tossetta, G.; Ciavattini, A.; Castellucci, M.; Di Renzo, G.C.; et al. Association between first trimester plasma htra1 level and subsequent preeclampsia: A possible early marker? *Pregnancy Hypertens.* **2019**, *18*, 58–62. [CrossRef]

47. Feng, S.; Cen, X.; Tan, R.; Wei, S.; Sun, L. The prognostic value of circulating tumor DNA in patients with melanoma: A systematic review and meta-analysis. *Transl. Oncol.* **2021**, *14*, 101072. [CrossRef]

48. Gault, A.; Anderson, A.E.; Plummer, R.; Stewart, C.; Pratt, A.G.; Rajan, N. Cutaneous immune-related ad-verse events in patients with melanoma treated with checkpoint inhibitors. *Br. J. Dermatol.* **2021**, *185*, 263–271. [CrossRef]

49. Chen, A.C.; Damian, D.L. Nicotinamide and the skin. *Australas. J. Dermatol.* **2014**, *55*, 169–175. [CrossRef] [PubMed]

50. Pissios, P. Nicotinamide N -Methyltransferase: More Than a Vitamin B3 Clearance Enzyme. *Trends Endocrinol. Metab.* **2017**, *28*, 340–353. [CrossRef] [PubMed]

51. Hanahan, D.; Weinberg, R.A. Hallmarks of Cancer: The Next Generation. *Cell* **2011**, *144*, 646–674. [CrossRef] [PubMed]

52. Hoeijmakers, J.H. DNA damage, aging, and cancer. *N. Engl. J. Med.* **2009**, *361*, 1475–1485. [CrossRef]

53. Nikas, I.P.; Paschou, S.A.; Ryu, H.S. The Role of Nicotinamide in Cancer Chemoprevention and Therapy. *Biomolecules* **2020**, *10*, 477. [CrossRef]

54. Park, J.; Halliday, G.M.; Surjana, D.; Damian, D.L. Nicotinamide prevents ultraviolet radiation-induced cellular energy loss. *Photochem. Photobiol.* **2010**, *86*, 942–948. [CrossRef]

55. Damian, D.L. Photoprotective effects of nicotinamide. *Photochem. Photobiol. Sci.* **2010**, *9*, 578–585. [CrossRef]

56. Thompson, B.C.; Halliday, G.M.; Damian, D.L. Nicotinamide enhances repair of arsenic and ultraviolet ra-diation-induced DNA damage in HaCaT keratinocytes and ex vivo human skin. *PLoS ONE* **2015**, *10*, e0117491. [CrossRef]

57. McLure, K.G.; Takagi, M.; Kastan, M.B. NAD + Modulates p53 DNA Binding Specificity and Function. *Mol. Cell. Biol.* **2004**, *24*, 9958–9967. [CrossRef] [PubMed]

58. Bitterman, K.J.; Anderson, R.M.; Cohen, H.Y.; Latorre-Esteves, M.; Sinclair, D. Inhibition of Silencing and Accelerated Aging by Nicotinamide, a Putative Negative Regulator of Yeast Sir2 and Human SIRT1. *J. Biol. Chem.* **2002**, *277*, 45099–45107. [CrossRef]

59. Campagna, R.; Mateuszuk, Ł.; Wojnar-Lason, K.; Kaczara, P.; Tworzydło, A.; Kij, A.; Bujok, R.; Mlynarski, J.; Wang, Y.; Sartini, D.; et al. Nicotinamide N-methyltransferase in endothelium protects against oxidant stress-induced endothelial injury. *Biochim. Biophys. Acta Bioenerg.* **2021**, *1868*, 119082. [CrossRef] [PubMed]

60. Vink, A.A.; Moodycliffe, A.M.; Shreedhar, V.; Ullrich, S.E.; Roza, L.; Yarosh, D.B.; Kripke, M.L. The inhibition of antigen-presenting activity of dendritic cells resulting from UV irradiation of murine skin is restored by in vitro photorepair of cyclobutane pyrimidine dimers. *Proc. Natl. Acad. Sci. USA* **1997**, *94*, 5255–5260. [CrossRef] [PubMed]

61. Gensler, H.L. Prevention of photoimmunosuppression and photocarcinogenesis by topical nicotinamide. *Nutr. Cancer* **1997**, *29*, 157–162. [CrossRef]

62. Yiasemides, E.; Sivapirabu, G.; Halliday, G.M.; Park, J.; Damian, D.L. Oral nicotinamide protects against ul-traviolet radiation-induced immunosuppression in humans. *Carcinogenesis* **2009**, *30*, 101–105. [CrossRef] [PubMed]

63. Chen, A.C.; Martin, A.J.; Choy, B.; Fernandez-Penas, P.; Dalziell, R.A.; McKenzie, C.A.; Scolyer, R.A.; Dhillon, H.M.; Vardy, J.L.; Kricker, A.; et al. A Phase 3 Randomized Trial of Nicotinamide for Skin-Cancer Chemo-prevention. *N. Engl. J. Med.* **2015**, *373*, 1618–1626. [CrossRef] [PubMed]

64. Thompson, B.C.; Surjana, D.; Halliday, G.M.; Damian, D.L. Nicotinamide enhances repair of ultraviolet ra-diation-induced DNA damage in primary melanocytes. *Exp. Dermatol.* **2014**, *23*, 509–511. [CrossRef] [PubMed]

65. Malesu, R.; Martin, A.J.; Lyons, J.G.; Scolyer, R.A.; Chen, A.C.; McKenzie, C.A.; Madore, J.; Halliday, G.M.; Damian, D.L. Nicotinamide for skin cancer chemoprevention: Effects of nicotinamide on melanoma in vitro and in vivo. *Photochem. Photobiol. Sci.* **2020**, *19*, 171–179. [CrossRef]

66. Aksoy, S.; Szumlanski, C.; Weinshilboum, R. Human liver nicotinamide N-methyltransferase. cDNA cloning, expression, and biochemical characterization. *J. Biol. Chem.* **1994**, *269*, 14835–14840. [CrossRef]

67. Peng, Y.; Sartini, D.; Pozzi, V.; Wilk, D.; Emanuelli, M.; Yee, V.C. Structural Basis of Substrate Recognition in Human Nicotinamide N-Methyltransferase. *Biochemistry* **2011**, *50*, 7800–7808. [CrossRef]

68. Mateuszuk, L.; Campagna, R.; Kutryb-Zajac, B.; Kus, K.; Slominska, E.M.; Smolenski, R.T.; Chlopicki, S. Re-versal of endothelial dysfunction by nicotinamide mononucleotide via extracellular conversion to nicotin-amide riboside. *Biochem. Pharmacol.* **2020**, *178*, 114019. [CrossRef]

69. Ulanovskaya, O.A.; Zuhl, A.M.; Cravatt, B.F. NNMT promotes epigenetic remodeling in cancer by creating a metabolic methyla-tion sink. *Nat. Chem. Biol.* **2013**, *9*, 300–306. [CrossRef]

70. Seta, R.; Mascitti, M.; Campagna, R.; Sartini, D.; Fumarola, S.; Santarelli, A.; Giuliani, M.; Cecati, M.; Muzio, L.L.; Emanuelli, M. Overexpression of nicotinamide N-methyltransferase in HSC-2 OSCC cell line: Effect on apoptosis and cell proliferation. *Clin. Oral Investig.* **2018**, *23*, 829–838. [CrossRef]

71. Xu, J.; Moatamed, F.; Caldwell, J.S.; Walker, J.R.; Kraiem, Z.; Taki, K.; Brent, G.A.; Hershman, J.M. Enhanced Expression of NicotinamideN-Methyltransferase in Human Papillary Thyroid Carcinoma Cells. *J. Clin. Endocrinol. Metab.* **2003**, *88*, 4990–4996. [CrossRef]

72. Sartini, D.; Seta, R.; Pozzi, V.; Morganti, S.; Rubini, C.; Zizzi, A.; Tomasetti, M.; Santarelli, L.; Emanuelli, M. Role of nicotinamide N-methyltransferase in non-small cell lung cancer: In vitro effect of shRNA-mediated gene silencing on tumourigenicity. *Biol. Chem.* **2015**, *396*, 225–234. [CrossRef]

73. Tomida, M.; Mikami, I.; Takeuchi, S.; Nishimura, H.; Akiyama, H. Serum levels of nicotinamide N-methyltransferase in patients with lung cancer. *J. Cancer Res. Clin. Oncol.* **2009**, *135*, 1223–1229. [CrossRef]

74. Lim, B.-H.; Cho, B.-I.; Na Kim, Y.; Kim, J.W.; Park, S.-T.; Lee, C.-W. Overexpression of nicotinamide N-methyltransferase in gastric cancer tissues and its potential post-translational modification. *Exp. Mol. Med.* **2006**, *38*, 455–465. [CrossRef]

75. Chen, C.; Wang, X.; Huang, X.; Yong, H.; Shen, J.; Tang, Q.; Zhu, J.; Ni, J.; Feng, Z. Nicotinamide N-methyltransferase: A potential biomarker for worse prognosis in gastric carcinoma. *Am. J. Cancer Res.* **2016**, *6*, 649–663.

76. Wu, M.; Hu, W.; Wang, G.; Yao, Y.; Yu, X.-F. Nicotinamide N-Methyltransferase Is a Prognostic Biomarker and Correlated With Immune Infiltrates in Gastric Cancer. *Front. Genet.* **2020**, *11*, 580299. [CrossRef] [PubMed]

77. Rogers, C.D.; Fukushima, N.; Sato, N.; Shi, C.; Prasad, N.; Hustinx, S.R.; Matsubayashi, H.; Canto, M.; Eshle-man, J.R.; Hruban, R.H.; et al. Differentiating pancreatic lesions by microarray and QPCR analysis of pan-creatic juice RNAs. *Cancer Biol. Ther.* **2006**, *5*, 1383–1389. [CrossRef] [PubMed]

78. Yu, T.; Wang, Y.T.; Chen, P.; Li, Y.H.; Chen, Y.X.; Zeng, H.; Yu, A.M.; Huang, M.; Bi, H.C. Effects of nicotina-mide N-methyltransferase on PANC-1 cells proliferation, metastatic potential and survival under metabol-ic stress. *Cell Physiol. Biochem.* **2015**, *35*, 710–721. [CrossRef] [PubMed]

79. Roessler, M.; Rollinger, W.; Palme, S.; Hagmann, M.L.; Berndt, P.; Engel, A.M.; Schneidinger, B.; Pfeffer, M.; Andres, H.; Karl, J.; et al. Identification of nicotinamide N-methyltransferase as a novel serum tumor mark-er for colorectal cancer. *Clin. Cancer Res.* **2005**, *11*, 6550–6557. [CrossRef] [PubMed]

80. Campagna, R.; Cecati, M.; Pozzi, V.; Fumarola, S.; Pompei, V.; Milanese, G.; Galosi, A.B.; Sartini, D.; Emanuelli, M. Involvement of transforming growth factor beta 1 in the transcriptional regulation of nicotinamide N-methyltransferase in clear cell renal cell carcinoma. *Cell. Mol. Biol.* **2018**, *64*, 51–55. [CrossRef]

81. Tang, S.-W.; Yang, T.-C.; Lin, W.-C.; Chang, W.-H.; Wang, C.-C.; Lai, M.-K.; Lin, J.-Y. Nicotinamide N -methyltransferase induces cellular invasion through activating matrix metalloproteinase-2 expression in clear cell renal cell carcinoma cells. *Carcinogenesis* **2010**, *32*, 138–145. [CrossRef]

82. Zhang, J.; Wang, Y.; Li, G.; Yu, H.; Xie, X. Down-Regulation of Nicotinamide N-methyltransferase Induces Apoptosis in Human Breast Cancer Cells via the Mitochondria-Mediated Pathway. *PLoS ONE* **2014**, *9*, e89202. [CrossRef]

83. Pozzi, V.; Di Ruscio, G.; Sartini, D.; Campagna, R.; Seta, R.; Fulvi, P.; Vici, A.; Milanese, G.; Brandoni, G.; Galosi, A.B.; et al. Clinical performance and utility of a NNMT-based urine test for bladder cancer. *Int. J. Biol. Markers* **2017**, *33*, 94–101. [CrossRef]

84. Tsuchiya, A.; Sakamoto, M.; Yasuda, J.; Chuma, M.; Ohta, T.; Ohki, M.; Yasugi, T.; Taketani, Y.; Hirohashi, S. Expression profiling in ovarian clear cell carcinoma: Identification of hepatocyte nuclear factor-1 beta as a molecular marker and a possible molecular target for therapy of ovarian clear cell carcinoma. *Am. J. Pathol.* **2003**, *163*, 2503–2512. [CrossRef]

85. Pozzi, V.; Salvolini, E.; Lucarini, G.; Salvucci, A.; Campagna, R.; Rubini, C.; Sartini, D.; Emanuelli, M. Cancer stem cell enrichment is associated with enhancement of nicotinamide N-methyltransferase expression. *IUBMB Life* **2020**, *72*, 1415–1425. [CrossRef] [PubMed]
86. Sartini, D.; Muzzonigro, G.; Milanese, G.; Pierella, F.; Rossi, V.; Emanuelli, M. Identification of Nicotinamide N-Methyltransferase as a Novel Tumor Marker for Renal Clear Cell Carcinoma. *J. Urol.* **2006**, *176*, 2248–2254. [CrossRef] [PubMed]
87. Sartini, D.; Santarelli, A.; Rossi, V.; Goteri, G.; Rubini, C.; Ciavarella, D.; Muzio, L.L.; Emanuelli, M. Nicotinamide N-Methyltransferase Upregulation Inversely Correlates with Lymph Node Metastasis in Oral Squamous Cell Carcinoma. *Mol. Med.* **2007**, *13*, 415–421. [CrossRef] [PubMed]
88. Emanuelli, M.; Santarelli, A.; Sartini, D.; Ciavarella, M.; Rossi, V.; Pozzi, V.; Rubini, C.; Muzio, L.L. Nicotinamide N-Methyltransferase upregulation correlates with tumour differentiation in oral squamous cell carcinoma. *Histol. Histopathol.* **2010**, *25*. [CrossRef]
89. Sartini, D.; Pozzi, V.; Renzi, E.; Morganti, S.; Rocchetti, R.; Rubini, C.; Santarelli, A.; Muzio, L.L.; Emanuelli, M. Analysis of tissue and salivary nicotinamide N-methyltransferase in oral squamous cell carcinoma: Basis for the development of a noninvasive diagnostic test for early-stage disease. *Biol. Chem.* **2012**, *393*, 505–511. [CrossRef]
90. Ganzetti, G.; Sartini, D.; Campanati, A.; Rubini, C.; Molinelli, E.; Brisigotti, V.; Cecati, M.; Pozzi, V.; Campagna, R.; Offidani, A.; et al. Nicotinamide N-methyltransferase: Potential involvement in cutaneous malignant melanoma. *Melanoma Res.* **2018**, *28*, 82–88. [CrossRef] [PubMed]
91. Mascitti, M.; Santarelli, A.; Sartini, D.; Rubini, C.; Colella, G.; Salvolini, E.; Ganzetti, G.; Offidani, A.; Emanuelli, M. Analysis of nicotinamide N-methyltransferase in oral malignant melanoma and potential prognostic significance. *Melanoma Res.* **2019**, *29*, 151–156. [CrossRef] [PubMed]
92. Campagna, R.; Salvolini, E.; Pompei, V.; Pozzi, V.; Salvucci, A.; Molinelli, E.; Brisigotti, V.; Sartini, D.; Campanati, A.; Offidani, A.; et al. Nicotinamide N-methyltransferase gene silencing enhances chemosensi-tivity of melanoma cell lines. *Pigment Cell Melanoma Res.* **2021**. [CrossRef]
93. Hah, Y.; Cho, H.Y.; Jo, S.Y.; Park, Y.S.; Heo, E.P.; Yoon, T. Nicotinamide N-methyltransferase induces the proliferation and invasion of squamous cell carcinoma cells. *Oncol. Rep.* **2019**, *42*, 1805–1814. [CrossRef] [PubMed]
94. Pompei, V.; Salvolini, E.; Rubini, C.; Lucarini, G.; Molinelli, E.; Brisigotti, V.; Pozzi, V.; Sartini, D.; Campanati, A.; Offidani, A.; et al. Nicotinamide N-methyltransferase in nonmelanoma skin cancers. *Eur. J. Clin. Investig.* **2019**, *49*, e13175. [CrossRef] [PubMed]
95. Sartini, D.; Pompei, V.; Lucarini, G.; Rubini, C.; Molinelli, E.; Brisigotti, V.; Salvolini, E.; Campanati, A.; Offidani, A.; Emanuelli, M. Differential expression of nicotinamide N-methyltransferase in cutaneous keratoacanthoma and squamous cell carcinoma: An immunohistochemical study. *J. Eur. Acad. Dermatol. Venereol.* **2019**, *34*, e121–e123. [CrossRef] [PubMed]
96. Kassem, H.; Sangar, V.; Cowan, R.; Clarke, N.; Margison, G.P. A potential role of heat shock proteins and nicotinamide N-methyl transferase in predicting response to radiation in bladder cancer. *Int. J. Cancer* **2002**, *101*, 454–460. [CrossRef]
97. D'Andrea, F.P.; Safwat, A.; Kassem, M.; Gautier, L.; Overgaard, J.; Horsman, M.R. Cancer stem cell overex-pression of nicotinamide N-methyltransferase enhances cellular radiation resistance. *Radiother. Oncol.* **2011**, *99*, 373–378. [CrossRef]
98. Neelakantan, H.; Wang, H.-Y.; Vance, V.; Hommel, J.D.; McHardy, S.F.; Watowich, S.J. Structure–Activity Relationship for Small Molecule Inhibitors of Nicotinamide N-Methyltransferase. *J. Med. Chem.* **2017**, *60*, 5015–5028. [CrossRef]
99. Gao, Y.; van Haren, M.J.; Moret, E.E.; Rood, J.J.M.; Sartini, D.; Salvucci, A.; Emanuelli, M.; Craveur, P.; Babault, N.; Jin, J.; et al. Bisubstrate Inhibitors of Nicotinamide N-Methyltransferase (NNMT) with Enhanced Activity. *J. Med. Chem.* **2019**, *62*, 6597–6614. [CrossRef]
100. Kannt, A.; Rajagopal, S.; Kadnur, S.V.; Suresh, J.; Bhamidipati, R.K.; Swaminathan, S.; Hallur, M.S.; Kristam, R.; Elvert, R.; Czech, J.; et al. A small molecule inhibitor of Nicotinamide N-methyltransferase for the treatment of metabolic disorders. *Sci. Rep.* **2018**, *8*, 3660. [CrossRef]
101. Policarpo, R.L.; Decultot, L.; May, E.; Kuzmič, P.; Carlson, S.; Huang, D.; Chu, V.; Wright, B.A.; Dhakshinamoorthy, S.; Kannt, A.; et al. High-Affinity Alkynyl Bisubstrate Inhibitors of Nicotinamide N-Methyltransferase (NNMT). *J. Med. Chem.* **2019**, *62*, 9837–9873. [CrossRef]
102. Ruf, S.; Hallur, M.S.; Anchan, N.K.; Swamy, I.N.; Murugesan, K.R.; Sarkar, S.; Narasimhulu, L.K.; Putta, V.P.R.K.; Shaik, S.; Chandrasekar, D.V.; et al. Novel nicotinamide analog as inhibitor of nicotinamide N-methyltransferase. *Bioorg. Med. Chem. Lett.* **2018**, *28*, 922–925. [CrossRef] [PubMed]
103. Lee, H.Y.; Suciu, R.M.; Horning, B.D.; Vinogradova, E.V.; Ulanovskaya, O.A.; Cravatt, B.F. Covalent inhibi-tors of nicotinamide N-methyltransferase (NNMT) provide evidence for target engagement challenges in situ. *Bioorg. Med. Chem. Lett.* **2018**, *28*, 2682–2687. [CrossRef]

Article

Novel Insights for Patients with Multiple Basal Cell Carcinomas and Tumors at High-Risk for Recurrence: Risk Factors, Clinical Morphology, and Dermatoscopy

Dimitrios Sgouros [1,*], Dimitrios Rigopoulos [2], Ioannis Panayiotides [3], Zoe Apalla [4], Dimitrios K. Arvanitis [1], Melpomeni Theofili [1], Sofia Theotokoglou [1], Anna Syrmali [1], Konstantinos Theodoropoulos [1], Georgia Pappa [1], Vasileia Damaskou [3], Alexander Stratigos [2] and Alexander Katoulis [1]

[1] 2nd Department of Dermatology-Venereology, "Attikon" General University Hospital, Medical School, National and Kapodistrian University of Athens, 12462 Athens, Greece; arvandi18@yahoo.com (D.K.A.); melpomenitheofili@hotmail.com (M.T.); theotokoglousofia@gmail.com (S.T.); annasyrmali@gmail.com (A.S.); theod28@gmail.com (K.T.); mdgeopap@med.uoa.gr (G.P.); akatoulis@med.uoa.gr (A.K.)

[2] 1st Department of Dermatology-Venereology, Andreas Sygros Hospital, Medical School, National and Kapodistrian University of Athens, 16121 Athens, Greece; drigopoul@med.uoa.gr (D.R.); alstrat@med.uoa.gr (A.S.)

[3] 2nd Department of Pathology, "Attikon" General University Hospital, Medical School, National and Kapodistrian University of Athens, 12462 Athens, Greece; ioagpan@med.uoa.gr (I.P.); damaskou@0320.syzefxis.gov.gr (V.D.)

[4] State Clinic of Dermatology, Hospital for Skin and Venereal Diseases, 54643 Thessaloniki, Greece; zapalla@auth.gr

* Correspondence: dsgouros@med.uoa.gr; Tel.: +30-21058-32396 or +30-69748-16025; Fax: +30-21058-32396

Citation: Sgouros, D.; Rigopoulos, D.; Panayiotides, I.; Apalla, Z.; Arvanitis, D.K.; Theofili, M.; Theotokoglou, S.; Syrmali, A.; Theodoropoulos, K.; Pappa, G.; et al. Novel Insights for Patients with Multiple Basal Cell Carcinomas and Tumors at High-Risk for Recurrence: Risk Factors, Clinical Morphology, and Dermatoscopy. *Cancers* **2021**, *13*, 3208. https://doi.org/10.3390/cancers13133208

Academic Editor: Claus Garbe

Received: 30 May 2021
Accepted: 24 June 2021
Published: 27 June 2021

Publisher's Note: MDPI stays neutral with regard to jurisdictional claims in published maps and institutional affiliations.

Simple Summary: We investigated 225 patients with 304 primary basal cell carcinomas (BCCs) and we conducted a retrospective, morphological, cohort study aimed at evaluating patients' demographics and tumors' clinical and dermatoscopic characteristics. Our main objectives were the detection of risk factors for multiple BCCs in individual patients and the description of clinical and dermatoscopic features of low and high risk for local recurrence tumors. The rising incidence of BCC and the occurrence of multiple tumors in individual patients poses BCC as a major issue for health systems. To the best of our knowledge, this is one of the first studies to attempt to unveil clinical and dermatoscopic features of low-/high-risk neoplasms beyond histopathology and take into equal account parameters, such as anatomic location and size of the lesion. We strongly support that profiling of multiple patients with BCCs and a thorough knowledge of high-risk tumors' clinico-dermatoscopic morphology could provide physicians with important information towards prevention of this neoplasm.

Abstract: Introduction: Basal cell carcinoma (BCC) quite frequently presents as multiple tumors in individual patients. Neoplasm's risk factors for local recurrence have a critical impact on therapeutic management. Objective: To detect risk factors for multiple BCCs (mBCC) in individual patients and to describe clinical and dermatoscopic features of low- and high-risk tumors. Materials & Methods: Our study included 225 patients with 304 surgically excised primary BCCs. All patients' medical history and demographics were recorded. Clinical and dermatoscopic images of BCCs were evaluated for predefined criteria and statistical analyses were performed. Results: Grade II-III sunburns before adulthood (OR 2.146, $p = 0.031$) and a personal history of BCC (OR 3.403, $p < 0.001$) were the major predisposing factors for mBCC. Clinically obvious white color (OR 3.168, $p < 0.001$) and dermatoscopic detection of white shiny lines (OR 2.085, $p = 0.025$) represented strongly prognostic variables of high-risk BCC. Similarly, extensive clinico-dermatoscopic ulceration (up to 9.2-fold) and nodular morphology (3.6-fold) raise the possibility for high-risk BCC. On the contrary, dermatoscopic evidence of blue-black coloration had a negative prognostic value for high-risk neoplasms (light OR 0.269, $p < 0.001$/partial OR 0.198, $p = 0.001$). Conclusions: Profiling of mBCC patients and a thorough knowledge of high-risk tumors' clinico-dermatoscopic morphology could provide physicians with important information towards prevention of this neoplasm.

Keywords: basal cell carcinoma; dermatoscopy; histopathology; skin cancer; diagnosis; non-melanoma skin cancer; prevention

1. Introduction

Basal cell carcinoma (BCC) represents the most common type of skin cancer and human malignancy overall [1]. Demographics, etiopathogenesis, risk factors, histopathology, and clinico-dermatoscopic presentation of this tumor are well-described in the current literature [2–4]. Despite its extremely rare metastatic potential, the rising incidence of this neoplasm poses a major health issue for patients and healthcare systems [5]. Surgical excision is the treatment of choice for skin cancers. However, there are several efficient, guideline-approved, non-surgical, therapeutic options for specific subtypes of BCC (e.g., superficial) [2,3,6]. Current evidence shows that the classification of BCC as low or high risk for local recurrence, based on several clinical and histopathological characteristics (e.g., lesion's maximum diameter, anatomic location, histologic subtype, etc.) can be decisive for the selection of treatment in everyday clinical practice [4,6].

Moreover, BCC quite frequently presents as multiple (>1) tumors in individual patients. In addition, a personal history of at least one BCC yields a 17-fold risk for a subsequent BCC [5]. Multiple BCC patient risk factors (e.g., sex, age, etc.) have been investigated in the literature [7–10]. Further studies on the clinical features of multiple tumors and patients' profiling could provide physicians with important knowledge towards prevention of this neoplasm.

Dermatoscopic examination is a non-invasive, safe, and patient-friendly procedure that enhances a physician's diagnostic accuracy. The use of this method for diagnosing BCC has been thoroughly investigated [11–14]. In addition to its diagnostic role, dermatoscopy can help as a follow-up measure for size reduction of locally advanced BCC under neo-adjuvant systemic treatments [15]. Of note is the correlation of currently accepted dermatoscopic criteria with histological subtypes of BCC [16–19].

Our study had a double primary objective, i.e., to detect risk factors for multiple BCC in individual patients and to describe clinical and dermatoscopic features of low and high-risk tumors in a population of 225 patients with 304 primary BCCs. As a secondary goal, we investigated the dermatoscopic findings in a subgroup analysis among solitary BCC and multiple BCCs, as well as high-risk tumors with histologically aggressive subtypes.

2. Materials and Methods

This was a retrospective, morphological, cohort study conducted by the 2nd Dermatology Department of the "ATTIKON" University Hospital of Athens (waiver decision by Ethics Committee 1248/19-1-2016). Informed consent was obtained from all subjects involved in the study.

The inclusion criterion was patients with a histopathological diagnosis of primary BCC that was surgically excised. The exclusion criteria were: (1) Gorlin–Goltz syndrome patients, (2) severely immunocompromised patients (i.e., patients under immunosuppressive treatment for autoimmune diseases or internal malignancies and HIV patients), (3) patients with locally recurrent tumors from prior treatments, and (4) patients that did not give their consent for data collection for the purposes of the study. The enrollment period was between January 2016 and January 2018. All selected patients had a thorough physical skin examination and a full report of their medical history. In addition, a detailed history of sun exposure habits, previous cutaneous diseases, and past treatments for skin malignancies were recorded. Clinical and dermatoscopic images of the suspicious lesions were both captured at the initial medical visit before surgical excision using a Nikon J1 camera (Tokyo, Japan) and a handheld Dermlite Hybrid II dermatoscope (3Gen Inc, San Juan Capistrano, CA, USA). All the clinical and dermatoscopic characteristics of tumors were retrospectively evaluated for predefined criteria (Supplementary Table S1) [20] by two investigators non-blinded to the final diagnosis (D.S. and A.K.). The cohort was divided into two groups of patients, i.e., those with solitary neoplasms and those with more than one synchronous tumors (solitary vs. multiple BCC). The tumors were classified into low- and high-risk tumors for recurrence based on three factors: (1) tumor size, (2) anatomic location, and (3) histological subtype (Table 1) [6]. Of note, we did not include the "clinical margins criterion" as a risk factor for tumor classification, since we planned to investigate it as an independent clinical feature. Statistical analysis with uni- and multivariate logistic regression was performed for demographic factors along with clinical and dermatoscopic features for the above-mentioned groups of patients and tumors.

Table 1. Risk factors for low and high risk for recurrence BCC [1].

Risk Factors	Low-Risk BCC	High-Risk BCC
Location/size	Trunk, extremities < 2 cm	Trunk, extremities ≥2 cm Cheeks, forehead, scalp, neck and pretibial any size "Mask areas" of face [2], genitalia, hands and feet
Histopathology	Nodular, superficial	Aggressive growth pattern [3]

[1] Any risk factor places the patient in the high-risk category; [2] center of face, eyelids, eyebrows, periorbital, nose, lips (cutaneous and vermilion), chin, mandible, pre- and post-auricular skin/sulci, temple, and ear; [3] infiltrative, basosquamous, morpheaform, micronodular, mixed, sclerosing/carcinosarcomatous features, perineural invasion. This table was adapted by Schmultz, C., Blitzblau, R., et al. Basal Cell Skin Cancer Version 2.2021 in NCCN Clinical Practice Guidelines in Oncology, available online at https://www.nccn.org/professionals/physician_gls/pdf/nmsc.pdf (accessed on 25 February 2021).

Statistical Analysis Methods

The Shapiro–Wilk and Shapiro–Francia tests were used for normality of distribution. Continuous variables following a normal distribution are presented as a mean ± standard deviation, whereas not normally distributed variables are presented as a median with interquartile range (25th and 75th percentiles). For categorical variables, the frequencies and percentages were used. Chi-squared and Fischer's exact tests were used for the comparison of categorical variables, while unpaired t-tests and Mann–Whitney U tests were applied depending on the distributions of the continuous variables. Univariate and multivariate logistic regression was also performed. All statistical calculations were based on a two-sided hypothesis, and a *p*-value of <0.05 was considered to be statistically significant. All statistical analyses were performed using Stata/IC version 15.1 (StataCorp, Lakeway Drive, Texas, USA).

3. Results

3.1. Solitary (sBCC) and Multiple BCCs (mBCC): Patients' Demographics

In total, 225 patients with 304 primary BCCs were included. There were 172 patients (76.4%) who presented with a solitary tumor and 53/225 patients (23.6%) were diagnosed with ≥2 tumors at the initial evaluation visit. The male sex prevailed in both groups of patients with 105/172 (61.1%) and 38/53 (71.2%) for sBCC and mBCC, respectively. The median age for the entire group of patients was 73 years. However, patients with mBCC (median age 75) were older than sBCC patients (median age 72.5). Skin exposure habits and chronic solar damage were more prominent in the group of patients with mBCC rather than the solitary neoplasms group. In specific, 24/53 patients (45.3%) reported occupational sun exposure, 19/53 patients (35.9%) had a history of at least one severe sunburn (≥grade II) during childhood-adolescence and 33/53 patients (62.3%) were diagnosed with actinic keratoses as compared with 58/172 (33.7%), 38/172 (22.1%) and 84/172 (48.8%) patients, respectively, for the sBCC group. A personal history of any type of skin cancer and history of at least one previous BCC were two independent risk factors, more prevalent in the group of multiple tumors (24/53 (45.3%) and 23/53 (43.4%) patients, respectively) as compared with the sBCC group (41/172 (23.8%) and 33/172 (19.2%) patients). All the aforementioned results can be seen in Table 2.

Table 2. Patients' demographics with solitary BCC and multiple BCCs (*n* = 225).

	Total (*n* = 225)	Solitary BCC (*n* = 172)	Multiple BCCs (*n* = 53)
Age, median years (range)	73 (28–94)	72.5 (28–91)	75 (37–94)
Sex, *n* (%)			
Males	143 (63.6)	105 (61)	38 (71.7)
Females	82 (36.4)	67 (39)	15 (28.3)
Fitzpatrick skin phototype, *n* (%)			
I	0	0	0
II	47 (20.9)	34 (19.8)	13 (24.5)
III	126 (56)	96 (55.8)	30 (56.6)
IV	52 (23.1)	42 (24.4)	10 (18.9)
Occupational sun exposure, *n* (%)	82 (36.4)	58 (33.7)	24 (45.2)
History of sunburns, *n* (%) *	57 (25.3)	38 (22.1)	19 (35.9)
Actinic keratosis, *n* (%)	117 (52)	84 (48.8)	33 (62.3)
Personal history of skin cancer, *n* (%)	65 (28.9)	41 (23.8)	24 (45.3)
Family history of skin cancer, *n* (%)	24 (10.7)	19 (11.1)	5 (9.4)
Personal history of BCC, *n* (%)	56 (24.9)	33 (19.2)	23 (43.4)

* Grade II/III sunburns < 18 years old.

Univariate logistic regression showed that a personal history of BCC (3.2-fold), a personal history of skin cancer (2.6-fold), sunburns grade II–III (<18 years old) (1.9-fold), and the presence of actinic keratosis (1.7-fold) were important risk factors for mBCC. However, the multivariate analysis revealed that severe sunburns during childhood-adolescence and a personal history of BCC were the two most critical risk factors for the development of ≥2 BCC in an individual patient with a 2.1-fold and 3.4-fold risk, respectively (Table 3).

Table 3. Uni- and multivariate logistic regression for multiple BCCs vs. solitary BCC.

Univariate	*p*-Value	OR	95% CIs
Age	0.453		
Sex	0.159		
Fitzpatrick skin phototype	0.611		
Occupational Sun Exposure	0.126		
Personal history of skin cancer	0.003	2.644	1.388–5.038
Family history of skin cancer	0.74		
Personal history of BCC	0.001	3.229	1.665–6.265
Actinic keratosis	0.089	1.729	0.92–3.248
History of sunburns	0.046	1.971	1.011–3.84
Multivariate	***p*-Value**	**OR**	**95% CIs**
Personal history of BCC	<0.001	3.403	1.732–6.685
History of sunburns	0.031	2.146	1.073–4.295

For the final model, a fitness of good control was performed based on Hosmer–Lemeshow criterion (*p*-value = 0.648).

3.2. Solitary BCC and Multiple BCCs: Tumors' Clinical and Histological Features

In total, 172 patients presented with one BCC and 53 patients had multiple lesions (132 overall). Specifically, 38 patients presented with two tumors (38/53, 71.7%) nine patients with three tumors (9/53, 17.1%), three patients with four tumors (3/53, 5.6%), and ≥5 tumors were detected in three patients (3/53, 5.6%). Regarding important factors of risk stratification for local recurrence (such as diameter, anatomic site, and histopathology) no striking difference was recorded between the two groups of patients. Specifically, sBCC comprised of 46/172 patients (26.7%) with low-risk tumors and 126/172 patients (73.3%) with high-risk tumors, while mBCC consisted of 46/132 patients (34.9%) with low-risk tumors and 86/132 patients (65.2%) with high-risk neoplasms. In both groups, the head/neck area was the most common anatomic location for the development of BCC (73.4%), followed by the trunk (19.1%) and extremities 7.6%). The median diameter of the neoplasms was 0.9 cm. Regarding histologic subtypes of BCC, our results showed a prevalence of types of indolent biologic behavior (i.e., nodular and superficial, 73.7%) rather than aggressive growth pattern forms (i.e., infiltrative, morpheaform, basosquamous, micro-nodular, and mixed, 26.3%). Specific results for the subgroups of sBCC and mBCC are shown in Table S2.

3.3. Low Risk versus High Risk for Local Recurrence BCC: Clinical Characteristics

Out of the 304 tumors that were included in the study, 92 tumors were classified as low-risk tumors and 212 tumors were classified as high-risk tumors based on three criteria, i.e., lesion's maximum diameter, anatomic location, and histopathology (Table 1). Clinical margins were investigated as a separate feature. Indeed, well-defined clinical borders were predominant (78/92, 84.8%) among the subgroup of low-risk BCC, as expected, although poorly defined clinical margins were not a prevalent characteristic in high-risk tumors either (82/212, 38.7%). In terms of ulceration, intact epidermis was evident in the majority of low-risk tumors (49/92, 53.3%), in contrast to high-risk tumors which exhibited prominent erosion/ulceration (155/212, 73.1%). Clinically, most of the high-risk tumors presented as nodular lesions (146/212, 68.9%) while 50% (46/92) of low-risk BCC had a nodular morphology. It is worthwhile mentioning that an important subset of low-risk tumors was flat (23/92, 25%) as compared with 5.7% (12/212) in high-risk BCC. Concerning coloration, pink was the most frequently observed color in both subgroups with a total proportion of 78.6% (239/304). A white color was more evident among high-risk tumors (54.7% versus 28.3%), while a blue-black color was more commonly encountered among low-risk BCC (51.1% versus 31.13%) (Table 4).

Table 4. Low-risk and high-risk tumors' clinical characteristics and multivariate analysis.

	Total (*n* = 304)	Low-Risk (*n* = 92)	High-Risk (*n* = 212)	Multivariate OR (95% CI/*p*-Value)
Margins, *n* (%)				
Well-defined	208 (68.4)	78 (84.8)	130 (61.3)	
Ill-defined	96 (31.6)	14 (15.2)	82 (38.7)	2.007 (0.952–4.23/0.067)
Ulceration, *n* (%)				
None	106 (34.9)	49 (53.3)	57 (26.9)	
Erosions	55 (18.1)	21 (22.8)	34 (16.)	
Prominent	108 (35.5)	20 (21.7)	88 (41.5)	2.533 (1.243–5.162/0.011)
>90%	35 (11.5)	2 (2.2)	33 (15.6)	9.241 (1.79–47.711/0.008)
Clinical presentation, *n* (%)				
Flat	35 (11.5)	23 (25)	12 (5.6)	
Elevated	77 (25.3)	23 (25)	54 (25.5)	2.384 (0.892–6.376/0.083)
Nodular	192 (63.2)	46 (50)	146 (68.9)	3.674 (1.502–8.988/0.004)
Colors, *n* (%)				
Pink color	239 (78.6)	75 (81.5)	164 (77.4)	
White color	142 (46.7)	26 (28.3)	116 (54.7)	3.682 (1.988–6.819/<0.001)
Blue-black color	113 (37.2)	47 (51.1)	66 (31.1)	0.193 (0.032–1.153/0.071)
Pigmentation intensity, *n* (%)				
None	183 (60.2)	44 (47.8)	139 (65.6)	
Light	44 (14.5)	20 (21.7)	24 (11.3)	
Partial	36 (11.8)	15 (16.3)	21 (9.9)	
Heavy	41 (13.5)	13 (14.2)	28 (13.2)	5.611 (0.771–40.82/0.088)

For the final model, a fitness of good control was performed based on Hosmer–Lemeshow criterion (*p*-value = 0.547).

Univariate logistic regression for the clinical features of high-risk versus low-risk BCC is presented in Supplementary Table S3. Multivariate logistic regression showed that extensive clinical ulceration (>90% of total lesion surface) and prominent ulceration yield a 9.2-fold and 2.5-fold probability for high-risk BCC, respectively. In the same context, white color and nodular morphology are strong prognostic factors for a high-risk tumor with an OR 3.6. On the contrary, clinical blue-black coloration is a negative prognostic factor for high-risk neoplasms (OR 0.2) (Table 4). A detailed analysis of multivariate logistic regression is presented in Table S4.

3.4. Low Risk versus High Risk for Local Recurrence BCC: Dermatoscopic Features

Vascular structures were the most striking dermatoscopic finding in both subgroups of low-risk and high-risk tumors (272/304, 89.5%). Arborizing vessels prevailed in both subgroups as well (247/304, 81.2%) followed by telangiectasias and glomerular vessels. Telangiectasias were more common among low-risk tumors (38% versus 22.2%) and glomerular vessels were more frequent in high-risk tumors (14.2% versus 4.4%). As expected, high-risk tumors were mostly eroded or ulcerated lesions (168/212, 79.3%), while dermatoscopic erosion/ulceration was also prevalent in low-risk tumors (57/92, 62%). In terms of pigmentation, the majority of high-risk tumors were non-pigmented (115/212, 54.3%), while 71.4% (66/92) of low-risk BCC had dermatoscopic signs of pigmentation. In specific, all types of pigmented structures were more frequently observed in low-risk tumors as compared with high-risk tumors. Finally, dermatoscopic clues for white coloration were more frequently observed among high-risk tumors; white shiny lines (46.2% versus 27.2%); multiple yellow-white globules (12.7% versus 7.6%); white circles and yellow clods (25.9% versus 7.6%) (Table 5).

Table 5. Low-risk and high-risk tumors' dermatoscopic features and multivariate analysis.

	Total (n = 304)	Low-Risk (n = 92)	High-Risk (n = 212)	Multivariate OR (95% CI/p-Value)
Vasculature, n (%)				
None	32 (10.5)	13 (14.1)	19 (9)	
Apparent (<50%)	219 (72.1)	63 (68.5)	156 (73.6)	
Prominent ($\geq$50%)	53 (17.4)	16 (17.4)	37 (17.4)	
Vessels, n (%)				
Arborizing	247 (81.3)	65 (70.7)	182 (85.9)	
Telangiectasias	82 (27)	35 (38.)	47 (22.2)	
Glomerular	34 (11.2)	4 (4.4)	30 (14.2)	3.314 (1.033–10.626/0.044)
Linear irregular	14 (4.6)	2 (2.2)	12 (5.7)	
Dotted	1 (0.3)	0	1 (0.5)	
Hairpin	3 (1)	0	3 (1.4)	
Polymorphous	14 (4.6)	0	14 (6.6)	
Pigmented structures, n (%)				
Blue-gray ovoid globules	127 (41.8)	46 (50)	81 (38.2)	
Multiple dots	90 (29.6)	40 (43.5)	50 (23.6)	
Spoke-wheel	24 (7.9)	15 (16.3)	9 (4.3)	
Leaf-like	28 (9.2)	18 (19.6)	10 (4.7)	
Concentric	16 (5.3)	10 (10.9)	6 (2.8)	
Pigmentation intensity, n (%)				
None	141 (46.4)	26 (28.2)	115 (54.2)	
Light (<10%)	77 (25.3)	33 (35.9)	44 (20.8)	0.269 (0.13–0.558/<0.001)
Partial (10–50%)	42 (13.8)	18 (19.6)	24 (11.3)	0.198 (0.078–0.5/0.001)
Heavy (>50%)	44 (14.5)	15 (16.3)	29 (13.7)	0.313 (0.105–0.934/0.037)
Pink-whitish background, n (%)	211 (69.4)	71 (77.2)	140 (66)	0.369 (0.158–0.862/0.021)
Diffuse white color, n (%)	11 (3.6)	1 (1.1)	10 (4.7)	
White shiny lines, n (%)	123 (40.5)	25 (27.2)	98 (46.2)	2.087 (1.097–3.971/0.025)
Multiple yellow-white globules, n (%)	34 (11.2)	7 (7.6)	27 (12.7)	
White circles and yellow clods, n (%)	62 (20.4)	7 (7.6)	55 (25.9)	
Ulceration, n (%)				
None	79 (26)	35 (38)	44 (20.8)	
Erosions	71 (23.3)	30 (32.6)	41 (19.3)	
Prominent	121 (39.8)	25 (27.2)	96 (45.3)	2.451 (1.198–5.014/0.014)
>90%	33 (10.9)	2 (2.2)	31 (14.6)	8.042 (1.637–39.505/0.01)

For the final model, a fitness of good control was performed based on Hosmer–Lemeshow criterion (p-value = 0.47).

Univariate logistic regression for the dermatoscopic findings of high-risk versus low-risk BCC is presented in Supplementary Table S5. The multivariate analysis revealed that extensive (8-fold) as well as prominent (2.4-fold) ulceration, glomerular vessels (3.3-fold), and white shiny linear structures (2-fold) are positive predictive factors for a high-risk BCC. On the contrary, pink-whitish background (0.37-fold) along with pigmentation of any extent (0.2–0.3-fold) represent negative prognostic factors for high-risk tumors (Table 5 and Table S6 and Figures 1 and 2).

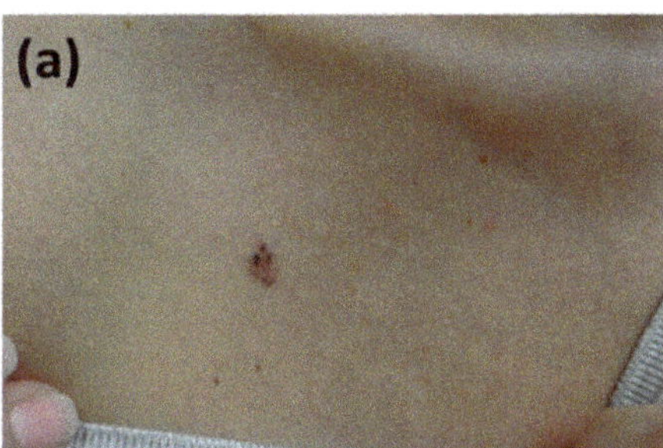
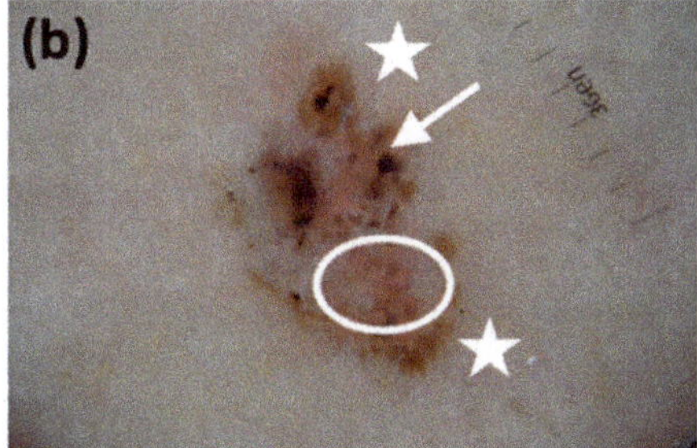

Figure 1. (**a**) A brown-black plaque on the chest of a 51-year-old female patient. The lesion has a maximum diameter of 0.6 cm. Histology set the diagnosis of a superficial BCC; (**b**) dermatoscopy confirmed our observations for low-risk neoplasms. Pigmentation was the striking feature in this tumor with leaf-like structures at the periphery (white asterisks), concentric structures (white circle), and a hint of telangiectasias (white arrow).

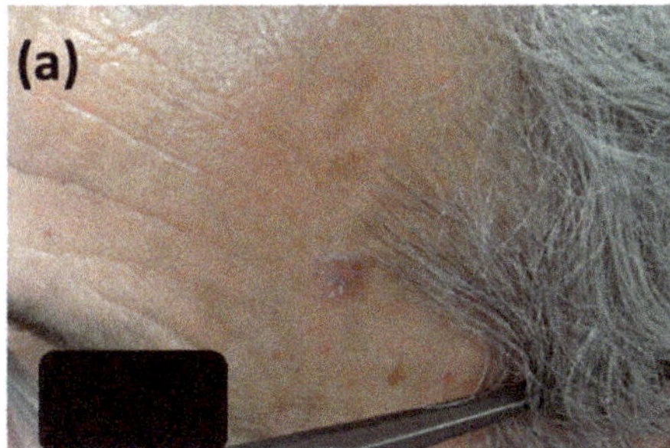
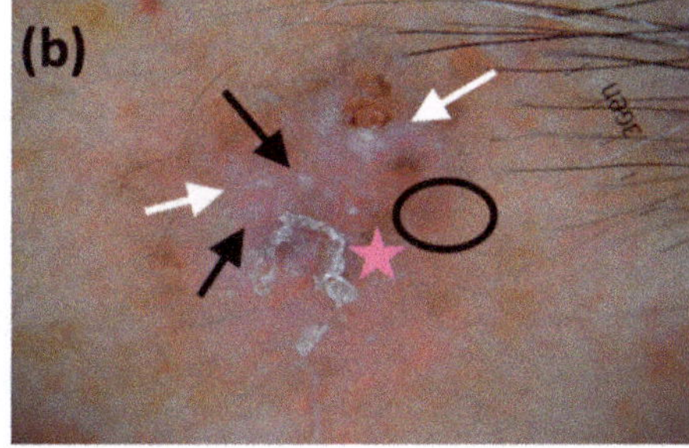

Figure 2. (**a**) A pinkish nodule on the left temple of a 66-year-old male patient with dark skin phototype and a maximum diameter of 1.3 cm, histopathologically diagnosed as a mixed BCC (nodular and metatypical). Due to anatomic location and size, the lesion was treated as a high-risk tumor for local recurrence; (**b**) dermatoscopic evaluation was in line with histology. A combination of arborizing (pink asterisk) and hairpin (black circle) vessels was evident. White, shiny linear structures (white arrows) were also obvious on the lesion's surface. Finally, white, perifollicular circles along with central yellow clods were dermatoscopically apparent (black arrows).

3.5. Dermatoscopic Features in Subgroup Analysis for Solitary BCC and Multiple BCCs and Aggressive Histologic Subtypes

Dermatoscopic features of BCC in the subgroups of solitary and multiple tumors can be seen in detail in Supplementary Table S5. No striking differences were detected in the frequencies of dermatoscopic findings between both subgroups of patients, except for "white features". In specific white shiny lines (51.7% versus 25.8%), multiple yellow-white globules (16.3% versus 4.6%) and white circles and yellow clods (24.41% versus 14.4%) were more commonly observed in sBCC rather than in the group of multiple tumors.

Supplementary Table S8 shows in detail the dermatoscopic features in aggressive histologic subtypes of BCC as compared with the group of high-risk tumors. The specific subtypes can also be seen in Supplementary Table S9. Of note, there are no significant differences in the dermatoscopic findings of histologically aggressive tumors as compared with high-risk BCC apart from a slight prevalence of "white structures" already observed in the subgroup analysis for solitary tumors (Table S9).

4. Discussion

4.1. Solitary versus Multiple BCCs

BCC morbidity represents a major issue for public health. Despite its extremely low mortality, the rising incidence of the tumor and the high occurrence of mBCC quantify the burden of disease comparable with esophageal, ovarian, or thyroid cancer, according to the WHO [21]. Profiling of patients with multiple tumors, identification of risk factors, and particular clinico-dermatoscopic features of mBCC could be valuable for the overall management of the disease.

Our study showed that grade II–III sunburns before adulthood (OR 2.146, $p = 0.031$) and a personal history of BCC (OR 3.403, $p < 0.001$) were the major predisposing factors for mBCC (Table 3). So far, the personal history of a previously treated BCC is considered to be a well-established risk factor for a subsequent BCC [10,22,23]. Chronic UV exposure has also been proven to be an important contributor for non-melanoma skin cancer [5]. However, there was no clear association of the sun exposure pattern with a single BCC or multiple BCCs [22]. According to our results, at least one severe sunburn during childhood or adolescence increases the risk for the development of mBCC later in life. Thus, preventive measures in early life could have a protective role against mBCC.

In our study, we observed that older patients at a slightly higher male/female ratio comprised the group of mBCC as compared with the sBCC group (Table 2). This finding was in accordance with the previously published literature [8,9,22]. The distribution of histological subtypes of the tumor in both groups was similar and corresponded to the average of the incidence of various forms of the neoplasm [1] (Table S2). This result does not confirm previously published data that superficial BCC is a more common subtype in patients with multiple tumors [7–10].

We also performed a subgroup analysis for the dermatoscopic features in sBCC and mBCC. (Table S7) No clear differences were detected between both groups, apart from a more frequent representation of white dermatoscopic structures (i.e., white shiny lines, white peri-follicular circles, multiple yellow and white globules) in the group with single BCC. White coloration in dermatoscopy of BCC is strongly correlated with collagen alteration, calcification, and thus deeper infiltration in dermis [24–27]. Our observation supports that the presence of multiple tumors is not necessarily associated with more aggressive BCC subtypes.

4.2. High-Risk versus Low-Risk BCC

Due to the extremely rare metastatic potential of BCC, the traditional classification for neoplasms is not applicable in this type of skin cancer. Thus, the tumors are categorized accordingly to the risk for local recurrence [3,6]. Our study evaluated BCCs as low/high risk based on three criteria (i.e., lesion's diameter, anatomic location, and histology) in order to investigate predefined criteria for risk stratification as independent variables (i.e., clinical margins). Moreover, the exclusion of other factors (i.e., locally recurrent tumor, mBCC syndromes, and severe immunosuppression) allowed us to detect characteristics of primary lesions without the statistical bias that may arise from patients' health status or physicians' previous topical treatments (Table 1).

Of note, 69.7% (212/304) of tumors were staged as high risk, although only 26.3% (80/304) of tumors had a histologic subtype with aggressive behavior. This finding highlights the importance of the size and the site of the lesion, as equally significant along with histopathology, for risk stratification, and therefore treatment selection (Table 4 and Table S2).

In terms of clinical morphology, low-risk BCC had mostly well-defined clinical margins (84.8%, 78/92), as expected. Surprisingly, high-risk tumors did not present with ill-defined borders as a prevalent feature (38.7%, 82/212) (Table 4). This result could be suggestive of the fact that a well-defined BCC is not always a low-risk tumor, and therefore other risk factors should be considered. Regarding clinically or dermatoscopically prominent ulceration and nodular presentation, we confirmed current evidence that both features typically characterize more aggressive subtypes of BCC [4,17–19,28–30] (Table 4, Table 5, Tables S4 and S6).

Coloration represents quite an interesting finding both clinically and dermatoscopically. Our results present clinically obvious white color (OR 3.168, $p < 0.001$) and dermatoscopically detected white shiny lines (OR 2.085, $p = 0.025$) as variables strongly prognostic of high-risk BCC. The latter is in accordance with the pre-existing literature for white dermatoscopic structures as features correlating with aggressive subtypes of BCC [17–19,24–30]. On the contrary, pigmentation due to melanin elements seems to be inversely associated with high-risk tumors. In terms of clinically apparent blue-black hue, the multivariate analysis showed a tendency for low-risk BCC (OR 0.193, $p = 0.071$). However, dermatoscopic evidence of blue-black coloration had a negative prognostic value for high-risk neoplasms (light OR 0.269, $p < 0.001$/partial OR 0.198, $p = 0.001$). This finding suggests that melanin may represent a positive predictive factor for BCC with a higher risk for local recurrence and supports recently published data on the hypothesis that well-differentiated non-aggressive BCC could preserve relatively more melanocytes [31] (Table 4, Table 5, Tables S4 and S6).

Concerning other dermatoscopic observations, we confirmed that arborizing vessels are typical for BCC and telangiectasias are mostly seen in low-risk tumors (Table 5). Of note, glomerular vessels in multivariate analysis were prognostic for high-risk BCC (OR 3.314, $p = 0.044$), another finding in line with the current literature [28,29]. Dermatoscopic evidence of ulceration also raised the possibility of a high-risk tumor (up to 8-fold) in accordance with the pre-existing published data [17–20,28–30] (Table 5).

Finally, we performed a subgroup comparative analysis investigating the occurrence of dermatoscopic variables between high-risk tumors ($n = 212$) and histologic subtypes of BCC with a more aggressive growth pattern ($n = 80$). The results are listed in Supplementary Table S8 and surprisingly we could not detect any statistically important variations among the two groups. Our observations support the significance of the current classification system for BCC in clinical practice and show that factors other than histopathology (i.e., tumor's diameter and anatomic location) have a critical impact on risk stratification, almost equal to histologic subtype. Thus, from a clinician's perspective, clinical morphology and dermatoscopic findings could efficiently negate unnecessary biopsies and prevent therapeutic pitfalls.

4.3. Limitations

Our study has certain limitations including the "non-blinded to diagnosis" investigators evaluating clinical and dermatoscopic findings, the lack of a control group, and certain personal history information that were unavailable (e.g., age of first BCC diagnosis) and are considered by literature significant risk factors for mBCC. Moreover, the current classification system of BCC does not take into account the different impacts of each risk factor and this is an additional limitation in our study. Finally, the visit-seeking attitude of patients with a personal history of BCC and the surveillance plan after diagnosis of an initial tumor might pose a bias.

5. Conclusions

In conclusion, a personal history of BCC and moderate-to-severe sunburns during childhood-adolescence are the two most important risk factors for the development of multiple BCC. The presence of multiple tumors does not seem to be related to more aggressive subtypes of the neoplasm. Concerning high-risk for recurrence BCC, ill-defined clinical margins is not an absolute criterion. Nodular morphology, clinical and dermatoscopic evidence of ulceration, and the color white, either clinically obvious or presented dermatoscopically in the form of white shiny lines, serve as strong predictors for high-risk BCC. On the contrary, pigmentation due to melanin represents a negative prognostic value for high-risk tumors. To the best of our knowledge, this is one of the first attempts to describe the dermatoscopic features of locally aggressive BCC beyond the histologic subtype. We strongly support the need for further studies to unveil important diagnostic clues for BCC.

Supplementary Materials: The following are available online at https://www.mdpi.com/article/ 10.3390/cancers13133208/s1, Table S1: Definition of dermatoscopic terms and criteria [4,19], Table S2: Solitary and multiple BCC tumors characteristics, n = 304, Table S3: Univariate logistic regression for the clinical features of high-risk vs. low-risk BCCs, Table S4: Multivariate logistic regression analysis for the clinical features of high-risk vs. low-risk BCCs, Table S5: Univariate logistic regression analysis for the dermatoscopic features of high-risk vs. low-risk BCCs, Table S6: Multivariate logistic regression for the dermatoscopic features of high-risk vs. low-risk BCCs, Table S7: Dermatoscopic findings in solitary and multiple BCC, Table S8: Dermatoscopic features in histologically aggressive subtypes of BCC (n = 80), Table S9: Aggressive histologic subtypes of BCC, n = 80 (%).

Author Contributions: Conceptualization, D.S., Z.A., and A.K.; formal analysis, D.S.; investigation, D.S., D.R., I.P., Z.A., M.T., S.T., A.S. (Anna Syrmali), K.T., G.P., V.D., A.S. (Alexander Stratigos), and A.K.; methodology, D.S., I.P., D.K.A., and A.K.; project administration, D.S.; supervision, D.R. and A.K.; writing—original draft, D.S.; writing—review and editing, D.S. and A.K. All authors have read and agreed to the published version of the manuscript.

Funding: This research received no external funding.

Institutional Review Board Statement: The study was conducted according to the guidelines of the Declaration of Helsinki and approved by the Ethics Committee of ATTIKON University Hospital of Athens (1248/19-1-2016).

Informed Consent Statement: Informed consent was obtained from all subjects involved in the study.

Data Availability Statement: Data sharing is not applicable to this article.

Conflicts of Interest: The authors declare no conflict of interest.

References

1. Messina, J.; Epstein EHJr Kossard, S.; McKenzie, C.; Patel, R.M.; Patterson, J.W.; Scolyer, R.A. Basal Cell Carcinoma. In *WHO Classification of Skin Tumors*; Elder, D.J., Massi, S., Scolyer, R.A., Willemze, R., Eds.; IARC: Lyon, France, 2018; Volume 1, pp. 26–34.
2. Trakatelli, M.; Morton, C.C.; Nagore, E.E.; Ulrich, C.; Del Marmol, V.; Peris, K.; Basset-Seguin, N.N. Update of the European guidelines for basal cell carcinoma management. *Eur. J. Dermatol.* **2014**, *24*, 312–329. [CrossRef] [PubMed]
3. Peris, K.; Fargnoli, M.C.; Garbe, C.; Kaufmann, R.; Bastholt, L.; Seguin, N.B.; Bataille, V.; Del Marmol, V.; Dummer, R.; Harwood, C.A.; et al. Diagnosis and treatment of basal cell carcinoma: European consensus–based interdisciplinary guidelines. *Eur. J. Cancer* **2019**, *118*, 10–34. [CrossRef] [PubMed]
4. Lallas, A.; Apalla, Z.; Argenziano, G.; Longo, C.; Moscarella, E.; Specchio, F.; Raucci, M.; Zalaudek, I. The dermatoscopic universe of basal cell carcinoma. *Dermatol. Pr. Concept.* **2014**, *4*. [CrossRef] [PubMed]
5. Verkouteren, J.; Ramdas, K.; Wakkee, M.; Nijsten, T. Epidemiology of basal cell carcinoma: Scholarly review. *Br. J. Dermatol.* **2017**, *177*, 359–372. [CrossRef] [PubMed]
6. Schmultz, C.; Blitzblau, R.; Aasi, S.Z.; Alam, M.; Andersen, J.; Baumann, B.C.; Bordeaux, J.; Jen, P.L.; Chin, R.; Contreras, C.M.; et al. Basal Cell Skin Cancer Version 2.2021 In NCCN clinical practice guidelines in Oncology. Available online: https://www.nccn.org/professionals/physician_gls/pdf/nmsc.pdf (accessed on 25 February 2021).
7. Kuo, K.Y.; Batra, P.; Cho, H.G.; Li, S.; Chahal, H.S.; Rieger, K.E.; Tang, J.Y.; Sarin, K.Y. Correlates of multiple basal cell carcinoma in a retrospective cohort study: Sex, histologic subtypes, and anatomic distribution. *J. Am. Acad. Dermatol.* **2017**, *77*, 233–234.e2. [CrossRef]
8. Adachi, K.; Yoshida, Y.; Noma, H.; Goto, H.; Yamamoto, O. Characteristics of multiple basal cell carcinomas: The first study on Japanese patients. *J. Dermatol.* **2018**, *45*, 1187–1190. [CrossRef]
9. Flohil, S.; Koljenovic, S.; De Haas, E.; Overbeek, L.; De Vries, E.; Nijsten, T. Cumulative risks and rates of subsequent basal cell carcinomas in the Netherlands. *Br. J. Dermatol.* **2011**, *165*, 874–881. [CrossRef] [PubMed]
10. Verkouteren, J.A.; Smedinga, H.; Steyerberg, E.W.; Hofman, A.; Nijsten, T. Predicting the Risk of a Second Basal Cell Carcinoma. *J. Investig. Dermatol.* **2015**, *135*, 2649–2656. [CrossRef]

11. Lallas, A.; Apalla, Z.; Ioannides, D.; Argenziano, G.; Castagnetti, F.; Moscarella, E.; Longo, C.; Palmieri, T.; Ramundo, D.; Zalaudek, I. Dermoscopy in the diagnosis and management of basal cell carcinoma. *Futur. Oncol.* **2015**, *11*, 2975–2984. [CrossRef] [PubMed]

12. Reiter, O.; Mimouni, I.; Gdalevich, M.; Marghoob, A.A.; Levi, A.; Hodak, E.; Leshem, Y.A. The diagnostic accuracy of dermoscopy for basal cell carcinoma: A systematic review and meta-analysis. *J. Am. Acad. Dermatol.* **2019**, *80*, 1380–1388. [CrossRef]

13. Reiter, O.; Mimouni, I.; Dusza, S.; Halpern, A.C.; Leshem, Y.A.; Marghoob, A.A. Dermoscopic features of basal cell carcinoma and its subtypes: A systematic review. *J. Am. Acad. Dermatol.* **2019**. [CrossRef] [PubMed]

14. Suppa, M.; Micantonio, T.T.; Di Stefani, A.; Soyer, H.P.; Chimenti, S.S.; Fargnoli, M.C.; Peris, K. Dermoscopic variability of basal cell carcinoma according to clinical type and anatomic location. *J. Eur. Acad. Dermatol. Venereol.* **2015**, *29*, 1732–1741. [CrossRef] [PubMed]

15. Conforti, C.; Corneli, P.; Harwood, C.; Zalaudek, I. Evolving Role of Systemic Therapies in Non-melanoma Skin Cancer. *Clin. Oncol.* **2019**, *31*, 759–768. [CrossRef] [PubMed]

16. Yélamos, O.; Braun, R.P.; Liopyris, K.; Wolner, Z.J.; Kerl, K.; Gerami, P.; Marghoob, A.A. Dermoscopy and dermatopathology correlates of cutaneous neoplasms. *J. Am. Acad. Dermatol.* **2019**, *80*, 341–363. [CrossRef] [PubMed]

17. Pampena, R.; Parisi, G.; Benati, M.; Borsari, S.; Lai, M.; Paolino, G.; Cesinaro, A.M.; Ciardo, S.; Farnetani, F.; Bassoli, S.; et al. Clinical and Dermoscopic Factors for the Identification of Aggressive Histologic Subtypes of Basal Cell Carcinoma. *Front. Oncol.* **2021**, *10*. [CrossRef]

18. Conforti, C.; Pizzichetta, M.; Vichi, S.; Toffolutti, F.; Serraino, D.; Di Meo, N.; Giuffrida, R.; Deinlein, T.; Giacomel, J.; Rosendahl, C.; et al. Sclerodermiform basal cell carcinomas vs. other histotypes: Analysis of specific demographic, clinical and dermatoscopic features. *J. Eur. Acad. Dermatol. Venereol.* **2021**, *35*, 79–87. [CrossRef]

19. Sgouros, D.; Apalla, Z.; Theofili, M.; Damaskou, V.; Kokkalis, G.; Kitsiou, E.; Lallas, A.; Kanelleas, A.; Stratigos, A.; Nikolaidou, C.; et al. How to spot a basosquamous carcinoma: A study on demographics, clinical—Dermatoscopic features and histopathologic correlations. *Eur. J. Dermatol.* **2021**. accepted publication.

20. Giacomel, J.; Zalaudek, I. Pink Lesions IN Dermoscopy. *Dermatol. Clin.* **2013**, *31*, 649–678. [CrossRef] [PubMed]

21. Global Burden of Disease Study 2013 Collaborators. Global, regional and national incidence, prevalence and years lived with disability for 301 acute and chronic diseases and injuries in 188 countries, 1990-2013: A systematic analysis for the Global Burden of Disease Study 2013. *Lancet* **2015**, *386*, 743–800. [CrossRef]

22. Kiiski, V.; de Vries, E.; Flohil, S.C.; Bijl, M.J.; Hofman, A.; Stricker, B.H.C.; Nijsten, T. Risk Factors for Single and Multiple Basal Cell Carcinomas. *Arch. Dermatol.* **2010**, *146*, 848–855. [CrossRef] [PubMed]

23. Rahimi, H.; Hallaji, Z.; Mirshams-Shahshahani, M. Comparison of risk factors of single basal cell carcinoma with multiple basal cell carcinomas. *Indian J. Dermatol.* **2011**, *56*, 398–402. [CrossRef] [PubMed]

24. Navarrete-Dechent, C.; Bajaj, S.; Marchetti, M.A.; Rabinovitz, H.S.; Dusza, S.W.; Marghoob, A.A. Association of Shiny White Blotches and Strands With Nonpigmented Basal Cell Carcinoma. *JAMA Dermatol.* **2016**, *152*, 546–552. [CrossRef]

25. Balagula, Y.; Braun, R.P.; Rabinovitz, H.S.; Dusza, S.W.; Scope, A.; Liebman, T.; Mordente, I.; Siamas, K.; Marghoob, A.A. The significance of crystalline/chrysalis structures in the diagnosis of melanocytic and nonmelanocytic lesions. *J. Am. Acad. Dermatol.* **2012**, *67*, 194.e1–194.e8. [CrossRef] [PubMed]

26. Liebman, T.; Rabinovitz, H.; Dusza, S.; Marghoob, A. White shiny structures: Dermoscopic features revealed under polarized light. *J. Eur. Acad. Dermatol. Venereol.* **2011**, *26*, 1493–1497. [CrossRef]

27. Navarrete-Dechent, C.; Liopyris, K.; Rishpon, A.; Marghoob, N.G.; Cordova, M.; Dusza, S.W.; Sahu, A.; Kose, K.; Oliviero, M.; Rabinovitz, H.; et al. Association of Multiple Aggregated Yellow-White Globules With Nonpigmented Basal Cell Carcinoma. *JAMA Dermatol.* **2020**, *156*, 882–890. [CrossRef] [PubMed]

28. Giacomel, J.; Lallas, A.; Argenziano, G.; Reggiani, C.; Piana, S.; Apalla, Z.; Ferrara, G.; Moscarella, E.; Longo, C.; Zalaudek, I. Dermoscopy of basosquamous carcinoma. *Br. J. Dermatol.* **2013**, *169*, 358–364. [CrossRef]

29. Akay, B.N.; Saral, S.; Heper, A.O.; Erdem, C.; Rosendahl, C. Basosquamous carcinoma: Dermoscopic clues to diagnosis. *J. Dermatol.* **2017**, *44*, 127–134. [CrossRef] [PubMed]

30. Popadic, M. Dermoscopy of aggressive basal cell carcinomas. *Indian J. Dermatol. Venereol. Leprol.* **2015**, *81*, 608–610. [CrossRef]

31. Moon, H.-R.; Park, T.J.; Ro, K.W.; Ryu, H.J.; Seo, S.H.; Son, S.W.; Kim, I.-H. Pigmentation of basal cell carcinoma is inversely associated with tumor aggressiveness in Asian patients. *J. Am. Acad. Dermatol.* **2019**, *80*, 1755–1757. [CrossRef]

MDPI

Commentary

Basosquamous Carcinoma: A Commentary

Christina Fotiadou *, Zoe Apalla and **Elizabeth Lazaridou**

Second Department of Dermatology—Venereology, Aristotle University, Medical School, 54124 Thessaloniki, Greece; zapalla@auth.gr (Z.A.); bethlaz@auth.gr (E.L.)
* Correspondence: cifotiad@auth.gr; Tel.: +30-2410553993

Simple Summary: Basosquamous carcinoma is a rare, aggressive non-melanoma skin cancer with features that lie between those of basal cell carcinoma and squamous cell carcinoma. A lot of controversy has been raised around the classification, pathogenesis, histologic morphology, biologic behavior, prognosis and management of this tumor. This is a narrative review based on articles published on PubMed in English language which had in their title the terms "basosquamous carcinoma" and/or "metatypical carcinoma of the skin". The aim of this review was to summarize and evaluate the latest data of the English literature regarding epidemiology, clinical presentation, dermoscopic and histopathologic characteristics, as well as the genetics and management of BSC to better characterize basosquamous skin lesions.

Abstract: Basosquamous carcinoma is a rare, aggressive non-melanoma skin cancer with features that lie between those of basal cell carcinoma and squamous cell carcinoma. A lot of controversy has been raised around the classification, pathogenesis, histologic morphology, biologic behavior, prognosis and management of this tumor. This is a narrative review based on an electronic search of articles published in PubMed in English language which had in their title the terms "basosquamous carcinoma" and/or "metatypical carcinoma of the skin". The aim of this review was to summarize and evaluate current data regarding epidemiology, clinical presentation, dermoscopic and histopathologic characteristics, as well as the genetics and management of BSC, in order to shed some more light onto this intriguing entity. As a conclusion, dermoscopy, deep incisional biopsies and immunohistologic techniques (Ber-EP4) should be applied in clinically suspicious lesions in order to achieve an early diagnosis and better prognosis of this tumor. Surgical treatments, including wide excision and Mohs' micrographic surgery, remain the treatment of choice. Finally, vismodegib, a Hedgehog pathway inhibitor, must be thoroughly investigated, with large controlled trials, since it may offer an alternative solution to irresectable or difficult-to-treat, locally advanced cases of basosquamous carcinoma.

Keywords: basosquamous carcinoma; metatypical basal cell carcinoma; diagnosis; treatment; biologic behavior; dermoscopy; histopathology; Mohs' micrographic surgery; genetics; vismodegib

Citation: Fotiadou, C.; Apalla, Z.; Lazaridou, E. Basosquamous Carcinoma: A Commentary. *Cancers* **2021**, *13*, 6146. https://doi.org/10.3390/cancers13236146

Academic Editors: Philip R. Cohen and Karl D. Lewis

Received: 19 October 2021
Accepted: 4 December 2021
Published: 6 December 2021

Publisher's Note: MDPI stays neutral with regard to jurisdictional claims in published maps and institutional affiliations.

1. Introduction

Basosquamous carcinoma (BSC) is a rare, relatively aggressive non-melanoma skin tumor which has raised a lot of controversy regarding its classification, its pathogenesis and its management since it was first described in the early 20th century. Multiple diagnostic and treatment challenges arise from the fact that BSC has a variable and non-characteristic clinical and histologic morphology seated between that of basal cell carcinoma (BCC) and squamous cell carcinoma (SCC) that is followed by a rather unpredictable biologic behavior.

This article aimed to summarize and evaluate all the latest data of the English language literature regarding epidemiology, clinical presentation, dermoscopic and histopathologic characteristics, as well as the genetics and management of BSC, in order to shed some more light to this intriguing entity.

2. Materials and Methods

A PubMed search of articles in English was conducted using the term BSC and/or metatypical BCC alone or with the following subheadings: classification, incidence, epidemiology, diagnosis, histopathology, dermoscopy, genetics, biologic behavior, treatment.

Inclusion criteria were: (1) case series or case reports of BSC; and (2) review articles, meta-analyses and systematic reviews on BSC;

Exclusion criteria were: (1) articles in a language other than English (i.e., French, German, Spanish, Chinese, etc.); and (2) articles about BSC of organs other than the skin (i.e., larynx, nasopharynx, lungs, anus, etc.);

Additionally, selected articles that were referenced in the included publications were used to support the discussion of our review.

3. Results

3.1. Background and Definition

The very first description of this tumor was made by Beadles back in 1894 and it was considered a type of rodent ulcer [1,2]. He described a lesion with features both of BCC and SCC which could not be clearly separated [1,2]. Again, in 1910, in a larger series of rodent ulcers, McCormack wrote about tumors with intermixed basaloid and squamous features [2,3]. For the first time, in 1928, the term BSC was used by Montgomery in order to describe 17 out of 119 carcinomas which he believed were transitional between basal and squamous cell carcinomas [4,5].

In the following years, the origin and the definition of this entity troubled pathologists [6,7]. Some of them considered these lesions as collisions of separate primary BCC and SCCs, or variants of BCC that form keratin while others believed them to be independent tumors with features of both BCC and SCC [2,8,9]. Meanwhile, in part of the literature, BSC were referred to as "metatypical carcinomas" [5,10,11].

In more recent years, the theory of summarization is gaining ground [12]. According to this, the BSC is derived from a BCC with genetic alterations that undergo squamous differentiation [7,11,12]. This theory is reflected in the most recent definition of BSC by World Health Organization (WHO) in the textbook "WHO classification of skin tumors" which stated that: "Basosquamous carcinoma is a term used to describe basal cell carcinomas that are associated with squamous differentiation" [13]. At the same time, the National Comprehensive Cancer Network (NCCN) states that basosquamous carcinomas have a metastatic capacity that is more similar to that of SCC than BCC [5,7,14].

3.2. Epidemiology

BCCs and SCCs are the most common non-melanoma skin cancers with a rising incidence in the general population, while BSC is considered a relatively rare entity [7]. Up to date, very few small studies have evaluated the specific epidemiologic characteristics of BSC [5,6,15]. According to previous review papers, the incidence of BSC ranges from 1.7 to 2.7% [5].

Indeed, Shuller et al. reported a BSC incidence of 1.2%, Martin et al. reported a BSC incidence of 1.4% while Bowman et al.—in a retrospective study of cases treated with Moh's micrographic surgery—found an incidence of 2.7% [6,16,17]. A newer retrospective study by Ciążyńska et al. revealed 180 lesions of BSC during a period of 20 years (1999–2019) which corresponded to 2.1% of all non-melanoma skin cancers (NMSCs) [7]. However, Gualdi et al., in a prospective study (2012–2015) which included 6042 NMSCs, reported a rate of 4.8% for BSCs, a percentage considerably higher than ever before [15].

3.3. Clinical and Demographic Characteristics

The clinical presentation of BSC is nonspecific and it does not present particular differences as compared to a common BCC (Figure 1a) [5,7]. The most common clinical scenario in BSC is a long-standing nodule that gradually becomes ulcerated. A similar clinical course was also described for metastatic BCCs [2,7,18]. Sun-exposed areas of the head and neck (82–97%), especially the perinasal area and ears, are the most frequent anatomic locations of this tumor [2,5–7,17–19]. However, though relatively fewer, BSCs were also observed in the trunk and extremities [2,5–7,17–19].

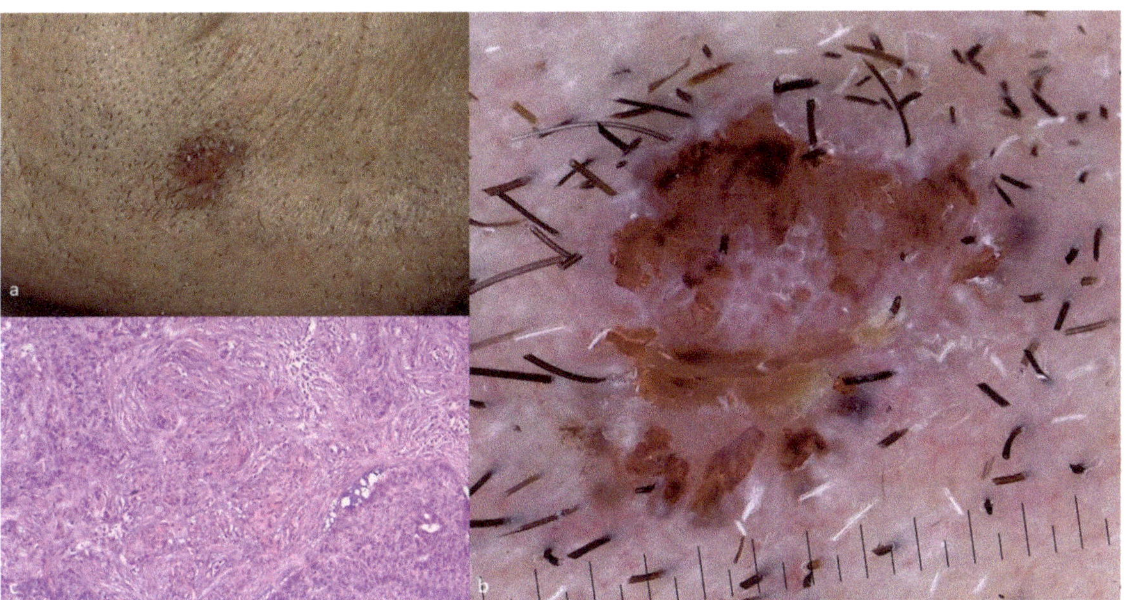

Figure 1. (**a**) Clinical image of a BSC; (**b**) dermoscopic image of a BSC revealing both a blue-grey ovoid nest and white circles; and (**c**) histologic image of a BSC (hematoxylin-eosin, 10×).

Fitzpatrick type I–II and high UVR exposure are considered potent risk factors for the development of a BSC [11]. The tumor usually develops in elderly individuals (34.4% of all BSCs are found in patients over 70 years of age) with a strong male preponderance [6,12,15,17].

3.4. Diagnosis of Basosquamous Carcinoma

The early and concise diagnosis of this tumor is crucial due to its potentially aggressive biologic behavior. However, it is a rather intriguing task because there is an absence of a standardized diagnosing protocol. The differential diagnosis may include entities such as viral wart, seborrheic keratosis (SK), hyperkeratotic actinic keratosis (AK), Bowen's disease (BD), BCC, invasive SCC, as well as amelanotic/hypomelanotic melanoma (AHM) [6,17,20,21].

3.4.1. Dermoscopy of BSC

Dermoscopy reveals features unrecognizable during naked-eye clinical examination and it is recommended for the early recognition of skin cancer. In this context, dermoscopy may raise the suspicion of BSC by the recognition of certain dermoscopic characteristics, seen in both types of differentiation, namely basaloid and squamoid (Figure 1b).

Unfortunately, due to the rarity of this entity, very few studies evaluating the dermoscopic features of BSC have been conducted until now [22,23]. In a study by Giacomel et al., the dermoscopic pattern of BSCs was characterized by features of both BCCs and SCCs, mirroring its complicated mixed histopathology [22]. In detail, the commonest dermoscopic criteria for BSCs were unfocused arborizing vessels, keratin masses, white structureless areas, scale, ulceration or blood crusts, white structures, blue-grey blotches and blood spots in keratin masses [22]. Finally, according to the authors, the most important dermoscopic clue that should raise suspicion for a possible BSC diagnosis is the simultaneous presence of at least one feature of both invasive SCC and BCC [22]. The study of Acay et al., in 2017, also demonstrated that BSC is dermoscopically characterized by BCC-related polymorphous or monomorphous vasculature, combined with findings of keratinization [23]. Specifically, a serpentine of branched vessels was the most common vascular finding while keratin masses, ulceration, and white structureless areas were the most common non-vascular features [23].

3.4.2. Histopathologic Features of BSC

Biopsy and histologic examination remain the gold standard diagnostic method for BSC (Figure 1c). The majority of the published literature on the subject, mostly including case series, retrospective studies, and review articles, describe the presence of both BCC and SCC histologic characteristics with a transition zone between them. However, there is a certain controversy regarding how these features are arranged within the lesions [2,5,11]. The transition zone is considered, by most authors, as a tissue which depicts a transitional stage of differentiation between BCC and SCC cells and not simply an area with atypical BCC cells [2,5,24]. The BCC component of a BSC usually contains basaloid cells with a small cytoplasm and large, uniform, pale, nuclei, whilst the SCC element consists of accumulations of polygonal squamous cells containing voluminous eosinophilic cytoplasm, larger open nuclei with prominent nucleoli and frequent mitosis [2,17,24,25]. These aggregates of squamoid cells are either found inside the basaloid islands or as the other authors describe, adjacent to them [2,5,23,24,26,27]. Two entities that may mimic and should be considered in the histologic differential diagnosis of BSC are a collision tumor and a keratinizing BCC. In the former case, the BCC and SCC areas are completely separated with no transition zone and in the latter case, there is abrupt keratinization in the center of a nodular BCC lesion without the intervening areas of squamoid cells [1,2,28]. Finally, the correct histologic diagnosis of a BSC can be jeopardized when the biopsy is superficial and not incisional. In this scenario, the lack of deep areas of the lesion in the sample, where the squamoid characteristics often lay, may result in the incorrect interpretation of the tumor as a classic BCC [1,2].

3.4.3. Immunohistologic Features of BSC

The histologic diagnosis of a BSC may be strengthened by the application of an immunohistochemical criterion such as human epithelial antigen (HEA) expression [20]. Despite the fact that no specific immunohistologic marker for BSCs exists, the Ber-EP4, an anti-HEA mouse monoclonal antibody, which is usually strongly positive in BCCs and the epithelial membrane antigen (EMA), which stains positive in SCCs, have proved to be helpful [2]. In a "typical" BSC, the BCC element is Ber-EP4, cytokeratin (AE1) and cytokeratin (AE3) positive, whereas the area of SCC is AE1, AE3, and CAM5.2 positive with variable staining for EMA [2,5,18]. Moreover, the transition zone, though not always present, is typically characterized by a gradual decline in Ber-EP4 staining as an indication of transitioning from the area of the basaloid to the area of squamous differentiation [2,5,29].

3.5. Genetics and Pathogenesis

The genetic origin of BSCs has not been fully clarified yet [5]. Although these tumors share histologic features of both BCCs and SCCs, the exact gene mutations that lead to their formation is still a matter of controversy [5,30]. On the other hand, the molecular background of BCC and SCC has been thoroughly studied. BCC derives from the over activation of the sonic hedgehog (HH) pathway which inhibits a transmembrane protein called PTCH or activates a transmembrane protein called SMO [31,32]. Other genetic drivers of BCC include PTEN, MYCN, PPP6C, GRIN2A, GLI1, CSMD3, DCC, PREX2, and APC [30,33,34]. SCC is characterized by a greater variety of gene mutations which include mutations in HRAS and disruptions of the TGFBR1, TGFBR2, NOTCH1, and NOTCH2 genes as well as additional mutations in CASP8, CDKN2A, NOTCH3, KRAS,

NRAS, PDK1, BAP1, AJUBA, KMT2D, MYH9, TRAF3, NSD1, CDH1, and TP63 [30,35–38]. A recent, very interesting study by Chiang tried to define the genomic alterations that characterize BSC by using the targeted sequencing of 1641 cancer genes from 20 BSCs, whole exome sequencing from 16 BCCs, and a mixture of previously published whole-exome and whole-genome datasets from 52 SCCs [30–39]. According to the findings of this study, the majority of BSCs had underlying PTCH1 and SMO mutations in addition to mutations in other known BCC drivers such as MYCN, PPP6C, GRIN2A, CSMD3, DCC, PREX2, APC, PTEN, and PIK3CA [30]. These data support the theory that the HH signaling pathway is the initial driver of BSC and that this tumor probably originates as a BCC that partially squamatizes through the accumulation of ARID1A mutations and RAS/MAPK pathway activation [30,39]. The most frequent gene mutations that characterize BCCs, SCCs, and BSCs are shown in Table 1.

Table 1. The most frequent gene mutations that characterize BCCs, SCCs and BSCs.

BCCs	SCCs	BSCs
PTCH	HRAS	PTCH
SMO	TGFBR1	SMO
PTEN	TGFBR2	MYCN
MYCN	NOTCH1	PPP6C
PPP6C	NOTCH2	GRIN2A
GRIN2A	CASP8	CSMD3
GLI1	CDKN2A	DCC
CSMD3	NOTCH3	PREX2
DCC	KRAS	APC
PREX2	NRAS	PTEN
APC	PDK1	PIK3CA
	BAP1	ARIDIA
	AJUBA	
	KMT2D	
	CDH1	

3.6. Biologic Behavior and Prognosis

According to many authors, in opposition to BCC, BSC is characterized by a more aggressive biologic behavior that is almost equal to that of SCC [2,5,18,25,40]. This "aggressiveness" is defined as a more dynamic topical growth of the tumor, as well as a greater potential for recurrences and metastases. As a consequence, the latter characteristics are associated with a worse prognosis as compared to the classic BCC [2,5].

Indeed, a relatively recent study by Zhu et al. including 19 patients with metastatic BSCs suggested an intermediate prognosis between that of BCC and SCC but more favorable than previous assumptions [40]. Volkestein et al. reported that the local recurrence rate of BSC—after wide surgical excision—may reach 45%, which is almost double that of BCC and SCC [18]. In another study, in which the physicians applied Mohs' micrographic surgery, the topical recurrence rate for BSC decreased to 4–9%, but remained higher than BCC (0.64%) and SCC (1.2%) [2,14,41,42]. Martin et al. concluded that based on their analysis, the most important predictive factors of topical recurrence for BSCs were male sex, positive surgical margins, lymphatic, and perineural invasion [17]. The metastatic potential of BSC ranges between 4 and 8.4%, which is closer to that of SCC [2,41]. A recent study by Ciążyńska et al. reported that 40% of patients diagnosed with BSC had a second skin neoplasm. This percentage is significantly higher to the corresponding 23% reported with other non-melanoma skin cancers (NMSCs) and multiple lesions [7]. In this context, they concluded that BSC patients are more prone to the development of new primary skin cancers and for this reason, they should be closely monitored [7].

3.7. Therapeutic Options and Management of BSC

To date, there are no established, standard therapeutic guidelines for the treatment of BSCs. The rarity of these tumors along with the absence of robust literature data on the subject is the most reasonable explanation for this situation. Nevertheless, several treatment options have been applied with a variety of outcomes. Superficial methods such as curettage and electrodesiccation have been used in the past but are not considered first line treatment due to their high recurrence rate.

3.7.1. Wide Surgical Excision

Although a very high recurrence rate, reaching 45%, has been reported after wide surgical excision, this method remains a first line treatment choice for many authors [2,5,12, 16–18,20,21,24,43]. Furthermore, it is suggested that surgical excision should be followed by the evaluation of a lymph node and distant metastases, and of course, close clinical follow up for recurrence and metastasis [2,6,17].

3.7.2. Mohs' Micrographic Surgery (MMS)

Based on the results of recent studies, MMS is considered the optimal surgical option for BSCs, since it is linked to lower recurrence rates compared to the wide surgical excision [17,24,25,44,45]. Analytically, Skaria et al. reported an 8.9% recurrence rate with MMS that is much lower than the 45% observed with wide surgical excision, but significantly higher when compared with the recurrence rates reported for BCCs and SCCs [44]. Allen et al. achieved an even lower recurrence rate (4.9%) [45]. However, according to Oldbury et al., there are several practical and financial issues to be addressed in order to officially support MMS as the first-line treatment of choice for BSCs: (a) inability to pre-operatively choose the right candidates for MMS (i.e., clinical diagnosis of BSC instead of BCC) in everyday clinical practice; (b) higher cost of MMS compared to surgical excision; (c) more time-consuming process; and (d) non-applicable in many centers worldwide [12].

3.7.3. Sentinel Lymph Node Biopsy (SLNB)

SLNB has been suggested by some authors as part of the management of BSC. However, there is open controversy regarding whether SLNB should be offered in BSCs for the early diagnosis of occult nodal metastasis, staging and treatment of subclinical node disease [17,27]. Kakagia et al., in a prospective study with 142 patients with BSC, concluded that tumor size >2 cm in addition to lymphatic and perineural invasion are significant determinants of SLN micrometastasis [46]. In the absence of palpable lymphadenopathy, wide resection and SLNB with long-term follow up is highly recommended in these patients [46]. However, a possible "preventive benefit" of early SLNB for the nodal spread and distant metastasis needs further prospective controlled studies, with a longer follow-up period, in order to be confirmed [46].

3.7.4. Radiotherapy–Chemotherapy

Adjuvant radiotherapy has been proposed by several authors for BSC, in the scenario of positive surgical margins and the inability to re-excise the tumor in order to achieve them, or in cases with local lymph node metastasis [2,12,17,47]. Although there are currently no reports in the literature about the use of radiotherapy in BSCs, some authors, based on their experience with aggressive and/or metastatic BCCs, suggest that treatment with radiotherapy either alone or in combination with surgery may be an appropriate option for the management of BSC, if standard surgical excision or MMS is not possible [12,48]. In a few cases of metastatic BSCs in the literature, palliative chemotherapy has been used (adriamycine, cisplatine) [11].

3.7.5. New Emerging Therapies

Vismodegib, a Hedgehog pathway inhibitor, is approved for the treatment of adults with metastatic BCC, or with locally advanced BCC that has recurred after surgery, or adults who are not candidates for surgery or radiation [31]. According to the available literature data, the risk of SCC development in individuals treated with vismodegib is a matter of controversy [49–51]. Taking into consideration the fact that BSC pathologically displays features of both BCC and SCC, treatment with vismodegib might bear the risk of the progression of the SCC component of the tumor. Indeed, reports of SCCs developing in patients under vismodegib have been published and a case–control study with 180 patients found an increased risk of SCC in patients under treatment with vismodegib [49]. On the other hand, the largest published cohort study with 1675 patients suggested that vismodegib is not associated with an increased risk of SCC while a systematic and a narrative review of the literature concluded that the existing evidence does not justify an association between vismodegib and SCC formation [31,51,52]. Moreover, there are three very recent case reports including four patients in total which provide preliminary evidence that vismodegib might be effective for difficult-to-treat BSC (Table 2) [53–55]. In all of them, the lesions did not "transform" into purely squamous tumors but in contrast they completely remitted and in two cases the remission was maintained for a very long period of time [53–55]. Although the effect of vismodegib on BSC treatment definitely requires further investigation in larger controlled studies, it could prove to be a future solution for locally advanced BSCs despite the presence of a squamous component in the tumors.

Table 2. Studies which showed a complete response of locally advanced basosquamous carcinoma with vismodegib.

Source	Study Design	Number of Patients	Results
McGrane J. et al. [53]	Case report	1	Marked improvement, stable after 28 months
Apalla Z. et al. [54]	Case series	2	Complete clinical response with long-term follow up (12 and 18 months, respectively
Sahuquillo-Torralba A. et al. [55]	Case report	1	Complete response after 7 months

Other treatments options such as the PD-L1 inhibitors pebrolizumab and cepilumab have been used for the treatment of advanced-metastatic BCCs and SCCs but still there are no literature data for their use in advanced BSC [56].

4. Conclusions

BSC remains a controversial entity, which belongs to the group of NMSCs, sharing characteristics of both BCC and SCC. The differential diagnosis between a BSC and a BCC cannot be made on the grounds of clinical examination. On the other hand, dermoscopy could be more helpful in the diagnosis by revealing features of both components (BCC, SCC) that are otherwise unrecognizable to the naked eye. Histology is the gold standard of diagnosis. Early recognition and correct histologic classification contribute to the optimization of the management of the tumor. Deep incisional biopsies and immunohistochemical techniques (Ber-EP4 staining) facilitate correct diagnosis. In the absence of standardized treatment protocols for BSC, prospective studies comparing various treatment options are needed, in order to reach a consensus regarding ideal management. Surgical treatments, including wide excision and Mohs' micrographic surgery, remain the treatments of choice for most clinicians. The addition of SLNB, radiation therapy, and imaging monitoring in suspicious cases (tumor size >2 cm, perineural and lymphatic invasion) remain matters of controversy. Radiotherapy could have a supportive role, postoperatively, when re-excision is not possible or not allowed by the patient. Finally, a new treatment prospective such as vismodegib must be thoroughly investigated, with large controlled trials, since it may offer an alternative solution to irresectable or difficult-to-treat locally advanced cases of BSC.

Author Contributions: Conceptualization, E.L., Z.A.; supervision, E.L., Z.A.; review and editing of the draft, E.L., Z.A., C.F.; investigation of the review, C.F., Z.A.; writing—original draft preparation, C.F. All authors have read and agreed to the published version of the manuscript.

Funding: This research received no external funding.

Conflicts of Interest: The authors declare no conflict of interest.

References

1. De Stefano, A.; Dispenza, F.; Petrucci, A.G.; Citraro, L.; Croce, A. Features of biopsy in diagnosis of metatypical basal cell carcinoma (basosquamous carcinoma) of head and neck. *Otolaryngol. Pol.* **2012**, *66*, 419–423. [CrossRef]
2. Tan, C.Z.; Rieger, K.E.; Sarin, K.Y. Basosquamous carcinoma: Controversy, advances, and future directions. *Dermatol. Surg.* **2017**, *43*, 23–31. [CrossRef]
3. MacCormac, H. The relation of rodent ulcer to squamous cell carcinoma of the skin. *Arch. Middlesex. Hosp.* **1910**, *19*, 172–183. [CrossRef]
4. Hamilton, M. Basal squamous cell epithelioma. *Arch. Dermatol. Syph.* **1928**, *18*, 50–73. [CrossRef]
5. Garcia, C.; Poletti, E.; Crowson, A.N. Basosquamous carcinoma. *J. Am. Acad. Dermatol.* **2009**, *60*, 137–143. [CrossRef]
6. Bowman, P.H.; Ratz, J.L.; Knoepp, T.G.; Barnes, C.J.; Finlay, R.E. Basosquamous carcinoma. *Dermatol. Surg.* **2003**, *29*, 830–832. [CrossRef] [PubMed]
7. Ciążyńska, M.; Sławińska, M.; Kamińska-Winciorek, G.; Lange, D.; Lewandowski, B.; Reich, A.; Pabianek, M.; Szczepaniak, K.; Hankiewicz, A.; Ułańska, M.; et al. Clinical and epidemiological analysis of basosquamous carcinoma: Results of the multicenter study. *Sci. Rep.* **2020**, *10*, 1–8. [CrossRef]
8. Burston, J.; Clay, R.D. THE PROBLEMS OF HISTOLOGICAL DIAGNOSIS IN BASO-SQUAMOUS CELL CARCINOMA OF THE SKIN. *J. Clin. Pathol.* **1959**, *12*, 73–79. [CrossRef]
9. de Faria, J. Basal cell carcinoma of the skin with areas of squamous cell carcinoma: A basosquamous cell carcinoma? *J. Clin. Pathol.* **1985**, *38*, 1273–1277. [CrossRef]
10. Tarallo, M.; Cigna, E.; Frati, R.; Delfino, S.; Innocenzi, D.; Fama, U.; Corbianco, A.; Scuderi, N. Metatypical basal cell carcinoma: A clinical review. *J. Exp. Clin. Cancer Res.* **2008**, *27*, 65. [CrossRef] [PubMed]
11. Shukla, S.; Khachemoune, A. Reappraising basosquamous carcinoma: A summary of histologic features, diagnosis, and treatment. *Arch. Dermatol. Res.* **2020**, *312*, 605–609. [CrossRef]
12. Oldbury, J.W.; Wain, R.; Abas, S.; Dobson, C.M.; Iyer, S.S. Basosquamous Carcinoma: A Single Centre Clinicopathological Evaluation and Proposal of an Evidence-Based Protocol. *J. Ski. Cancer* **2018**, *2018*, 1–7. [CrossRef]
13. LeBoit, P.E.; International Agency for Research on Cancer; World Health Organization; International Academy of Pathology; European Organization for Research on Treatment of Cancer; UniversitätsSpital Zürich. *Pathology and Genetics of Skin Tumours*; IARC Press: Lyon, France, 2006.
14. Wermker, K.; Roknic, N.; Goessling, K.; Klein, M.; Schulze, H.-J.; Hallermann, C. Basosquamous Carcinoma of the Head and Neck: Clinical and Histologic Characteristics and Their Impact on Disease Progression. *Neoplasia* **2015**, *17*, 301–305. [CrossRef]
15. Gualdi, G.; Soglia, S.; Fusano, M.; Monari, P.; Giuliani, F.; Porreca, A.; Di Nicola, M.; Calzavara-Pinton, P.; Amerio, P. Characterization of Basosquamous Carcinoma. A distinct type of keratinizing tumour. *Acta Derm. Venereol.* **2021**, *101*, adv00353. [CrossRef] [PubMed]
16. Schuller, D.E.; Berg, J.W.; Sherman, G.; Krause, C.J. Cutaneous Basosquamous Carcinoma of the Head and Neck: A Comparative Analysis. *Otolaryngol. Neck Surg.* **1979**, *87*, 420–427. [CrossRef]
17. Martin, R.C.; Edwards, M.J.; Cawte, T.G.; Sewell, C.L.; McMasters, K.M. Basosquamous carcinoma: Analysis of prognostic factors influencing recurrence. *Cancer* **2000**, *88*, 1365–1369. [CrossRef]
18. Volkenstein, S.; Wohlschlaeger, J.; Liebau, J.; Arens, A.; Lehnerdt, G.; Jahnke, K.; Neumann, A. Basosquamous carcinoma-A rare but aggressive skin malignancy. *J. Plast. Reconstr. Aesthetic Surg.* **2010**, *63*, e304–e306. [CrossRef] [PubMed]
19. Borel, D.M. Cutaneous basosquamous carcinoma. Review of the literature and report of 35 cases. *Arch. Pathol.* **1973**, *95*, 293–297. [PubMed]
20. Mougel, F.; Kanitakis, J.; Faure, M.; Euvrard, S. Basosquamous cell carcinoma in organ transplant patients: A clinicopathologic study. *J. Am. Acad. Dermatol.* **2012**, *66*, e151–e157. [CrossRef]
21. Betti, R.; Crosti, C.; Ghiozzi, S.; Cerri, A.; Moneghini, L.; Menni, S. Basosquamous cell carcinoma: A survey of 76 patients and a comparative analysis of basal cell carcinomas and squamous cell carcinomas. *Eur. J. Dermatol. EJD* **2013**, *23*, 83–86. [CrossRef]
22. Giacomel, J.; Lallas, A.; Argenziano, G.; Reggiani, C.; Piana, S.; Apalla, Z.; Ferrara, G.; Moscarella, E.; Longo, C.; Zalaudek, I. Dermoscopy of basosquamous carcinoma. *Br. J. Dermatol.* **2013**, *169*, 358–364. [CrossRef] [PubMed]
23. Akay, B.N.; Saral, S.; Heper, A.O.; Erdem, C.; Rosendahl, C. Basosquamous carcinoma: Dermoscopic clues to diagnosis. *J. Dermatol.* **2016**, *44*, 127–134. [CrossRef] [PubMed]
24. Leibovitch, I.; Huilgol, S.C.; Selva, D.; Richards, S.; Paver, R. Basosquamous carcinoma: Treatment with Mohs micrographic surgery. *Cancer* **2005**, *104*, 170–175. [CrossRef] [PubMed]
25. Costantino, D.; Lowe, L.; Brown, D.L. Basosquamous carcinoma—an under-recognized, high-risk cutaneous neoplasm: Case study and review of the literature. *J. Plast. Reconstr. Aesthetic Surg.* **2006**, *59*, 424–428. [CrossRef]

26. Anand, R.L.; Collins, D.; Chapman, A. Basosquamous carcinoma: Appearance and reality. *Oxf. Med. Case Rep.* **2017**, *2017*, omw095. [CrossRef]
27. Jankovic, I.; Kovacevic, P.; Visnjic, M.; Jankovic, D.; Binic, I.; Jankovic, A.; Ilic, I. Application of sentinel lymph node biopsy in cutaneous basosquamous carcinoma. *Ann. Dermatol.* **2011**, *23* Suppl. 1, S123–S126. [CrossRef]
28. Malloney, M.L. What is basosquamous carcinoma? *Dermatol. Surg.* **2000**, *26*, 505–506. [CrossRef]
29. Jones, M.S.; Helm, K.F.; Maloney, M.E. The Immunohistochemical Characteristics of the Basosquamous Cell Carcinoma. *Dermatol. Surg.* **1997**, *23*, 181–184. [CrossRef]
30. Chiang, A.; Tan, C.Z.; Kuonen, F.; Hodgkinson, L.M.; Chiang, F.; Cho, R.J.; South, A.P.; Tang, J.Y.; Chang, A.L.S.; Rieger, K.E. Genetic mutations underlying phenotypic plasticity in basosquamous carcinoma. *J. Investig. Dermatol.* **2019**, *139*, 2263–2271.e5. [CrossRef]
31. Apalla, Z.; Papageorgiou, C.; Lallas, A.; Sotiriou, E.; Lazaridou, E.; Vakirlis, E.; Kyrgidis, A.; Ioannides, D. Spotlight on vismodegib in the treatment of basal cell carcinoma: An evidence-based review of its place in therapy. *Clin. Cosmet. Investig. Dermatol.* **2017**, *ume 10*, 171–177. [CrossRef]
32. Marzuka, A.G.; Book, S.E. Basal Cell Carcinoma: Pathogenesis, Epidemiology, Clinical Features, Diagnosis, Histopathology, and Management. *Yale J. Boil. Med.* **2015**, *88*, 167–179.
33. Bonilla, X.; Parmentier, L.; King, B.; Bezrukov, F.; Kaya, G.; Zoete, V.; Seplyarskiy, V.B.; Sharpe, H.J.; McKee, T.; Letourneau, A.; et al. Genomic analysis identifies new drivers and progression pathways in skin basal cell carcinoma. *Nat. Genet.* **2016**, *48*, 398–406. [CrossRef] [PubMed]
34. Jayaraman, S.S.; Rayhan, D.J.; Hazany, S.; Kolodney, M.S. Mutational landscape of basal cell carcinomas by whole-exome sequencing. *J. Invest. Dermatol.* **2014**, *134*, 213–220. [CrossRef] [PubMed]
35. Cammareri, P.; Rose, A.M.; Vincent, D.F.; Wang, J.; Nagano, A.; Libertini, S.; Ridgway, R.A.; Athineos, D.; Coates, P.J.; McHugh, A.; et al. Inactivation of TGFbeta receptors in stem cells drives cutaneous squamous cell carcinoma. *Nat. Commun.* **2016**, *7*, 12493. [CrossRef]
36. South, A.P.; Purdie, K.J.; Watt, S.A.; Haldenby, S.; Breems, N.Y.D.; Dimon, M.; Arron, S.; Kluk, M.J.; Aster, J.C.; McHugh, A.; et al. NOTCH1 Mutations Occur Early during Cutaneous Squamous Cell Carcinogenesis. *J. Investig. Dermatol.* **2014**, *134*, 2630–2638. [CrossRef] [PubMed]
37. Pickering, C.R.; Zhou, J.H.; Lee, J.J.; Drummond, J.A.; Peng, S.A.; Saade, R.E.; Tsai, K.Y.; Curry, J.L.; Tetzlaff, M.T.; Lai, S.Y.; et al. Mutational Landscape of Aggressive Cutaneous Squamous Cell Carcinoma. *Clin. Cancer Res.* **2014**, *20*, 6582–6592. [CrossRef] [PubMed]
38. Schwaederle, M.; Elkin, S.K.; Tomson, B.N.; Carter, J.L.; Kurzrock, R. Squamousness: Next-generation sequencing reveals shared molecular features across squamous tumor types. *Cell Cycle* **2015**, *14*, 2355–2361. [CrossRef] [PubMed]
39. Tarapore, E.; Atwood, S.X. Defining the genetics of Basoaquamous Carcinoma. *J. Invest. Dermatol.* **2019**, *139*, 2258–2260. [CrossRef] [PubMed]
40. Zhu, G.A.; Danial, C.; Liu, A.; Li, S.; Chang, A.L.S. Overall and progression-free survival in metastatic basosquamous cancer: A case series. *J. Am. Acad. Dermatol.* **2014**, *70*, 1145–1146. [CrossRef]
41. Alam, M.; Desai, S.; Nodzenski, M.; Dubina, M.; Kim, N.; Martini, M.; Fife, D.; Reid, D.; Pirigyi, M.; Poon, E.; et al. Active ascertainment of recurrence rate after treatment of primary basal cell carcinoma (BCC). *J. Am. Acad. Dermatol.* **2015**, *73*, 323–325. [CrossRef]
42. Belkin, D.; Carucci, J.A. Mohs Surgery for Squamous Cell Carcinoma. *Dermatol. Clin.* **2011**, *29*, 161–174. [CrossRef] [PubMed]
43. Kececi, Y.; Argon, A.; Kebat, T.; Sir, E.; Gungor, M.; Vardar, E. Basosquamous carcinoma: Is it an aggressive tumor? *J. Plas. Surg. Hand Surg.* **2015**, *49*, 107–111. [CrossRef]
44. Skaria, A. Recurrence of Basosquamous Carcinoma after Mohs Micrographic Surgery. *Dermatology* **2010**, *221*, 352–355. [CrossRef]
45. Allen, K.J.; Cappel, M.A.; Killian, J.M.; Brewer, J.D. Basosquamous carcinoma and metatypical basal cell carcinoma: A review of treatment with Mohs micrographic surgery. *Int. J. Dermatol.* **2014**, *53*, 1395–1403. [CrossRef]
46. Kagakia, D.; Zapandioti, P.; Tryspiannis, G.; Grekou, A.; Tsoutsos, D. Sentinel lymph node metastasis in primary cutaneous basosquamous carcinoma. A cross-sectional study. *J. Surg. Oncol.* **2018**, *117*, 1752–1758. [CrossRef] [PubMed]
47. Deganello, A.; Gitti, G.; Struijs, B.; Paiar, F.; Gallo, O. Palliative combined treatment for unresectable cutaneous basosquamous cell carcinoma of the head and neck. *Acta Otorhinolaryngol. Ital.* **2013**, *33*, 353–356. [PubMed]
48. Farmer, E.R.; Helwig, E.B. Metastatic basal cell carcinoma: A clinicopathologic study of seventeen cases. *Cancer* **1980**, *46*, 748–757. [CrossRef]
49. Mohan, S.V.; Chang, J.; Li, S.; Henry, S.; Wood, D.J.; Chang, A.L.S. Increased risk of cutaneous squamous cell carcinoma after vismodegib therapy for basal cell carcinoma. *JAMA Dermatol.* **2016**, *152*, 527–532. [CrossRef] [PubMed]
50. Puig, S.; Sampogna, F.; Tejera-Vaquerizo, A. Study on the risk of cutaneous squamous cell carcinoma after vismodegib therapy for basal cell carcinoma: Not a case-control study. *JAMA Dermatol.* **2016**, *152*, 1172–1173. [CrossRef] [PubMed]
51. Bhutani, T.; Abrouk, M.; Sima, C.S.; Sadetsky, N.; Hou, J.; Caro, I.; Chren, M.-M.; Arron, S. Risk of cutaneous squamous cell carcinoma after treatment of basal cell carcinoma with vismodegib. *J. Am. Acad. Dermatol.* **2017**, *77*, 713–718. [CrossRef] [PubMed]
52. Jacobsen, A.A.; Aldahan, A.S.; Hughes, O.B.; Shah, V.V.; Strasswimmer, J. Hedgehog pathway inhibitor therapy for locally advanced and metastatic basal cell carcinoma: A systematic review and pooled analysis of interventional studies. *JAMA Dermatol.* **2016**, *152*, 816–824. [CrossRef]

53. McGrane, J.; Carswell, S.; Talbot, T. Metastatic spinal cord compression from basal cell carcinoma of the skin treated with surgical decompression and vismodegib: Case report and review of Hedgehog signalling pathway inhibition in advanced basal cell carcinoma. *Clin. Exp. Dermatol.* **2017**, *42*, 80–83. [CrossRef] [PubMed]
54. Apalla, Z.; Giakouvis, V.; Gavros, Z.; Lazaridou, E.; Sotiriou, E.; Bobos, M.; Vakirlis, E.; Ioannides, D.; Lallas, A. Complete response of locally advanced basosquamous carcinoma to vismodegib in two patients. *Eur. J. Dermatol. EJD* **2019**, *29*, 102–104. [PubMed]
55. Sahuquillo-Torralba, A.; Llavador-Ros, M.; Caballero-Daroqui, J.; Botella-Estrada, R. Complete response of a locally advanced basosquamous carcinoma with vismodegib treatment. *Indian J. Dermatol. Venereol. Leprol.* **2019**, *85*, 549–552. [CrossRef]
56. Shalhout, S.Z.; Emerick, K.S.; Kaufman, H.L.; Miller, D.M. Immunotherapy for Non-melanoma Skin Cancer. *Curr. Oncol. Rep.* **2021**, *23*, 1–10. [CrossRef] [PubMed]

Article

Characterizing Malignant Melanoma Clinically Resembling Seborrheic Keratosis Using Deep Knowledge Transfer

Panagiota Spyridonos [1,*], George Gaitanis [2], Aristidis Likas [3] and Ioannis Bassukas [2,*]

[1] Department of Medical Physics, Faculty of Medicine, School of Health Sciences, University of Ioannina, 45110 Ioannina, Greece
[2] Department of Skin and Venereal Diseases, Faculty of Medicine, School of Health Sciences, University of Ioannina, 45110 Ioannina, Greece; ggaitan@uoi.gr
[3] Department of Computer Science & Engineering, School of Engineering, University of Ioannina, 45110 Ioannina, Greece; arly@cs.uoi.gr
* Correspondence: pspyrid@uoi.gr (P.S.); ibassuka@uoi.gr (I.B.)

Simple Summary: Malignant melanomas (MMs) with aypical clinical presentation constitute a diagnostic pitfall, and false negatives carry the risk of a diagnostic delay and improper disease management. Among the most common, challenging presentation forms of MMs are those that clinically resemble seborrheic keratosis (SK). On the other hand, SK may mimic melanoma, producing 'false positive overdiagnosis' and leading to needless excisions. The evolving efficiency of deep learning algorithms in image recognition and the availability of large image databases have accelerated the development of advanced computer-aided systems for melanoma detection. In the present study, we used image data from the International Skin Image Collaboration archive to explore the capacity of deep knowledge transfer in the challenging diagnostic task of the atypical skin tumors of MM and SK.

Citation: Spyridonos, P.; Gaitanis, G.; Likas, A.; Bassukas, I. Characterizing Malignant Melanoma Clinically Resembling Seborrheic Keratosis Using Deep Knowledge Transfer. *Cancers* **2021**, *13*, 6300. https://doi.org/10.3390/cancers13246300

Academic Editor: Aimilios Lallas

Received: 25 October 2021
Accepted: 13 December 2021
Published: 15 December 2021

Publisher's Note: MDPI stays neutral with regard to jurisdictional claims in published maps and institutional affiliations.

Abstract: Malignant melanomas resembling seborrheic keratosis (SK-like MMs) are atypical, challenging to diagnose melanoma cases that carry the risk of delayed diagnosis and inadequate treatment. On the other hand, SK may mimic melanoma, producing a 'false positive' with unnecessary lesion excisions. The present study proposes a computer-based approach using dermoscopy images for the characterization of SK-like MMs. Dermoscopic images were retrieved from the International Skin Imaging Collaboration archive. Exploiting image embeddings from pretrained convolutional network VGG16, we trained a support vector machine (SVM) classification model on a data set of 667 images. SVM optimal hyperparameter selection was carried out using the Bayesian optimization method. The classifier was tested on an independent data set of 311 images with atypical appearance: MMs had an absence of pigmented network and had an existence of milia-like cysts. SK lacked milia-like cysts and had a pigmented network. Atypical MMs were characterized with a sensitivity and specificity of 78.6% and 84.5%, respectively. The advent of deep learning in image recognition has attracted the interest of computer science towards improved skin lesion diagnosis. Open-source, public access archives of skin images empower further the implementation and validation of computer-based systems that might contribute significantly to complex clinical diagnostic problems such as the characterization of SK-like MMs.

Keywords: melanoma; seborrheic keratosis; SK-like MM; deep learning; knowledge transfer

1. Introduction

Malignant melanomas (MMs) with atypical clinical presentation constitute a diagnostic pitfall, and false negatives carry the risk of a diagnostic delay and improper disease management [1,2]. Among the most common, challenging presentation forms of MMs are those that clinically resemble seborrheic keratosis (seborrheic keratosis-like MMs, SK-like MMs) [3].

SK is one of the most frequently diagnosed benign skin tumors in everyday clinical practice. It is a hallmark of aged, chronically sun-exposed skin of older individuals, with well-characterized, in most cases, diagnostic clinical features. The patients are usually alarmed about the sometimes rapidly growing exophytic lesions; however, in most cases, they can be assured that their growths are benign simply based on the clinical examination and without the need for histologic confirmation. Moreover, in many clinically doubtful cases, an additional dermoscopic assessment of the suspect lesion enables a clear-cut diagnosis of the condition based on a series of well-elaborated, typical dermoscopic features [4]. However, none of the SK dermoscopic findings is specific to SK [4], as they can be observed in other skin tumors, including malignant ones, among which are also distinct MM cases (SK-like MMs) [5].

The true incidence of SK-like MM is largely unknown since many of these lesions are misdiagnosed as SK on the basis of the clinical and dermoscopic examination and are not biopsied at this stage [3]. Izikson et al. [6], in a retrospective study covering ten years (1992 to 2001), retrieved 9204 pathology reports of material admitted with the clinical differential diagnosis SK. Melanoma was confirmed by histological examination in 61 of these cases (0.66%).

SK-like melanoma shares clinical and dermoscopic features of SK and melanoma, making the diagnosis challenging. A somewhat regular shape and the presence of benign dermoscopic patterns suggestive of an SK lead to underestimating the true malignant nature of this type of lesion. This ambiguity in the diagnosis was highlighted in a study by Carrera et al. [7] in which 54 dermatologists with about ten years of clinical practice clinically misdiagnosed 40% of 134 SK-like melanomas as benign lesions. An additional dermoscopic evaluation could improve the overall diagnostic accuracy from 60.9 to 68.1%, i.e., not more than by about 20%. Additionally, in the largest dermoscopic study of SK-like melanomas to date, the dermoscopy score and the seven-point checklist score showed benignity range with values 4.2 and 2 [5]. In the same study, Carrera et al. found that the most helpful criteria in correctly diagnosing SK-like MMs, despite the presence of other SK features, were the identification of blue–white veil, streaks, and a pigmented network [5]. Noninvasive optical methods, such as reflectance confocal microscopy (RCM) and optical coherence tomography can be employed to improve accuracy in melanoma diagnosis [8–10]. However, in SK-like MMs, the application has been limited due to frequent clinical, dermoscopic misdiagnosis [3].

The diagnostic grey zone between SK and MM becomes even broader as SK mimicking melanomas (MMs-like SK) have also been reported, with an increased risk of false MM diagnoses [11–14]. Dermoscopy of typical SK is characterized by milia-like cysts, comedo-like openings, and brain-like and finger-like structures [4]. However, pigmented SK can sometimes present dermoscopic patterns that mimic melanocytic lesions, the most frequent of which is the so-called false pigmented network. Dermoscopic evaluation of 402 lesions indicated that pigmented SK could show at least one of the criteria most predictive of melanocytic proliferations [11].

Recent studies have highlighted the contribution of RCM in characterizing MM-like SK. Farnetani et al. [15] retrospectively evaluated RCM images of atypical SK lesions suspicious of MM at dermoscopy to identify a diagnostic approach able to minimize surgical biopsies or excisions. They assessed 111 facial lesions with histological SK diagnosis. By dermoscopy, most lesions (n = 83 lesions, 75%) were classified as melanocytic-like. With RCM, only 16% were classified as suspicious of malignancy, with the remaining 84% considered 'SK-like'. The presence of RCM features associated with typical SK, the rare presence of melanoma-associated features, and the absence of medusa head-like structures seem to be the most sensitive indicators for atypical SK facial lesions.

In another retrospective study, Pezzine et al. [16], applied RCM to analyze excised skin lesions with a ≥ 1 score of the revisited seven-point dermoscopy checklist [17]. Their objective was to evaluate the agreement of RCM classification and histological diagnoses and the reliability of well-known RCM criteria for SK in identifying SK with atypical dermoscopy presentation. An excellent agreement (97%) was confirmed for RCM and histopathologic examination for SK with atypical dermoscopy presentation, allowing an effective noninvasive differential diagnosis. More importantly, RCM features in this group of atypical lesions were similar to those described for typical SK cases.

Recently, computer-aided diagnosis (CAD) systems are increasingly combined with various noninvasive imaging techniques to encompass advanced image processing and enable the application of artificial intelligence (AI) methods to improve diagnostic accuracy [18–20]. In the field of quantitative noninvasive optical techniques, Bozsànyi et al. [21] assessed the usefulness of spectral reflectance and autofluorescence measurements of MM and SK for their accurate differentiation. Using image analysis, they have extracted quantitative autofluorescence intensity measures and created a multiparameter descriptor—the SK index. High values of SK index (resulting from high fluorescence intensity values and the number of highly autofluorescent particles detected in the lesion area) were associated with SK lesions and were mainly caused by the milia-like cysts and comedo-like opening, which are primarily filled with keratin. On the other hand, compared with SK, the melanomas exhibited significantly lower intensity values. The authors used a threshold value of SK index and discriminated SK ($n = 319$) from MM ($n = 161$) with a sensitivity of 91.9% and specificity of 57.0%. It is worth noting that their data set included six image sets of MM-like SK and 52 image sets of SK-like MM; however, they did not clarify the clinical or dermoscopic atypia criteria of these latter cases.

In the same context, Wang et al. [22] developed a support vector machine (SVM) classification model fed with speckle patterns estimated from image histogram of copolarized and cross-polarized speckle images and a depolarization ratio image D to differentiate between MM and SK. Using a data set of 143 patients (MM $n = 37$, SK $n = 106$), they could discriminate SK from MM with this approach with a sensitivity of 87.63% and a specificity of 85.74%.

The increasing worldwide integration of dermoscopy in clinical dermatology practice [23,24], the evolving efficiency of deep learning algorithms in image recognition, and the availability of extensive image archives have greatly accelerated the development of advanced CAD systems for melanoma detection [25–30]. Earlier efforts were mainly concentrated on discriminating benign melanocytic lesions from MM. However, with the availability of large image datasets, the interest has shifted towards a more sophisticated categorization of skin tumors. Today, the largest, publicly available dataset of dermoscopic images is the International Skin Image Collaboration (ISIC) archive [31]. ISIC promotes CAD-based research by sponsoring annual related challenges for the computer science community in association with leading computer vision conferences. Thus in recognition of the immense clinical impact of differentiating between MM and SK, in 2017 ISIC released a focused dataset with a three-task challenge: lesion segmentation, visual dermoscopic features detection, and lesion discrimination firstly between melanoma vs. nevus and seborrheic keratosis (malignant vs. benign lesions), and secondly between seborrheic keratosis vs. nevus and melanoma (nonmelanocytic vs. melanocytic lesions) [32].

In the present study, we used image data from the ISIC archive to investigate the discrimination efficiency of image embeddings derived from pretrained convolutional network VGG16 to differentiate between MM and SK in the challenging diagnostic task of the preinvasive diagnosis of SK-like MMs. To the best of our knowledge, this study is the first effort exploring the capacity of deep knowledge transfer in refined complexity diagnostic tasks of clinically atypical skin tumors.

2. Materials and Methods

2.1. Data Set Description

Our data set comprised 978 dermoscopic images (malignant melanoma, MM, $n = 550$; seborrheic keratosis, SK, $n = 428$) retrieved from the International Skin Image Collaboration archive [31]. Patients' metadata are summarized in Table 1.

Table 1. Patient metadata: Gender and age of the patients.

Patient Characteristics	MM	SK
Female	240	195
Male	248	230
Undefined	62	3
Mean Age	60.8	64
Median Age	65	65
Standard Deviation (SD) of Age	15.9	13.3

MM: malignant melanoma, SK: seborrheic keratosis.

The clinical diagnosis of all MM cases and of 310 SK cases (72.4%) was confirmed by histological examination.

A large part of the images came from ISIC 2017 challenge [32]. This database provides ground truth lesion images with annotation of the lesion area and the dermoscopic patterns. To enhance our training set, we retrieved 200 additional images (n = 100 MM, n = 100 SK; the BCN_20000 dataset, Hospital Clínic de Barcelona) from the ISIC archive. For the remaining images (BCN_2000 dataset), the lesion area was annotated manually by our experts. The study did not include images in which hair (or another type of noise such as bubbles) substantially corrupted the lesion area. The image resolution in the dataset ranged from 639 × 602 to 6720 × 4461 pixels.

To train our system, we used n = 349 cases of MM and n = 318 cases of SK. The inclusion criteria of dermoscopic images in the test set (MM n = 201, SK n = 110) were the presence of at least one atypical dermoscopy pattern. For MM, this is the absence of pigmented network or the presence of milia-like cysts (or both). On the other hand, atypical SK lacked milia-like cysts or had a pigmented network (or both) (Figures 1 and 2).

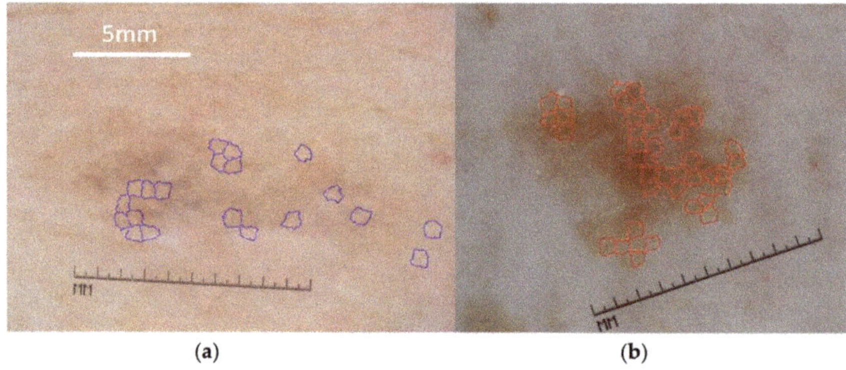

(a)   (b)

Figure 1. Atypical cases: (**a**) MM with milia-like cysts (annotated) (**b**) SK with a pigmented network (annotated). Scale bar = 5mm applies to both panels. Images in the figure were adapted to a uniform magnification (compare same lengths of the original integrated dermatoscope scale) Figures are available online [31].

2.2. Feature Extraction Using Deep Knowledge Transfer

The objective of machine learning in CAD systems is to extract patterns from images and use these patterns to make diagnostic predictions. These patterns are feature vector representations of input images, also called embeddings. From the deep learning perspective, using pretrained embeddings to encode images into feature vectors is known as transfer learning [33]. A typical example is to repurpose pretrained embeddings trained on a large corpus of millions of images [34] for a large-scale classification task to implement a classification model for a different classification task, with much fewer data available.

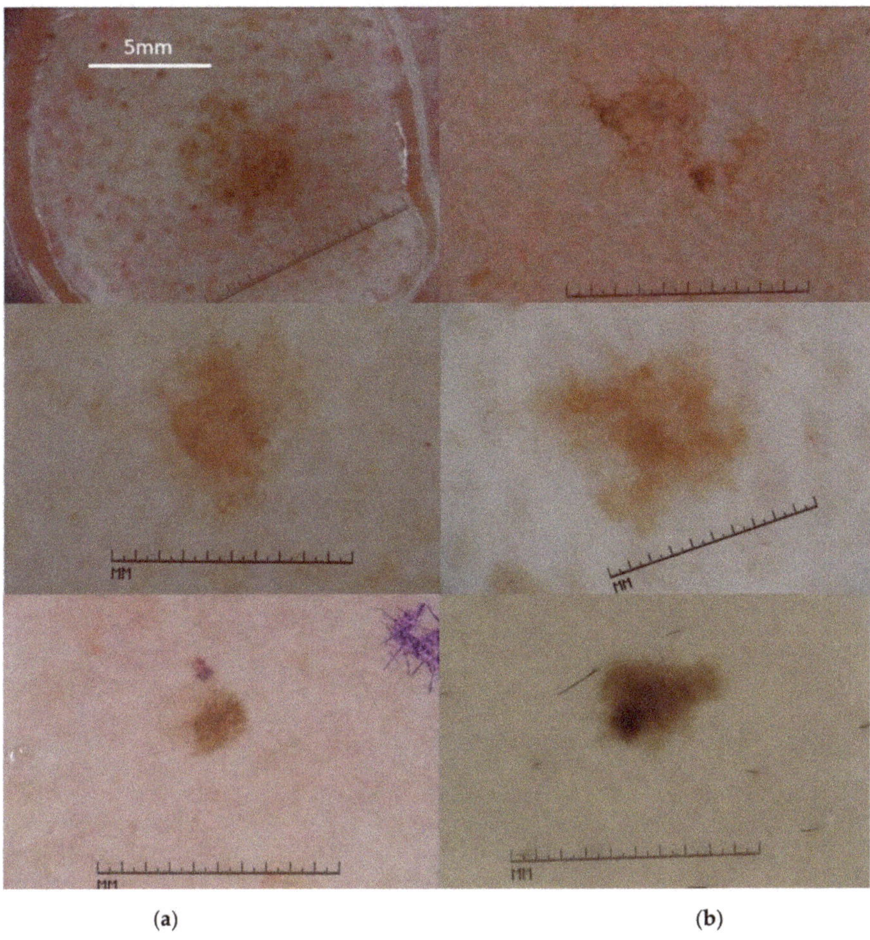

(a)   (b)

Figure 2. Examples of MM (**a**) and SK (**b**) cases. Pairs (left-right) of selected cases are displayed to highlight the distinct overlap of the morphological features. Scale bar = 5mm applies to all panels. All images in the figure were adapted to a uniform magnification (compare same lengths of the original integrated dermatoscope scale) (Figures are available online [31]).

Several studies have indicated that embeddings extracted from deep convolutional neural networks (CNNs) are powerful for various visual recognition tasks [35–37]. Their outstanding performance as image representation learners grew the trend of utilizing them as optimized feature generators for skin lesion classification [38–43]. Our work, aligned with previous research evidence, explores the efficiency of the pretrained CNN, namely the VGG16 [44] as the starting point, for the generation of image embeddings in order to discriminate between cases of atypical MM and atypical SK.

As a conventional deep CNN, VGG16 is a 16-layer architecture that consists of convolutional and fully connected parts. VGG16 pretrained on ImageNet is a classifier architecture for distinguishing a large number of object classes [34]. This goal is achieved gradually by learning image representations in a hierarchical order (Figure 3).

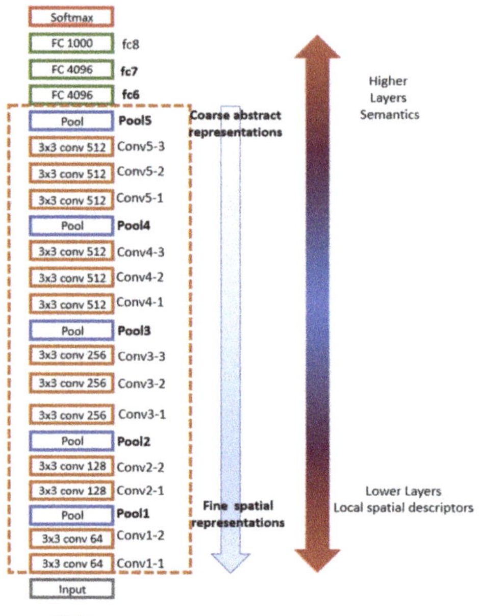

Figure 3. VGG16 architecture and the image representation hierarchies.

Top layers capture more abstract and high-level semantic features. They are robust at distinguishing objects of different classes (i.e., flowers, dogs, etc.) even at significant appearance changes or in the presence of a noisy background. Still, they are less discriminative to objects of the same category (i.e., differentiate between different species of flowers). Moreover, several studies confirmed that the fully connected layers of the CNN, whose role is primarily that of classification, tend to exhibit relatively worse generalization ability and transferability [45]. In contrast, the lower convolutional layers provide more detailed spatial representations. They are more helpful to localize fine-grained details and distinguish a target object from its distracters (other objects with similar appearance, i.e., distinguish between bird species). However, they are less robust to appearance changes. The convolutional layers, acting progressively from fine, spatial to coarse, abstract representations generally transfer well [33,37,45,46] to diverse classification tasks. Based on this evidence, in the present work, we aimed to find the optimal transition point in the convolutional layers to mine high-capacity image representations for the challenging diagnostic task of SK-like MMs characterization.

We exploited image representations from the layers "pool2–pool5". For comparison purposes, we also extracted the fully connected layers' "FC6", "FC7" feature maps so that we can contrast the behavior of the convolutional and fully connected layers (Figure 3).

Finally, the efficiency of VGG16 representations was compared with hierarchical feature embeddings from the ResNet50 convolutional network [47]. Image encoding from fine spatial to coarse abstract, was explored using the layers ReLU_10, ReLU_22, ReLU_40, and ReLU_49.

The image representation of a convolutional layer (activation) forms a tensor of HxWxd, consisting of d feature maps of size H × W. Each feature map is flattened using global average pooling to produce a d-dimensional feature vector. Table 2 summarizes the different VGG16 and ResNet50 layers' representations and their resulting feature vectors for an input image of 224 × 224 × 3 pixels.

Table 2. VGG16 and ResNet50 pretrained image representations and their corresponding d-dimensions feature vectors by global averaging. CNN: convolutional neural network.

CNN	Layer	Imager Representation (Activation)	Feature Vector Dimension (d)
VGG16	Pool2	$56 \times 56 \times 128$	128
	Pool3	$28 \times 28 \times 256$	256
	Pool4	$14 \times 14 \times 512$	512
	Pool5	$7 \times 7 \times 512$	512
	FC6	$1 \times 1 \times 4096$	4096
	FC7	$1 \times 1 \times 4096$	4096
ResNet50	ReLU_10	$56 \times 56 \times 256$	256
	ReLU_22	$28 \times 28 \times 512$	512
	ReLU_40	$14 \times 14 \times 1024$	1024
	ReLU_49	$7 \times 7 \times 2048$	2048

2.3. Implementation and Evaluation of the Diagnostic Model

The extracted deep feature vectors (Table 2) were used to train different binary SVM classifiers. SVM is the classifier of choice for assessing representations from pretrained CNNs [35,36]. For all SVM models, optimal hyperparameter selection (Box Constraint, Kernel function, Kernel scale, Polynomial order) was carried out using the Bayesian optimization method [48] that minimizes k-fold (k = 5) cross-validation classifier error. For each model, the accuracy performance was evaluated in an independent data set of challenging cases of MM and SK in terms of sensitivity, specificity, and overall accuracy:

$$Sensitivity = \frac{TP}{TP + FN} \tag{1}$$

$$Specificity = \frac{TN}{TN + FP} \tag{2}$$

$$Accuracy = \frac{TP + TN}{TP + FP + TN + FN} \tag{3}$$

where TN is the number of SK correctly identified, FN is the number of MM incorrectly identified as SK, TP is the number of MM correctly identified, and FP is the number of SK incorrectly identified as MM.

The models' accuracies were assessed with the McNemar test to detect whether the misclassification rates between any of the two models were statistically significant or not [49,50].

2.4. Image Preprocessing

Before being used as input to pretrained CNNs, all images were preprocessed following a standard pipeline of color normalization, cropping, and resizing (Figure 4). To achieve a color constancy in the whole data set, we used the Grey world color constancy method [51], initially used by [52] and followed by many researchers in automated skin classification works [53–55]. Finally, the exact lesion dimensions were used to crop the images as proposed in [55].

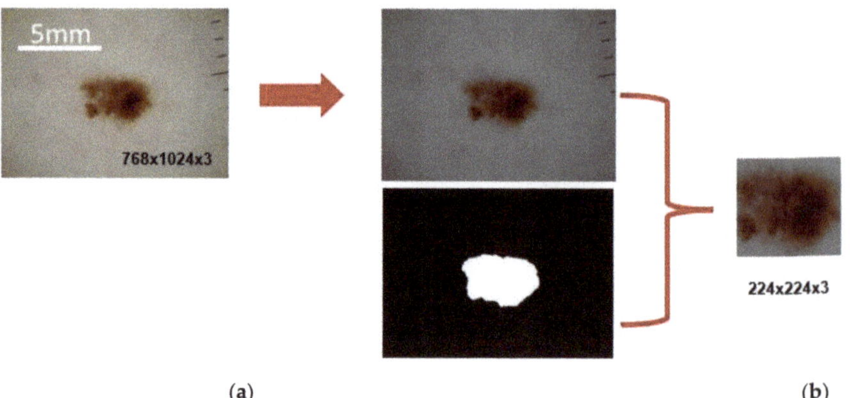

Figure 4. Image preprocessing example. (**a**) Each image is color normalized and combined with the lesion image mask to produce (**b**) the final lesion-cropped and adequately resized input to the CNN model. Scale bar = 5mm. (Figure available online [31]).

3. Results

Bayesian optimization was run for 100 iterations, and different image embeddings from pretrained VGG16 and ResNet50 layers resulted in different classification models, with noticeable differences in test classification accuracies (Table 3).

Table 3. SVM classification models performance using different image representations. Bold annotation highlights the best performance yielded by VGG16 and ResNet50, respectively.

CNN	Layer	SVM Model	Sensitivity (%)	Specificity (%)	Accuracy (%)
	Pool2	Polynomial	56.7	86.4	67.2
	Pool3	**Gaussian**	**78.6**	**84.5**	**80.7**
VGG16	Pool4		68.2	90.9	75.2
	Poo5		59.2	85.4	68.5
	FC6	Linear	57.2	86.4	67.5
	FC7		62.2	82.7	69.4
	ReLU_10	Polynomial	68.1	86.4	74.6
ResNet50	**ReLU_22**	**Gaussian**	**76.1**	**85.4**	**79.4**
	ReLU_40		70.6	89.1	77.2
	ReLU_49	Linear	62.7	86.4	71.1

The SVM model with a gaussian kernel using feature vectors from the 'pool3' layer exhibited the best overall accuracy of 80.7% (251/311 cases) and a sensitivity and specificity of 78.6% (158/201 cases) and 84.5% (93/110 cases), respectively. The highest specificity, 90.9% (100/110 cases), was achieved by a linear SVM classifier and features from the convolutional layer 'pool4'. Considering the ResNet50 approach, there was also the SVM model with a gaussian kernel using feature vectors from the 'ReLU_22' layer that exhibited the best overall accuracy of 79.4% with a sensitivity and specificity of 76.1% and 85.4%, respectively.

More detailed comparison results are illustrated in Table 4, where the statistical significance (McNemar test) of the differences in the observed accuracies is displayed. Considering the VGG16 embeddings, layer 'pool3' produced significantly better sensitivity and overall accuracy with more than a 99.9% confidence level. The 'pool4' layer outperformed the sensitivity and overall accuracy of pool5 and FC7 layers with a confidence of more than 95% and those of layers pool2 and FC6 with a confidence of more than 99.9%. The fully connected layer FC7 outperformed the FC6 layer in sensitivity with more than 95% confidence.

Table 4. Cross-comparison of the classifiers' accuracies (McNemar test). The arrowheads point to the classifier with the highest accuracy, and the lines denote comparable accuracies. The overall accuracy, sensitivity, and specificity results are denoted with dark, red, and blue colors. For example, comparing the performance of layers' representations FC6 and FC7, the FC7 layer exhibited statistically higher sensitivity with a confidence level of more than 95%. Only *p*-values of significantly different outcomes are displayed.

	Pool2	Pool4	Pool5	FC6	FC7	ReLU_22
Pool3	$p < 0.001$	$p < 0.001$	$p < 0.001$	$p < 0.001$	$p < 0.001$	
Pool2	-	$p < 0.001$				$p < 0.001$
Pool4	-	-	$p < 0.05$	$p < 0.001$	$p < 0.05$	$p < 0.001$
Pool5	-	-	-			$p < 0.001$
FC6	-	-	-	-	$p < 0.05$	$p < 0.001$
FC7	-	-	-	-	-	$p < 0.001$

Overall, all the representations resulted in comparable levels of specificity.

4. Discussion

The importance of the timely diagnosis of difficult to recognize melanomas that can clinically resemble benign tumors, such as the SK-like MMs, has been emphasized in previous studies [3,5,7,55,56]. Carrera et al. have indicated specific dermoscopic criteria for correctly identifying such challenging SK-like MM cases [5]. On the other hand, given their larger numbers and significant dermoscopic variability, SK may, at times, mimic melanoma contributing to the clinical MM overdiagnosis [14,15]. RCM may help diagnose challenging cases [3], and recent studies have highlighted the ability of RCM patterns to identify SK with atypical dermoscopy presentation [15,16]. However, there is a lack of related RCM studies focusing on SK-like MM [3]. Moreover, these later studies [5,15,16] have unilaterally highlighted the diagnostic accuracy of dermoscopic and RCM features. The dermoscopic features that assist experts in characterizing SK-like MM have not been employed to assess atypical cases of SK, and the specific RCM patterns were not evaluated in SK-like MM cases. Moreover, the use of RCM is time-consuming, and the increased cost of the equipment restricts the wide availability of this technology.

Today, with the rapid advancement of deep learning methods and the publicly available data sets, dermoscopic images almost monopolize the research interest of CAD skin lesion systems. Numerous breakthrough studies, mainly from the field of computer science, have demonstrated high (expert-level) accuracy in melanoma detection. These high accuracy rates are either related to binary classification tasks as benign vs. malignant or multidifferential diagnosis tasks. In this study, we explored the potential of deep knowledge transfer to approach the challenging 'grey zone' of atypical cases of MM and SK. Studying the different image representation transfer results from a well-known VGG16 architecture and following a standard workflow, we achieved a sensitivity of 78.6% and a specificity of 84.5% using the convolutional layer 'pool3' as a feature extractor. Our results confirm that meaningful feature reuse is concentrated at the convolutional layers rather than at higher, fully connected layers [33,36]. We have also tested the ResNet50 network, and we have verified the existence of the optimal transition from fine spatial to coarse semantic features through the deeper convolutional blocks of ResNet. However, since the discriminating image embeddings are located at the middle layers, the middle-level image embeddings from ResNet50 are of comparable capacity to that of middle-level VGG16.

Moreover, a meta-analysis of 70 studies on CAD systems, published between 2002 and 2018 [19], gave a melanoma sensitivity of 0.74 (95% CI, 0.66–0.80) and a specificity of 0.84 (95% CI, 0.79–0.88), indicating that we have tackled the challenging discrimination of SK-like MMs with comparable accuracies.

In future work, aggregating methods to combine embeddings from middle convolutional layers of the same network or different networks in a global, dense image representation might further boost the system's accuracy. However, the availability of annotated and high-quality image data remains the key contributor to improving accuracy.

Our present contribution is thus twofold: Firstly, the comprehensive evaluation of the transferability of features from different layers of pretrained VGG16 and ResNet50 unveiled the excellent efficiency and generalization properties of the middle-level convolutional layers. Secondly, we targeted a challenging diagnostic task where key dermoscopic patterns of either condition are shared between benign and malignant lesions. It is worth noting that the herein proposed CAD system is aligned with the recent technological advances in smartphone-based teledermatology that promise to enhance diagnostic efficacy at the clinical level [57].

The main limitation of this study is that the feature extraction from pretrained image embeddings is acting more like a black box. The exploited image patterns generate little human interpretable evidence of lesion diagnosis. The effectiveness of this algorithm in prediagnosed cases is within the scopes of a future prospective study.

5. Conclusions

Deep learning has boosted the efficiency of CAD systems significantly. With the publicly available data collections, the computer science community has now the opportunity to test the accuracy of these systems in melanoma diagnosis. Moreover, when these systems clearly focus on a specific diagnostic task and are trained and tested sufficiently, they may support dermatologists in challenging diagnostic tasks.

Author Contributions: Conceptualization, P.S., I.B. and G.G.; methodology and software implementation, P.S. and A.L.; writing—original draft preparation, P.S.; writing—review and editing, I.B., G.G. and A.L. All authors have read and agreed to the published version of the manuscript.

Funding: The authors did not receive support from any organization for the submitted work.

Institutional Review Board Statement: Ethical review and approval were waived for this study due to the fact that the patient images were retrieved from a publicly accessible database.

Informed Consent Statement: Patient consent was waived due to the fact that the employed figures are retrieved from a public database.

Data Availability Statement: Publicly available datasets were analyzed in this study. This data can be found here: https://challenge.isic-archive.com/data/ 2017 challenge. Last accessed date 9 December 2021.

Conflicts of Interest: The authors declare no conflict of interest.

References

1. Conic, R.; Cabrera, C.I.; Khorana, A.; Gastman, B.R. Determination of the Impact of Melanoma Surgical Timing on Survival Using the National Cancer Database. *J. Am. Acad. Dermatol.* **2017**, *78*, 40–46.e7. [CrossRef] [PubMed]
2. Naik, P.P. Cutaneous Malignant Melanoma: A Review of Early Diagnosis and Management. *World J. Oncol.* **2021**, *12*, 7–19. [CrossRef]
3. Janowska, A.; Oranges, T.; Iannone, M.; Romanelli, M.; Dini, V. Seborrheic Keratosis-Like Melanoma: A Diagnostic Challenge. *Melanoma Res.* **2021**, *31*, 407–412. [CrossRef] [PubMed]
4. Moscarella, E.; Brancaccio, G.; Briatico, G.; Ronchi, A.; Piana, S.; Argenziano, G. Differential Diagnosis and Management on Seborrheic Keratosis in Elderly Patients. *Clin. Cosmet. Investig. Dermatol.* **2021**, *14*, 395–406. [CrossRef]

5.   Carrera, C.; Segura, S.; Aguilera, P.; Scalvenzi, M.; Longo, C.; Barreiro-Capurro, A.; Broganelli, P.; Cavicchini, S.; Llambrich, A.; Zaballos, P.; et al. Dermoscopic Clues for Diagnosing Melanomas That Resemble Seborrheic Keratosis. *JAMA Dermatol.* **2017**, *153*, 544–551. [CrossRef]

6.   Izikson, L.; Sober, A.J.; Mihm, M.C.; Zembowicz, A. Prevalence of Melanoma Clinically Resembling Seborrheic Keratosis: Analysis of 9204 Cases. *Arch. Dermatol.* **2002**, *138*, 1562–1566. [CrossRef]

7.   Carrera, C.; Segura, S.; Aguilera, P.; Takigami, C.M.; Gomes, A.; Barreiro, A.; Scalvenzi, M.; Longo, C.; Cavicchini, S.; Thomas, L.; et al. Dermoscopy Improves the Diagnostic Accuracy of Melanomas Clinically Resembling Seborrheic Keratosis: Cross-Sectional Study of the Ability to Detect Seborrheic Keratosis-Like Melanomas by a Group of Dermatologists with Varying Degrees of Experience. *Dermatology* **2017**, *233*, 471–479. [CrossRef]

8.   Xiong, Y.-Q.; Ma, S.; Li, X.; Zhong, X.; Duan, C.; Chen, Q. A meta-analysis of reflectance confocal microscopy for the diagnosis of malignant skin tumours. *J. Eur. Acad. Dermatol. Venereol.* **2016**, *30*, 1295–1302. [CrossRef]

9.   Lan, J.; Wen, J.; Cao, S.; Yin, T.; Jiang, B.; Lou, Y.; Zhu, J.; An, X.; Suo, H.; Li, D.; et al. The Diagnostic Accuracy of Dermoscopy and Reflectance Confocal Microscopy for Amelanotic/Hypomelanotic Melanoma: A Systematic Review and Meta-Analysis. *Br. J. Dermatol.* **2020**, *183*, 210–219. [CrossRef] [PubMed]

10.   Blundo, A.; Cignoni, A.; Banfi, T.; Ciuti, G. Comparative Analysis of Diagnostic Techniques for Melanoma Detection: A Systematic Review of Diagnostic Test Accuracy Studies and Meta-Analysis. *Front. Med.* **2021**, *8*, 637069. [CrossRef]

11.   De Giorgi, V.; Massi, D.; Stante, M.; Carli, P. False "Melanocytic" Parameters Shown by Pigmented Seborrheic Keratoses: A Finding Which is not Uncommon in Dermoscopy. *Dermatol. Surg.* **2002**, *28*, 776–779. [CrossRef] [PubMed]

12.   Scope, A.; Benvenuto-Andrade, C.; Agero, A.L.C.; Marghoob, A.A. Nonmelanocytic Lesions Defying the Two-Step Dermoscopy Algorithm. *Dermatol. Surg.* **2006**, *32*, 1398–1406. [CrossRef]

13.   Lin, J.; Han, S.; Cui, L.; Song, Z.; Gao, M.; Yang, G.; Fu, Y.; Liu, X. Evaluation of Dermoscopic Algorithm for Seborrhoeic Keratosis: A Prospective Study in 412 Patients. *J. Eur. Acad. Dermatol. Venereol.* **2013**, *28*, 957–962. [CrossRef]

14.   Squillace, L.M.; Cappello, C.M.; Longo, E.C.; Moscarella, R.E.; Alfano, G.; Argenziano, G. Unusual Dermoscopic Patterns of Seborrheic Keratosis. *Dermatology* **2016**, *232*, 198–202. [CrossRef]

15.   Farnetani, F.; Pedroni, G.; Lippolis, N.; Giovani, M.; Ciardo, S.; Chester, J.; Kaleci, S.; Pezzini, C.; Cantisani, C.; Dattola, A.; et al. Facial Seborrheic Keratosis With Unusual Dermoscopic Patterns Can Be Differentiated From Other Skin Malignancies By In Vivo Reflectance Confocal Microscopy. *J. Eur. Acad. Dermatol. Venereol.* **2021**, *35*, e784–e787. [CrossRef] [PubMed]

16.   Pezzini, C.; Mandel, V.D.; Persechino, F.; Ciardo, S.; Kaleci, S.; Chester, J.; De Carvalho, N.; Persechino, G.; Pellacani, G.; Farnetani, F. Seborrheic Keratoses Mimicking Melanoma Unveiled by In Vivo Reflectance Confocal Microscopy. *Ski. Res. Technol.* **2018**, *24*, 285–293. [CrossRef] [PubMed]

17.   Argenziano, G.; Catricalà, C.; Ardigo, M.; Buccini, P.; De Simone, P.; Eibenschutz, L.; Ferrari, A.; Mariani, G.; Silipo, V.; Sperduti, I.; et al. Seven-Point Checklist of Dermoscopy Revisited. *Br. J. Dermatol.* **2010**, *164*, 785–790. [CrossRef]

18.   Ferrante di Ruffano, L.; Takwoingi, Y.; Dinnes, J.; Chuchu, N.; Bayliss, S.E.; Davenport, C.; Matin, R.N.; Godfrey, K.; O'Sullivan, C.; Gulati, A.; et al. Computer-Assisted Diagnosis Techniques (Dermoscopy and Spectroscopy-Based) for Diagnosing Skin Cancer in Adults. *Cochrane Database Syst. Rev.* **2018**, *2018*, CD013186. [CrossRef]

19.   Dick, V.; Sinz, C.; Mittlböck, M.; Kittler, H.; Tschandl, P. Accuracy of Computer-Aided Diagnosis of Melanoma: A Meta-Analysis. *JAMA Dermatol.* **2019**, *155*, 1291–1299. [CrossRef]

20.   Maiti, A.; Chatterjee, B.; Ashour, A.S.; Dey, N. Computer-Aided Diagnosis of Melanoma: A Review of Existing Knowledge and Strategies. *Curr. Med. Imaging Former. Curr. Med. Imaging Rev.* **2020**, *16*, 835–854. [CrossRef]

21.   Bozsányi, S.; Farkas, K.; Bánvölgyi, A.; Lőrincz, K.; Fésűs, L.; Anker, P.; Zakariás, S.; Jobbágy, A.; Lihacova, I.; Lihachev, A.; et al. Quantitative Multispectral Imaging Differentiates Melanoma from Seborrheic Keratosis. *Diagnostics* **2021**, *11*, 1315. [CrossRef]

22.   Wang, Y.; Cai, J.; Louie, D.C.; Lui, H.; Lee, T.K.; Wang, Z.J. Classifying Melanoma and Seborrheic Keratosis Automatically with Polarization Speckle Imaging. In Proceedings of the 2019 Global Conference on Signal and Information Processing, Ottawa, ON, Canada, 11–14 November 2019; pp. 1–4. [CrossRef]

23.   Bahadoran, P.; Malvehy, J. Dermoscopy in Europe: Coming of Age. *Br. J. Dermatol.* **2016**, *175*, 1132–1133. [CrossRef] [PubMed]

24.   Piliouras, P.; Buettner, P.; Soyer, H.P. Dermoscopy Use in the Next Generation: A Survey of Australian Dermatology Trainees. *Australas. J. Dermatol.* **2013**, *55*, 49–52. [CrossRef] [PubMed]

25.   Celebi, M.E.; Codella, N.; Halpern, A. Dermoscopy Image Analysis: Overview and Future Directions. *IEEE J. Biomed. Health Inform.* **2019**, *23*, 474–478. [CrossRef]

26.   Naeem, A.; Farooq, M.S.; Khelifi, A.; Abid, A. Malignant Melanoma Classification Using Deep Learning: Datasets, Performance Measurements, Challenges and Opportunities. *IEEE Access* **2020**, *8*, 110575–110597. [CrossRef]

27.   Adegun, A.; Viriri, S. Deep Learning Techniques for Skin Lesion Analysis and Melanoma Cancer Detection: A Survey of State-Of-The-Art. *Artif. Intell. Rev.* **2020**, *54*, 811–841. [CrossRef]

28.   Li, L.-F.; Wang, X.; Hu, W.J.; Xiong, N.N.; Du, Y.X.; Li, B.S. Deep Learning in Skin Disease Image Recognition: A Review. *IEEE Access* **2020**, *8*, 208264–208280. [CrossRef]

29.   Baig, R.; Bibi, M.; Hamid, A.; Kausar, S.; Khalid, S. Deep Learning Approaches Towards Skin Lesion Segmentation and Classification from Dermoscopic Images—A Review. *Curr. Med. Imaging Former. Curr. Med. Imaging Rev.* **2020**, *16*, 513–533. [CrossRef]

30. Kassem, M.; Hosny, K.; Damaševičius, R.; Eltoukhy, M. Machine Learning and Deep Learning Methods for Skin Lesion Classification and Diagnosis: A Systematic Review. *Diagnostics* **2021**, *11*, 1390. [CrossRef]
31. The International Skin Imaging Collaboration. Available online: https://www.isic-archive.com (accessed on 24 October 2021).
32. Codella, N.C.F.; Gutman, D.; Celebi, M.E.; Helba, B.; Marchetti, M.A.; Dusza, S.W.; Kalloo, A.; Liopyris, K.; Mishra, N.; Kittler, H.; et al. Skin Lesion Analysis toward Melanoma Detection: A Challenge. *arXiv* **2017**, arXiv:1710.05006v3, 168–172. Available online: https://arxiv.org/abs/1710.05006v3 (accessed on 22 September 2021).
33. Raghu, M.; Zhang, C.; Kleinberg, J.; Bengio, S. Transfusion: Understanding Transfer Learning for Medical Imaging. *arXiv* **2019**, arXiv:1902.07208. Available online: http://arxiv.org/abs/1902.07208 (accessed on 4 January 2021).
34. ImageNet. Available online: https://image-net.org/ (accessed on 27 September 2021).
35. Sharif Razavian, A.; Azizpour, H.; Sullivan, J.; Carlsson, S. CNN Features Off-The-Shelf: An Astounding Baseline For Recognition. In Proceedings of the IEEE Conference on Computer Vision and Pattern Recognition Workshops, Columbus, OH, USA, 23–28 June 2014; pp. 512–519. [CrossRef]
36. Girshick, R.; Donahue, J.; Darrell, T.; Malik, J. Rich Feature Hierarchies for Accurate Object Detection and Semantic Segmentation. In Proceedings of the IEEE Conference on Computer Vision and Pattern Recognition (CVPR), Columbus, OH, USA, 23–28 June 2014; pp. 580–587. [CrossRef]
37. Garcia-Gasulla, D.; Parés, F.; Vilalta, A.; Moreno, J.; Ayguadé, E.; Labarta, J.; Cortés, U.; Suzumura, T. On the Behavior of Convolutional Nets for Feature Extraction. *J. Artif. Intell. Res.* **2018**, *61*, 563–592. [CrossRef]
38. Codella, N.; Cai, J.; Abedini, M.; Garnavi, R.; Halpern, A.; Smith, J.R. Deep Learning, Sparse Coding, and SVM for Melanoma Recognition in Dermoscopy Images. In Proceedings of the 6th International Workshop on Machine Learning in Medical Imaging, Munich, Germany, 5–9 October 2015; pp. 118–126. [CrossRef]
39. Majtner, T.; Yildirim-Yayilgan, S.; Hardeberg, J.Y. Optimised Deep Learning Features for Improved Melanoma Detection. *Multimed Tools Appl.* **2018**, *78*, 11883–11903. [CrossRef]
40. Devassy, B.M.; Yildirim-Yayilgan, S.; Hardeberg, J.Y. The Impact of Replacing Complex Hand-Crafted Features with Standard Features for Melanoma Classification Using Both Hand-Crafted and Deep Features. *Adv. Intell. Syst. Comput.* **2018**, 150–159. [CrossRef]
41. Mahbod, A.; Schaefer, G.; Wang, C.; Ecker, R.; Ellinge, I. Skin Lesion Classification Using Hybrid Deep Neural Networks. In Proceedings of the ICASSP 2019—2019 IEEE International Conference on Acoustics, Speech and Signal Processing (ICASSP), Brighton, UK, 12–17 May 2019; pp. 1229–1233. [CrossRef]
42. Liu, L.; Mou, L.; Zhu, X.X.; Mandal, M. Automatic Skin Lesion Classification Based on Mid-Level Feature Learning. *Comput. Med. Imaging Graph.* **2020**, *84*, 101765. [CrossRef] [PubMed]
43. Yildirim-Yayilgan, S.; Arifaj, B.; Rahimpour, M.; Hardeberg, J.Y.; Ahmedi, L. Pre-trained CNN Based Deep Features with Hand-Crafted Features and Patient Data for Skin Lesion Classification. *Commun. Comput. Inf. Sci.* **2021**, *1382*, 151–162. [CrossRef]
44. Simonyan, K.; Zisserman, A. Very Deep Convolutional Networks for Large-Scale Image Recognition. In Proceedings of the International Conference on Learning Representations, San Diego, CA, USA, 7–9 May 2015; pp. 1–4. [CrossRef]
45. Liu, L.; Chen, J.; Fieguth, P.; Zhao, G.; Chellappa, R.; Pietikäinen, M. From BoW to CNN: Two Decades of Texture Representation for Texture Classification. *Int. J. Comput. Vis.* **2018**, *127*, 74–109. [CrossRef]
46. Tan, C.; Sun, F.; Kong, T.; Zhang, W.; Yang, C.; Liu, C. A Survey on Deep Transfer Learning. In Proceedings of the 27th International Conference on Artificial Neural Networks, Rhodes, Greece, 4–7 October 2018; pp. 270–279. [CrossRef]
47. He, K.M.; Zhang, X.Y.; Ren, S.Q.; Sun, J. Deep residual learning for image recognition. In Proceedings of the IEEE Conference on Computer Vision and Pattern Recognition (CVPR), Seattle, WA, USA, 27–30 June 2016; pp. 770–778. [CrossRef]
48. Gelbart, M.A.; Snoek, J.; Adams, R.P. Bayesian Optimization with Unknown Constraints. *arXiv* **2014**, arXiv:1403.5607, 250–259. Available online: https://arxiv.org/abs/1403.5607v1 (accessed on 26 September 2021).
49. Dietterich, T.G. Approximate Statistical Tests for Comparing Supervised Classification Learning Algorithms. *Neural Comput.* **1998**, *10*, 1895–1923. [CrossRef]
50. Bostanci, B.; Bostanci, E. An Evaluation of Classification Algorithms Using Mc Nemar's Test. In Proceedings of the Seventh International Conference on Bio-Inspired Computing: Theories and Applications, ABV-Indian Institute of Information Technology and Management Gwalior (ABV-IIITM Gwalior), Madhya Pradesh, India, 14–16 December 2012; pp. 15–26. [CrossRef]
51. Gijsenij, A.; Gevers, T.; van de Weijer, J. Computational Color Constancy: Survey and Experiments. *IEEE Trans. Image Process.* **2011**, *20*, 2475–2489. [CrossRef]
52. Barata, A.F.; Celebi, M.E.; Marques, J.S. Improving Dermoscopy Image Classification Using Color Constancy. *IEEE J. Biomed. Health Inform.* **2015**, *19*, 1146–1152. [CrossRef]
53. Mahbod, A.; Schaefer, G.; Ellinger, I.; Ecker, R.; Pitiot, A.; Wang, C. Fusing Fine-Tuned Deep Features for Skin Lesion Classification. *Comput. Med. Imaging Graph.* **2018**, *71*, 19–29. [CrossRef]
54. Zhang, J.; Xie, Y.; Xia, Y.; Shen, C. Attention Residual Learning for Skin Lesion Classification. *IEEE Trans. Med. Imaging* **2019**, *38*, 2092–2103. [CrossRef] [PubMed]
55. Mahbod, A.; Tschandl, P.; Langs, G.; Ecker, R.; Ellinger, I. The Effects of Skin Lesion Segmentation on the Performance of Dermatoscopic Image Classification. *Comput. Methods Programs Biomed.* **2020**, *197*, 105725. [CrossRef] [PubMed]

56. Papageorgiou, V.; Apalla, Z.; Sotiriou, E.; Papageorgiu, C.; Lazaridou, E.; Vakirlis, S.; Ioannides, D.; Lallas, A. The Limitations of Dermoscopy: False-Positive and False-Negative Tumours. *J. Eur. Acad. Dermatol. Venereol.* **2018**, *32*, 879–888. [CrossRef]
57. Wang, Y.; Wang, J.; Zhang, W.; Zhan, Y.; Guo, S.; Zheng, Q.; Wang, X. A Survey on Deploying Mobile Deep Learning Applications: A Systemic and Technical Perspective. *Digit. Commun. Netw.* **2021**, in press. [CrossRef]

Article

In-Depth Characterisation of Real-World Advanced Melanoma Patients Receiving Immunotherapies and/or Targeted Therapies: A Case Series

Saira Sanjida [1,2,†], Brigid Betz-Stablein [2,*,†], Victoria Atkinson [3], Monika Janda [1], Ramez Barsoum [2], Harrison Aljian Edwards [2], Frank Chiu [2], My Co Tran [2], H Peter Soyer [2,4] and Helmut Schaider [2,4]

[1] Centre for Health Services Research, Faculty of Medicine, The University of Queensland, Woolloongabba, QLD 4102, Australia; s.sanjida@uq.edu.au (S.S.); m.janda@uq.edu.au (M.J.)
[2] The University of Queensland Diamantina Institute, Dermatology Research Centre, The University of Queensland, Woolloongabba, QLD 4102, Australia; r.barsoum@uq.edu.au (R.B.); harrison.edwards@uq.net.au (H.A.E.); f.chiu@uq.edu.au (F.C.); myco.tran@uq.edu.au (M.C.T.); p.soyer@uq.edu.au (H.P.S.); h.schaider@uq.edu.au (H.S.)
[3] Cancer Care Services, Princess Alexandra Hospital, Woolloongabba, QLD 4102, Australia; victoria.atkinson@health.qld.gov.au
[4] Dermatology Department, Princess Alexandra Hospital, Woolloongabba, QLD 4102, Australia
[*] Correspondence: b.betzstablein@uq.edu.au; Tel.: +61-7-34437399
[†] These authors contributed equally to this work.

Citation: Sanjida, S.; Betz-Stablein, B.; Atkinson, V.; Janda, M.; Barsoum, R.; Edwards, H.A.; Chiu, F.; Tran, M.C.; Soyer, H.P.; Schaider, H. In-Depth Characterisation of Real-World Advanced Melanoma Patients Receiving Immunotherapies and/or Targeted Therapies: A Case Series. *Cancers* **2022**, *14*, 2801. https://doi.org/10.3390/cancers14112801

Academic Editor: Adam C. Berger

Received: 18 May 2022
Accepted: 2 June 2022
Published: 4 June 2022

Publisher's Note: MDPI stays neutral with regard to jurisdictional claims in published maps and institutional affiliations.

Simple Summary: Immunotherapies and targeted therapies have led to improved melanoma survival in clinical trial settings. While this is also true in real-world settings, it is much less studied. Clinical trials have strict inclusion criteria and, therefore, a relatively homogenous set of participants. However, this is not the case in real-world settings, and the differences in characteristics are rarely described in association with survival. In this 3D total-body photography imaging study, we describe the characteristics and clinical outcomes of 41 study participants who received immuno- and/or targeted therapies for metastatic melanoma in a real-world setting. After a median of 39 months follow-up, 59% ($n = 24/41$) of the participants were alive. Our sample size was too small to detect significant differences between patient characteristics; however, despite the majority of our participants having multiple comorbidities, survival was similar to previous reported clinical trials and other real-world settings.

Abstract: Immunotherapies and targeted therapies have shown significant benefits for melanoma survival in the clinical trial setting. Much less is known about the characteristics and associated outcomes of those receiving such therapies in real-world settings. This study describes the characteristics of patients with advanced melanoma receiving immuno- and/or targeted therapies in a real-world setting. This prospective cohort study enrolled participants aged >18 years, diagnosed with advanced melanoma and currently undergoing immuno- and/or targeted therapies outside a clinical trial for follow-up with three-dimensional (3D) total-body imaging. Participants ($n = 41$) had a mean age of 62 years (range 29–86), 26 (63%) were male and the majority ($n = 26$, 63%) had ≥2 comorbidities. After a median of 39 months (range 1–52) follow-up, 59% ($n = 24/41$) of participants were alive. Despite multiple co-morbidities, the survival of participants with advanced melanoma treated using immuno- and/or targeted therapies was similar or better in our real-world setting compared to those treated in clinical trials using similar therapies. Larger studies powered to evaluate phenotypic and socio-economic characteristics, as well as specific comorbidities associated with survival in a real-world setting, are required to help determine those who will most benefit from immuno- and/or targeted therapies.

Keywords: advanced melanoma; immunotherapy; targeted therapy; survival; real-world setting

1. Introduction

Australia and New Zealand have the highest incidence rates of melanoma in the world [1,2]. The risk of developing melanoma is associated with multiple characteristics, including socio-demographic (e.g., age, country), clinical (e.g., melanoma history), environmental (e.g., sun exposure), behavioural (e.g., use of sun protection) and phenotypical (e.g., hair, skin colour) [3–5]. When diagnosed early, the 5-year survival rate is typically greater than 90% [6]. However, until recently, there was a reduced survival rate of 18% for those diagnosed with advanced metastatic melanoma [3].

Over the past decade (2010–2020), there have been significant improvements in the treatment options for advanced melanoma, particularly through the use of immunotherapy (e.g., programmed cell death 1, PD-1; cytotoxic T-lymphocyte-associated protein 4, CLTA-4) and targeted therapy (e.g., mitogen-activated protein kinase, MEK; B-Raf proto-oncogene serine/threonine kinase, BRAF), resulting in improved survival [7]. These therapies are currently considered as the standard of care and the first-line treatment for advanced stage melanoma, given the rapid treatment response, efficacy against brain metastases and long-term treatment benefits [8]. A recent review [9] reported that 5 year survival of metastatic melanoma patients in clinical trials was 1 in 2 for those receiving a combination of immunotherapy [10] and more than 1 in 3 when receiving a combination of BRAF/MAPK kinase targeted therapy [11] or single-agent PD-1 blockade [12]. This is a stark improvement to the <5% survival of metastatic melanoma patients 10 years ago [13]. However, clinical trials follow strict inclusion and exclusion criteria where age, health and accessibility criteria may not reflect the true heterogeneity of people treated for metastatic melanoma in the real-world.

Several studies have been conducted in real-world settings to determine the survival outcomes after immuno- and/or targeted therapies. A Dutch study evaluated 1004 advanced melanoma patients treated between 2014–2017 who were ineligible for clinical trials and found poorer overall survival (8.8 vs. 23 months). The study found poorer survival was associated with Eastern Cooperative Oncology Group Performance Score ≥ 2, brain metastases and lactate dehydrogenase > 500 U/L [14]. However, smaller studies have reported survival rates more similar to those of clinical trials, including a US study ($n = 484$) with an overall survival rate of 20.7 months [15] and a Slovenian study ($n = 116$) with overall survival of 33 months in advanced cutaneous melanoma patients [16]. A German real-world study showed, for advanced melanoma patients with mutant BRAF (V600E/K), a higher overall survival of 29 months for those receiving PD-1 compared with 12 months for those receiving dual MAPKi [17]. Many of these studies required participants to be receiving a specific immunotherapy, whereas there is sparse published research reflecting the real-world scenario where participants often receive multiple treatments (immuno- and/or targeted therapies), and in combinations that change over time. This prospective cohort study was designed to use 3D total-body photography to observe changes in naevi over time and provided the opportunity to comprehensively describe the characteristics, treatment and survival of a series of advanced-stage melanoma patients receiving immuno- and/or targeted therapies in a real-world setting in Queensland, Australia.

2. Methods and Materials

2.1. Study Setting and Participant Recruitment

This prospective cohort study enrolled participants with advanced melanoma at the Dermatology Outpatients Department at the Princess Alexandra Hospital, Queensland, Australia, between June 2016 and December 2017. Participants were eligible for inclusion if they were >18 years of age, diagnosed with stage III–IV melanoma and were then undergoing treatment with immuno- and/or targeted therapy. Concurrent enrolment in a clinical trial during 12 months' follow-up excluded participants from participating in this observational study. Participants were actively followed for 12 months with 3D total-body imaging to study the natural history of naevi in participants undergoing immuno- and/or targeted therapy. Inactive observation ended on 31 October 2020 where further treatment and survival outcome data were collected from electronic medical records.

2.2. Data Collection

Socio-demographic, environmental, phenotypic and clinical characteristics were collected by clinical research assistants using standard questionnaires. The socio-demographic characteristics collected including gender, age, private health insurance and highest educational attainment. Participants were asked about their sun exposure patterns, such as whether their occupations after leaving school and their sport and leisure activities were indoors or outdoors. The participants' natural hair colour at age 21 and severity of freckling were recorded. Freckling density was assessed on three body sites, the face, the dorsum of the right arm and the shoulders, and rated for level of freckling by clinic staff from 1 for 'none' to 4 for 'severe'. The composite of these was given an overall freckling score [18] and re-categorised into none/light (1–3), mild (4–6), medium (7–9) and severe (10–12). Body mass index (BMI) was calculated and categorised according to World Health Organization (WHO) [19]. Three-dimensional total-body images along with dermoscopic images of all naevi > 5 mm were collected every 4 months. For those with longitudinal follow-up, change was evaluated across all available sequential dermoscopic (range 2–4) images by a dermatology registrar and confirmed by a senior dermatologist. Change was defined as a change > 15% in size, shape, colour, profile or naevus dermoscopic pattern and categorised as decreasing, stable or increasing. Clinical staff collected melanoma history and comorbidity information from electronic medical records and included details of melanoma diagnosis (such as primary melanoma location, multiple primary melanomas diagnosed), time since diagnosis (the first and the most recent melanoma if applicable) and details of treatment received during study timeline.

2.3. Statistical Analyses

Descriptive statistics were used to determine the frequency and distribution of participants' socio-demographic, clinical, environmental, behavioural and phenotypic characteristics. Continuous variables were summarised as means with standard deviation (SD) or medians (range) as appropriate, with categorical variables as frequency and proportion. Survival status was calculated as of the censored date (OS) of 31 October 2020. Follow-up time was calculated from the time between baseline visit to censored date according to the survival status of participants. Median overall survival time could not be calculated as > 50% of participants were still alive at the end of the observation period. All statistical analyses were performed in SPSS 27.0 (Statistical Package for the Social Sciences software, IBM SPSS Statistics for Windows, Chicago, IL, USA) or R [20].

3. Results

Forty-three participants were enrolled in the study between June 2016 and December 2017, and 2 were excluded as no record was found of the participants receiving immuno- and/or targeted therapies. Therefore, 41 participants were followed for a median of 39 months (range 1–52 months).

3.1. Baseline Characteristics

The participants had a median age of 65 years (range 29–86), and the majority were male (n = 26, 63%) (Table 1). The minority had private health insurance (n = 10, 29%), and just over half of participants had a post-high school qualification (n = 23, 56%). Approximately half of participants (n = 23, 56%) worked mainly indoors, with most participants spending their leisure time outdoors (n = 28, 68%). About half reported dark brown or black hair at age 21 (n = 21, 51%), with innate skin colour classified as fair in the majority (n = 31, 76%). Approximately half of the participants had no or only mild freckling on their arms, shoulders and faces (n = 18, 44%), while the remainder had medium to severe freckling (n = 23, 56%). Median naevus count > 5 mm was 8 (range 1–60.) Based on

BMI, one participant was underweight, 15 (37%) were considered normal weight, 10 (24%) overweight and 15 (37%) obese.

Table 1. Baseline sociodemographic and clinical characteristics (frequency, column percentages) of participants with advanced melanoma by survival status (frequency, row percentages).

	n = 41 (%)	Alive *n* = 24 (59%)	Deceased *n* = 17 (41%)
Socio-demographic characteristics, *n* (%)			
Age [1]: Mean (SD)	62 [14]	61 [16]	63 [13]
Category			
≤65 years	21 (51)	12 (57)	9 (43)
>65 years	20 (49)	12 (60)	8 (40)
Gender			
Male	26 (63)	14 (54)	12 (46)
Female	15 (37)	10 (67)	5 (33)
Private health insurance			
Yes	10 (24)	8 (80)	2 (20)
No	24 (59)	13 (54)	11 (46)
Not Reported	7 (17)	3 (43)	4 (57)
Education			
Higher school or less	18 (44)	9 (50)	9 (50)
Post-high school qualification	23 (56)	15 (65)	8 (35)
Environmental exposure, *n* (%)			
Occupations since leaving school			
Mainly indoors	18 (44)	12 (67)	6 (33)
Mainly outdoors/both indoors and outdoors	23 (56)	12 (52)	11 (48)
Overall sports and leisure activity			
Mainly indoors/both indoors and outdoors	13 (32)	9 (69)	4 (31)
Mainly outdoors	28 (68)	15 (54)	13 (46)
Phenotype characteristics, *n* (%)			
Naevi Count (>5 mm)			
Median (range)	8 (1–60)	8 (1–59)	9 (3–60)
Natural hair colour at age 21			
Red/auburn/blonde/light brown	21 (51)	16 (76)	5 (24)
Dark brown/black	20 (49)	8 (40)	12 (60)
Innate skin colour			
Fair	31 (76)	19 (61)	12 (39)
Medium or olive	10 (24)	5 (50)	5 (50)
Facultative skin colour			
Fair	16 (39)	11 (69)	5 (31)
Medium or olive	25 (61)	13 (52)	12 (48)
Freckling score			
Nil/mild (0–6)	18 (44)	11 (61)	7 (39)
Mild/severe (7–12)	23 (56)	13 (57)	10 (43)
Body mass index (kg/m^2)			
Normal (≤24.9) [2]	16 (39)	10 (63)	6 (37)
Overweight (25.0–29.9)	10 (24)	4 (40)	6 (60)
Obese (≥30.0)	15 (37)	10 (67)	5 (33)

Table 1. *Cont.*

	n = 41 (%)	Alive *n* = 24 (59%)	Deceased *n* = 17 (41%)
Clinical characteristics			
Number of comorbidities			
None	5 (12)	3 (60)	2 (40)
1	10 (25)	5 (50)	5 (50)
2 or more	26 (63)	16 (62)	10 (38)
Comorbidities [3]			
Hypertension			
Yes	14 (34)	7 (50)	7 (50)
No	27 (66)	17 (37)	10 (63)
Hypercholesterolemia or hyperlipidemia			
Yes	10 (22)	6 (60)	4 (40)
No	31 (78)	18 (58)	13 (42)
Cardiovascular disease			
Yes	8 (20)	4 (50)	4 (50)
No	33 (80)	20 (61)	13 (39)
Diabetes mellitus			
Yes	8 (20)	3 (38)	5 (62)
No	33 (80)	21 (64)	12 (36)
Melanoma history			
Number of primary melanomas			
1	22 (54)	14 (64)	8 (36)
2–7	17 (41)	10 (59)	7 (41)
Time since diagnosis of most recent primary melanoma (*n*= 36) [4]			
≤5 years	21 (58)	12 (57)	9 (43)
5 to ≤10 years	11 (30)	8 (73)	3 (27)
≥11 years	4 (12)	3 (75)	1 (8)
Time since diagnosis metastatic melanoma			
≤1 year	16 (39)	0	16 (100)
2 to 3 years	13 (32)	12 (92)	1 (8)
≥4 years	12 (29)	12 (100)	0
Melanoma stage			
Stage III	5 (12)	3 (60)	2 (40)
Stage IV	36 (88)	21 (58)	15 (42)
Brain metastasis			
Yes	9 (22)	3 (33)	6 (67)
No	32 (78)	21 (66)	11 (34)

NR: Not reported, SD: standard deviation. [1] Median: 65 years: range: 29 to 86 years; [2] includes 1 underweight (BMI = 16.8); [3] percentages do not add up to 100 as participants could have multiple co-morbidities; [4] Median: 5 years: range: 1 to 24 years

3.2. Comorbidities

The majority of participants had multiple comorbidities (*n* = 26, 63%) (Table 1). The most common comorbidity was hypertension (*n*= 14, 34%), followed by hypercholesterolemia or hyperlipidemia (*n* = 10, 22%). A considerable number of participants also reported cardiovascular disease (*n* = 8, 19%), diabetes mellitus (*n* = 8, 19%), cancer other than skin cancer (*n* = 4, 8%) and liver disease (*n* = 4, 9%).

3.3. Melanoma History

About half of participants (*n* = 22, 54%) had only one previous melanoma, while 17 participants (*n* = 41%) had multiple prior melanomas (range 2–7) (Table 1). The median time since diagnosis of the first and most recent primary melanoma was 6 years (2–24) and 5 years (1–24), respectively. The most recent primaries were located primarily on the chest/abdomen (*n* = 12, 29%), followed by the head and neck (*n* = 10, 24%), lower limbs

(n = 9, 22%), back (n = 4, 10%) and upper limbs (n = 3, 7%). Three participants (7%) had unknown primaries. Most participants had stage IV metastatic melanoma (n = 36, 88%), while the remaining 5 (12%) had stage III melanoma with only lymph metastases reported. Twenty-two per cent (n = 9) reported brain metastases.

3.4. Treatments Received for Advanced Melanoma

Around two-thirds of participants (n = 24, 59%) received two or more therapies (Table 2). PD-1 blocker was the first line of treatment for the majority of patients (63%, n = 26), followed by the combination of BRAF and MEK inhibitors (n = 9, 22%). The most common immuno- and/or targeted therapies prescribed for the treatment of advanced melanoma in this case series were PD-1 blocker (pembrolizumab, n = 29), BRAF inhibitor (dabrafenib, n = 14) and MEK inhibitor (trametinib n = 14). Treatment regimens were commonly adjusted due to progressive disease (n = 19), toxicity or other adverse effects (n = 8). At the censoring date (October 2020), five participants were still receiving treatment (Table 2).

Table 2. Immunotherapy and/or targeted therapy delivered to the participants with advanced melanoma at any time throughout follow-up. Percentages do not add up to 100 as participants could receive multiple treatments over the follow-up period.

Type of Treatment Delivered	n = 41 (%)	Alive, n = 24 (%)	Deceased, n = 17 (%)
Immunotherapy at any time			
PD-1 blocker			
Pembrolizumab	29 (71)	17 (71)	12 (71)
Nivolumab	13 (32)	6 (25)	7 (41)
Atezolizumab	1 (2)	0	1 (6)
CTLA-4 blocker			
Ipilimumab	13 (32)	7 (29)	6 (35)
Targeted therapy at any time			
MEK inhibitor			
Trametinib	14 (34)	8 (33)	6 (35)
BRAF inhibitor			
Dabrafenib	14 (34)	8 (33)	6 (35)
Vemurafenib	3 (7)	-	3 (18)
MEK inhibitor			
Cobimetinib	2 (5)	-	2 (11)
Total number of melanoma treatments received during follow-up			
One	17 (41)	1 (4)	6 (35)
Two or more	24 (59)	13 (54)	11 (65)
Combination of immune- and targeted therapy at one time	3 (7)	2 (8)	1 (6)
Continuation of treatments until October 2020	5 (12)	5 (21)	-

Line of treatment * (June 2016–October 2020)	PD-1 blocker	PD-1 and CTLA-4 blockers combination	BRAF and MEK inhibitors combination	Other [¶]
First-line treatment (n = 41)	26 (63)	4 (10)	9 (22)	2 (5)
Second-line treatment (n = 19)	4 (21)	6 (31)	4 (21)	5 (26)
Third-line and following treatment (n = 3)	1 (33)	1 (34)	1 (33)	0

[¶] Included ipilimumab, dabrafenib and trametinib alone. * The line of treatment was determined from the delivery of immuno- and/or targeted therapies to the participants with advanced melanoma after enrolment in this study according to date. Percentages presented by row.

3.5. Overall Survival

After a median follow-up time of 39 months, 24 (58%) participants were alive. The sample size did not allow testing for significant differences between those alive and those deceased; however, here we summarise characteristics showing a difference in survival at proportions greater than 10% (Table 1). A slightly higher proportion of females survived compared with males (67% vs. 54%). A higher proportion of those with private health insurance (80%) survived, compared with those without private health insurance (54%) and those that did not report whether they had insurance (43%). Those who worked mainly indoors as opposed to mainly outdoors or both indoors and outdoors showed higher rate of survival (67% vs. 52%). Similarly, a higher proportion of those who spent their leisure time 'mainly indoors' or 'both indoors and outdoors' survived opposed to those who spent their leisure time 'mainly outdoors' (69% vs. 54%). Participants with red/auburn/blonde/light brown hair showed higher levels of survival compared to those with dark brown or black hair (76% vs. 40%). A higher proportion of those with fair innate and fair facultative skin colour survived than those with medium to olive skin tones (respectively, 61% vs. 50%, 69% vs. 52%). With regards to BMI, a higher proportion classified as obese (67%) survived compared to those who were considered normal (63%) or overweight (40%). While the number of co-morbidities showed similar levels of survival, those with hypertension ($n = 16$) showed a higher level of survival compared to those without (50% vs. 37%), and those with diabetes showed a lower rate of survival (38% vs. 64%). Similar rates of survival were seen in those with stage III and stage IV melanoma (60% vs. 62%); however, it should be noted that only five (12%) participants included in the study were stage III. A higher proportion of participants without brain metastasis survived compared to participants with brain metastasis (69% vs. 33%).

3.6. Naevus Change

Of the 41 participants enrolled in this study, 28 (68%) completed a minimum of two imaging visits, with a median follow-up time of 11 months (range 3–19). Demographics of this subset are presented in Supplementary Table S1. Overall, 387 naevi > 5 mm were imaged, with each participant having a median of 7 naevi (range 1–59) (Table 3). The majority of naevi ($n = 259$, 67%) did not change, with 114 (29%) decreasing and 15 (4%) increasing. Per person, this corresponded to a median of 6 (range 0–51) stable and 1 (range 0–26) changing naevi, with 0 increasing (range 0–5) and 1 (range 0–26) decreasing. Within each individual, the mean proportion of changing naevi > 5 mm was 34% (SD 36). Given the small sample and innate bias, those who completed the longitudinal follow-up were healthier and/or responded better to treatment, we did not compare change by survival status; however, numbers are provided in Table 3 for future meta-analyses.

Table 3. Naevus change >5 mm for each patient.

	$n = 28$ (%)	Alive $n = 21$ (75%)	Deceased $n = 7$ (25%)
Total Body Naevus Count (>5 mm)			
Median (range)	7 (1–59)	7 (1–59)	7 (4–45)
Number of Stable Naevi			
Median (range)	6 (0–51)	4 (0–51)	7 (0–19)
Number Naevi Changing			
Median (range)	1 (0–26)	1 (0–24)	3 (0–26)
Number of Increasing Naevi			
Median (range)	0 (0–5)	0 (0–5)	0 (0–4)
Number of Decreasing Naevi			
Median (range)	1 (0–26)	1 (0–19)	1 (0–26)
Proportion of Stable Naevi			
Mean (SD)	66 (37)	65 (37)	66 (39)
Proportion of Changing Naevi			
Mean (SD)	34 (37)	35 (37)	34 (39)
Proportion of Increasing Naevi			
Mean (SD)	4 (15)	2 (4)	12 (30)
Proportion of Decreasing Naevi			
Mean (SD)	30 (34)	33 (37)	22 (25)

4. Discussion

This prospective cohort study originally designed to study the natural history of naevi in individuals receiving immune- and/or targeted therapies for metastatic melanoma allowed for the comprehensive description of characteristics, treatment and survival in an under-studied population in a real-world setting. At the end of the median follow-up of 39 months, 58% (n = 24/41) of participants remained alive. This is similar or higher than the proportions reported in clinical trial settings [10].

Treatment response and survival in patients with metastatic melanoma is likely a complex interaction between phenotypic, genetic and sociodemographic characteristics. While this study was not originally designed, and therefore powered, to test for significance between survival and such variables, it may point towards characteristics that should be investigated in larger studies. The majority of variables, including age, showed similar proportions of patients both alive and deceased (within $\pm$ 10%). Contradicting results have been seen with respect to age, with some showing increased survival in those under 60 [14] and 64 [21], respectively, while others show no difference in overall survival [15,22–24]. In our participants, a slightly higher proportion of females remained alive at the end of our study. While in men, lower awareness and self-detection of melanoma, less frequent skin monitoring and higher sun exposure are all possible reasons for both late diagnosis of melanoma and reduced survival in males [25], when tested, no significant difference in survival was seen between sexes in other real-world studies [14,15,22,26,27]. Few studies have investigated phenotypic, sun behaviour-related characteristics or socio-economic status. A French study (n = 87) has explored survival outcomes for real-world data with participant characteristics since the new melanoma immuno- and/or targeted therapies have become available and showed no significant association between number of naevi, phototype and sun exposure characteristics [23]. However, we saw slightly higher proportions of survival in those with lower occupational and leisure sun exposure and those with lighter facultative and innate skin tone. This difference could easily be explained by the different measures of sun exposure and phototype or merely be an artefact due to the small sample size. Our study also showed a higher proportion of those with private health insurance survived compared to those who did not report having private health insurance, but as this is contrary to an earlier Australian study of PBS-subsidised ipilimumab [22], it may merely reflect our sample size.

In contrast to the clinical trial setting, the majority of our participants had more than one comorbidity. A recent retrospective Dutch study of 2216 metastatic melanoma patients $\geq$ 65 years under immuno- and/or targeted therapies showed no association with age, sex or number of comorbidities with respect to response or survival after a median follow-up of 0.7 years [26]. A smaller Italian study (n = 174) of metastatic melanoma patients $\geq$ 75 years receiving Anti-PD1 antibodies also showed no association with the number of comorbidities relating to either progression-free or overall survival [27]. While several studies have considered the effects of the number of comorbidities, few have looked at the effects of specific comorbidities. While we cannot draw any conclusions given the limited sample size of our study, our study indicates that specific comorbidities may have opposing effects, with potential associations with both improved and decreased survival. Interestingly, a recent review [28] summarised several studies where obesity was shown to be associated with improved overall survival of metastatic melanoma patients on immunotherapies. Although we did not have the power to test for a significant association, we did see a higher proportion of those classified as obese survive. In two studies, this association was stronger [29] or seen only in males [30]. Those with brain metastasis had a lower proportion survive in our study, which is consistent with other real-world studies of those receiving immuno- and/or targeted therapies [14,15,24,26].

With respect to naevus change in those undergoing immuno- and/or targeted therapies, several case reports have described hypopigmentation of naevi, regression/involution of naevi with and without the halo effect and, in a few cases, appearance of new naevi [31–38]. An Italian study of 11 patients receiving dabrafenib observed the appearance of new naevi in 4/11 patients, and hyperkeratosis/hyperpigmentation in 6/11 patients [39]. An Australian study compared naevus changes in patients ($n = 40$) receiving four different combinations of immuno- and targeted therapies to controls ($n = 10$), and observed naevus darkening most often under BRAF inhibitor therapy, with naevus hypopigmentation more common in those under a combination of dabrafenib and trametinib [40]. There was also lesion lightening observed in those receiving anti-PD1 therapies with and without ipilimumab. In some cases, regression of naevi has been observed alongside regression of both primary [31] and metastatic melanomas [33,36,37], and it has been suggested that regression of many naevi may be a prognostic marker, highlighting the importance of including dermatologist follow-up in clinical trials and the real-world setting [33,35,40]. Our study supports the above findings, largely showing regressing naevi but also a small number of increasing naevi.

While our sample size was small ($n = 41$), our participants were followed prospectively and our study was not limited to a specific treatment. The small sample size did limit statistical power and, therefore, we were unable to test for statistical significance. A limitation of this study is that health-conscious participants were more likely to participate in this study, allowing the potential for health awareness bias [41]. This may explain the longer survival times observed in this study or, more likely, reflects the innate self-selection bias in that patients with a short life expectancy are less likely to be referred to a dermatology department for participation in research studies. In addition, as this study was not designed to evaluate survival, other important characteristics such as LDH and frailty score were not collected. Nonetheless, this prospective case series shows that, despite a higher rate of co-morbidities, participants in a real-world setting can have similar survival times compared to those in the clinical trial setting. However, given our sample size, even descriptive statistics can be misleading. Therefore, we suggest this study not be used to draw conclusions but rather to inform future data collection in real-world clinical studies. Specifically, collecting data such as demographic and phenotypic (including naevus change) characteristics, as well as specific comorbidities, in larger survival studies powered for both univariate and multivariable analysis may help determine those who will most benefit from immune- and/or targeted therapy treatment.

Supplementary Materials: The following supporting information can be downloaded at: https://www.mdpi.com/article/10.3390/cancers14112801/s1, Table S1: Baseline sociodemographic and clinical characteristics (Fre-quency, column percentages) of participants with and without longitudinal follow up (Frequency, row percentages).

Author Contributions: H.P.S. and H.S. contributed to the study design. V.A., H.A.E., M.C.T. and R.B. contributed to sample recruitment and data collection. S.S. was involved in data cleaning, analysis and drafting and writing of the manuscript. B.B.-S. was involved in data collection, analysis and writing of the manuscript. H.P.S., H.S., V.A., H.A.E., M.C.T., R.B., S.S., B.B.-S., M.J. and F.C. contributed to the interpretation and critical revision of the manuscript. All authors have read and agreed to the published version of the manuscript.

Funding: This research was conducted with the support of the Centre of Research Excellence for the Study of Naevi funded by the National Health and Medical Research Council (NHMRC) (grant ID: APP1099021). M.J. is funded by a TRIP Fellowship (APP1151021). H.P.S. is funded by the Medical Research Future Fund—Next Generation Clinical Researcher's Program Practitioner Fellowship (APP1137127).

Institutional Review Board Statement: This study was approved by the University of Queensland (ethics approval number: 2016000429) and the Princess Alexandra Hospital (ethics approval number: HREC/16/QPAH/037) human research ethics committee.

Informed Consent Statement: Informed consent was obtained from all subjects involved in the study.

Data Availability Statement: Due to privacy and ethical concerns, the data that support the findings of this study are available on request from the corresponding author.

Acknowledgments: The authors would like to acknowledge the staff at the Clinical Research Facility, Fleur Kong, and the participants without whom this study could not have been undertaken. We thank Caitlin Horsham for her editorial support.

Conflicts of Interest: H.P.S. is a shareholder of MoleMap NZ Limited and e-derm consult GmbH and undertakes regular teledermatological reporting for both companies. H.P.S. is a Medical Consultant for Canfield Scientific Inc., MoleMap Australia Pty Ltd., Blaze Bioscience Inc. and Revenio Research Oy and a medical advisor for First Derm.

References

1. Arnold, M.; de Vries, E.; Whiteman, D.C.; Jemal, A.; Bray, F.; Parkin, D.M.; Soerjomataram, I. Global burden of cutaneous melanoma attributable to ultraviolet radiation in 2012. *Int. J. Cancer* **2018**, *143*, 1305–1314. [CrossRef] [PubMed]
2. Whiteman, D.C.; Green, A.C.; Olsen, C. The Growing Burden of Invasive Melanoma: Projections of Incidence Rates and Numbers of New Cases in Six Susceptible Populations through 2031. *J. Investig. Dermatol.* **2016**, *136*, 1161–1171. [CrossRef] [PubMed]
3. Johnson, M.M.; Leachman, S.A.; Aspinwall, L.G.; Cranmer, L.D.; Curiel-Lewandrowski, C.; Sondak, V.K.; Stemwedel, C.E.; Swetter, S.M.; Vetto, J.; Bowles, T.; et al. Skin cancer screening: Recommendations for data-driven screening guidelines and a review of the US Preventive Services Task Force controversy. *Melanoma Manag.* **2017**, *4*, 13–37. [CrossRef]
4. Olsen, C.M.; Pandeya, N.; Thompson, B.S.; Dusingize, J.C.; Green, A.C.; Neale, R.E.; Whiteman, D.C.; for the QSkin Study. Association between Phenotypic Characteristics and Melanoma in a Large Prospective Cohort Study. *J. Investig. Dermatol.* **2019**, *139*, 665–672. [CrossRef] [PubMed]
5. Rastrelli, M.; Tropea, S.; Rossi, C.R.; Alaibac, M. Melanoma: Epidemiology, risk factors, pathogenesis, diagnosis and classification. *In Vivo* **2014**, *28*, 1005–1011. [PubMed]
6. National Cancer Institute. Cancer Stat Facts: Melanoma of the Skin 2019. Available online: https://seer.cancer.gov/statfacts/html/melan.html (accessed on 22 October 2019).
7. Dummer, R.; Schadendorf, D.; Ascierto, P.A.; Larkin, J.; Lebbe, C.; Hauschild, A. Integrating first-line treatment options into clinical practice: What's new in advanced melanoma? *Melanoma Res.* **2015**, *25*, 461–469. [CrossRef]
8. Luke, J.J.; Flaherty, K.T.; Ribas, A.; Long, G.V. Targeted agents and immunotherapies: Optimizing outcomes in melanoma. *Nat. Rev. Clin. Oncol.* **2017**, *14*, 463–482. [CrossRef]
9. Jenkins, R.W.; Fisher, D.E. Treatment of Advanced Melanoma in 2020 and Beyond. *J. Investig. Dermatol.* **2020**, *141*, 23–31. [CrossRef]
10. Larkin, J.; Chiarion-Sileni, V.; Gonzalez, R.; Grob, J.-J.; Rutkowski, P.; Lao, C.D.; Cowey, C.L.; Schadendorf, D.; Wagstaff, J.; Dummer, R.; et al. Five-Year Survival with Combined Nivolumab and Ipilimumab in Advanced Melanoma. *N. Engl. J. Med.* **2019**, *381*, 1535–1546. [CrossRef]
11. Robert, C.; Grob, J.J.; Stroyakovskiy, D.; Karaszewska, B.; Hauschild, A.; Levchenko, E.; Chiarion Sileni, V.; Schachter, J.; Garbe, C.; Bondarenko, I.; et al. Five-Year Outcomes with Dabrafenib plus Trametinib in Metastatic Melanoma. *N. Engl. J. Med.* **2019**, *381*, 626–636. [CrossRef]
12. Hamid, O.; Robert, C.; Daud, A.; Hodi, F.S.; Hwu, W.J.; Kefford, R.; Wolchok, J.D.; Hersey, P.; Joseph, R.; Weber, J.S.; et al. Five-year survival outcomes for patients with advanced melanoma treated with pembrolizumab in KEYNOTE-001. *Ann. Oncol.* **2019**, *30*, 582. [CrossRef] [PubMed]
13. Dickson, P.V.; Gershenwald, J.E. Staging and Prognosis of Cutaneous Melanoma. *Surg. Oncol. Clin. N. Am.* **2011**, *20*, 1–17. [CrossRef] [PubMed]
14. Van Zeijl, M.C.T.; Ismail, R.K.; de Wreede, L.C.; van den Eertwegh, A.J.M.; de Boer, A.; van Dartel, M.; Hilarius, D.L.; Aarts, M.J.; van den Berkmortel, F.W.; Boers-Sonderen, M.J.; et al. Real-world outcomes of advanced melanoma patients not represented in phase III trials. *Int. J. Cancer* **2020**, *147*, 3461–3470. [CrossRef] [PubMed]
15. Cowey, C.L.; Liu, F.X.; Boyd, M.; Aguilar, K.M.; Krepler, C. Real-world treatment patterns and clinical outcomes among patients with advanced melanoma A retrospective, community oncology-based cohort study (A STROBE-compliant article). *Medicine* **2019**, *98*, e16328. [CrossRef]
16. Hribernik, N.; Boc, M.; Ocvirk, J.; Knez-Arbeiter, J.; Mesti, T.; Ignjatovic, M.; Rebersek, M. Retrospective analysis of treatment-naive Slovenian patients with metastatic melanoma treated with pembrolizumab—real-world experience. *Radiol. Oncol.* **2020**, *54*, 119–127. [CrossRef]
17. Schilling, B.; Martens, A.; Foppen, M.H.G.; Gebhardt, C.; Hassel, J.C.; Rozeman, E.A.; Gesierich, A.; Gutzmer, R.; Kähler, K.C.; Livingstone, E.; et al. First-line therapy-stratified survival in BRAF-mutant melanoma: A retrospective multicenter analysis. *Cancer Immunol. Immunother.* **2019**, *68*, 765–772. [CrossRef]
18. Duffy, D.; Box, N.; Chen, W.; Palmer, J.S.; Montgomery, G.; James, M.R.; Hayward, N.K.; Martin, N.; Sturm, R.A. Interactive effects of MC1R and OCA2 on melanoma risk phenotypes. *Hum. Mol. Genet.* **2003**, *13*, 447–461. [CrossRef]
19. World Health Organization. Body Mass Index 2019. Available online: http://www.euro.who.int/en/health-topics/disease-prevention/nutrition/a-healthy-lifestyle/body-mass-index-bmi (accessed on 30 October 2019).
20. RC Team. *R: A Language and Environment for Statistical Computing*; R Foundation for Statistical Computing: Vienna, Austria, 2019.

21. Moser, J.C.; Chen, D.; Hu-Lieskovan, S.; Grossmann, K.F.; Patel, S.; Colonna, S.V.; Ying, J.; Hyngstrom, J.R. Real-world survival of patients with advanced BRAF V600 mutated melanoma treated with front-line BRAF/MEK inhibitors, anti-PD-1 antibodies, or nivolumab/ipilimumab. *Cancer Med.* **2019**, *8*, 7637–7643. [CrossRef]

22. Kim, H.; Comey, S.; Hausler, K.; Cook, G. A real world example of coverage with evidence development in Australia—Ipilimumab for the treatment of metastatic melanoma. *J. Pharm. Policy Pract.* **2018**, *11*, 4. [CrossRef]

23. Bocquet-Tremoureux, S.; Scharbarg, E.; Nguyen, J.M.; Varey, E.; Quereux, G.; Saint-Jean, M.; Peuvrel, L.; Khammari, A.; Dreno, B. Efficacy and safety of nivolumab in metastatic melanoma: Real-world practice. *Eur. J. Dermatol. EJD* **2019**, *29*, 315–321.

24. Howell, A.V.; Gebregziabher, M.; Thiers, B.H.; Paulos, C.M.; Wrangle, J.M.; Hunt, K.J.; Wallace, K. Immune checkpoint inhibitors retain effectiveness in older patients with cutaneous metastatic melanoma. *J. Geriatr. Oncol.* **2020**, *12*, 394–401. [CrossRef] [PubMed]

25. Joosse, A.; de Vries, E.; Eckel, R.; Nijsten, T.; Eggermont, A.M.; Hölzel, D.; Coebergh, J.W.W.; Engel, J. Gender Differences in Melanoma Survival: Female Patients Have a Decreased Risk of Metastasis. *J. Investig. Dermatol.* **2011**, *131*, 719–726. [CrossRef] [PubMed]

26. De Glas, N.; Bastiaannet, E.; Bos, F.V.D.; Mooijaart, S.; van der Veldt, A.; Suijkerbuijk, K.; Aarts, M.; Berkmortel, F.V.D.; Blank, C.; Boers-Sonderen, M.; et al. Toxicity, Response and Survival in Older Patients with Metastatic Melanoma Treated with Checkpoint Inhibitors. *Cancers* **2021**, *13*, 2826. [CrossRef] [PubMed]

27. Ridolfi, L.; De Rosa, F.; Petracci, E.; Tanda, E.T.; Marra, E.; Pigozzo, J.; Marconcini, R.; Guida, M.; Cappellini, G.C.A.; Gallizzi, G.; et al. Anti-PD1 antibodies in patients aged ≥75 years with metastatic melanoma: A retrospective multicentre study. *J. Geriatr. Oncol.* **2020**, *11*, 515–522. [CrossRef]

28. Smith, L.K.; Arabi, S.; Lelliott, E.J.; McArthur, G.A.; Sheppard, K.E. Obesity and the impact on cutaneous melanoma: Friend or foe? *Cancers* **2020**, *12*, 1583. [CrossRef]

29. McQuade, J.L.; Daniel, C.R.; Hess, K.R.; Mak, C.; Wang, D.Y.; Rai, R.R.; Park, J.J.; Haydu, L.E.; Spencer, C.; Wongchenko, M.; et al. Association of body-mass index and outcomes in patients with metastatic melanoma treated with targeted therapy, immunotherapy, or chemotherapy: A retrospective, multicohort analysis. *Lancet Oncol.* **2018**, *19*, 310–322. [CrossRef]

30. Naik, G.S.; Waikar, S.S.; Johnson, A.E.W.; Buchbinder, E.I.; Haq, R.; Hodi, F.S.; Schoenfeld, J.D.; Ott, P.A. Complex inter-relationship of body mass index, gender and serum creatinine on survival: Exploring the obesity paradox in melanoma patients treated with checkpoint inhibition. *J. Immunother. Cancer* **2019**, *7*, 89. [CrossRef] [PubMed]

31. Grigore, L.E.; Ungureanu, L.; Bejinariu, N.; Seceac, C.; Vasilovici, A.; Senila, S.C.; Candrea, E.; Fechete, O.; Cosgarea, R. Complete regression of primary melanoma associated with nevi involution under BRAF inhibitors: A case report and review of the literature. *Oncol. Lett.* **2019**, *17*, 4176–4182. [CrossRef]

32. McClenahan, P.; Lin, L.L.; Tan, J.M.; Flewell-Smith, R.; Schaider, H.; Jagirdar, K.; Atkinson, V.; Lambie, D.; Prow, T.W.; Sturm, R.A. BRAFV600E mutation status of involuting and stable nevi in dabrafenib therapy with or without trametinib. *JAMA Dermatol.* **2014**, *150*, 1079–1082. [CrossRef]

33. Grassi, S.; Corsetti, P.; Moliterni, E.; Calvieri, S. Regression of benign melanocytic nevi ipilimumab-induced in an adult patient affected by metastatic melanoma: Is it a sign of response to therapy? *Ital. J. Dermatol. Venerol.* **2021**, *156* (Suppl. 1 to No. 6), 54–55.

34. Wolner, Z.; Marghoob, A.; Pulitzer, M.; Postow, M.; Marchetti, M. A case report of disappearing pigmented skin lesions associated with pembrolizumab treatment for metastatic melanoma. *Br. J. Dermatol.* **2017**, *178*, 265–269. [CrossRef] [PubMed]

35. Farinazzo, E.; Zelin, E.; Agozzino, M.; Papa, G.; Pizzichetta, M.A.; di Meo, N.; Zalaudek, I. Regression of nevi, vitiligo-like depigmentation and halo phenomenon may indicate response to immunotherapy and targeted therapy in melanoma. *Melanoma Res.* **2021**, *31*, 582–585. [CrossRef] [PubMed]

36. Plaquevent, M.; Greliak, A.; Pinard, C.; Duval-Modeste, A.-B.; Joly, P. Simultaneous long-lasting regression of multiple nevi and melanoma metastases after ipilimumab therapy. *Melanoma Res.* **2019**, *29*, 311–312. [CrossRef]

37. Libon, F.; Arrese, J.E.; Rorive, A.; Nikkels, A.F. Ipilimumab induces simultaneous regression of melanocytic naevi and melanoma metastases. *Clin. Exp. Dermatol.* **2013**, *38*, 276–279. [CrossRef] [PubMed]

38. Schwager, Z.; Laird, M.E.; Latkowski, J.-A. Regression of pigmented lesions in a patient with metastatic melanoma treated with immunotherapy. *JAAD Case Rep.* **2018**, *4*, 421–423. [CrossRef] [PubMed]

39. Dika, E.; Lambertini, M.; Fanti, P.A.; Piraccini, B.M.; Gurioli, C.; Ravaioli, G.M.; Chessa, M.A.; Gradassi, A.T.; Melotti, B.; Sperandi, F.; et al. Sequential monitoring of pigmented lesions during dabrafenib treatment: A prospective study and a literature overview. *G. Ital. Dermatol. E Venereol.* **2019**, *154*, 170–176. [CrossRef]

40. Zhao, C.Y.; Hwang, S.J.E.; Wakade, D.; Carlos, G.; Anforth, R.; Fernández-Peñas, P. Melanocytic lesion evolution patterns with targeted therapies and immunotherapies for advanced metastatic melanoma: An observational study. *Australas. J. Dermatol.* **2017**, *58*, 292–298. [CrossRef]

41. Tripepi, G.; Jager, K.J.; Dekker, F.W.; Zoccali, C. Selection bias and information bias in clinical research. *Nephron Clin. Pract.* **2010**, *115*, c94–c99. [CrossRef]

Article

Timeline of Adverse Events during Immune Checkpoint Inhibitors for Advanced Melanoma and Their Impacts on Survival

Lorena Villa-Crespo [1,†], Sebastian Podlipnik [1,†], Natalia Anglada [2], Clara Izquierdo [2], Priscila Giavedoni [1], Pablo Iglesias [1], Mireia Dominguez [1], Francisco Aya [3], Ana Arance [3], Josep Malvehy [1,2,4], Susana Puig [1,2,4] and Cristina Carrera [1,2,4,*]

[1]   Melanoma Group, Institut d'Investigacions Biomediques August Pi I Sunyer (IDIBAPS), 08036 Barcelona, Spain; lorevc_63@hotmail.com (L.V.-C.); podlinik@clinic.cat (S.P.); pgiavedo@clinic.cat (P.G.); piglesia@clinic.cat (P.I.); midoming@clinic.cat (M.D.); jmalvehy@clinic.cat (J.M.); spuig@clinic.cat (S.P.)
[2]   Medicine Department, Medicine Faculty, Campus Clínic, University of Barcelona, 08036 Barcelona, Spain; nanglada@gmail.com (N.A.); cizquierdo95@gmail.com (C.I.)
[3]   Medical Oncology Department, Hospital Clinic of Barcelona, University of Barcelona, 08036 Barcelona, Spain; faya@clinic.cat (F.A.); amarance@clinic.cat (A.A.)
[4]   Biomedical Research Networking Center on Rare Diseases (CIBERER), ISCIII, 08036 Barcelona, Spain
*   Correspondence: ccarrera@clinic.cat; Tel.: +34-93-2275400
†   These authors contributed equally to this work.

Simple Summary: A cohort of 153 melanoma patients treated with immune checkpoint inhibitors as first line therapy were studied, to specifically describe the timelines of all adverse events by the target organ, demonstrating a different profile of appearance over time. Interestingly, the survival benefit of presenting immune-related adverse events was demonstrated only for dermatological events, but a multivariate analysis found that this benefit is no longer significant after adjusting for the duration of therapy and the baseline stage of disease. In our opinion, the apparent good response marker of adverse events needs to be analyzed, taking into account the time in therapy and other prognostic markers, such as disease burden.

Abstract: Immune-related adverse events (irAEs) are frequent and could be associated with improved response to immune checkpoint inhibitors (ICIs). A prospective cohort of advanced melanoma patients receiving ICI as first-line therapy was retrospectively reviewed (January 2011–February 2019). A total of 116 of 153 patients presented with at least one irAE (75.8%). The most frequent irAEs were dermatological (derm irAEs, 50%), asthenia (38%), and gastrointestinal (29%). Most irAEs appeared within the first 90 days, while 11.2% appeared after discontinuation of the therapy. Mild grade 1–2 derm irAEs tended to appear within the first 2 months of therapy with a median time of 65.5 days (IQR 26-139.25), while grade 3–4 derm irAEs appeared later (median 114 days; IQR 69-218) and could be detected at any time during therapy. Only derm irAE occurrence was related to improved survival (HR 6.46). Patients presenting derm irAEs showed better 5-year overall survival compared to those with no derm irAEs (53.1% versus 24.9%; $p < 0.001$). However, the difference was not significant when adjusting for the duration of therapy. In conclusion: the timeline of immune-related-AEs differs according to the organ involved. The (apparent) improved survival of patients who present derm AEs during immunotherapy could be partially explained by longer times under treatment.

Keywords: immunotherapy; melanoma; dermatological adverse events; immune-related adverse events; immune checkpoint inhibitors; dermatological drug reactions; survival; outcome

Citation: Villa-Crespo, L.; Podlipnik, S.; Anglada, N.; Izquierdo, C.; Giavedoni, P.; Iglesias, P.; Dominguez, M.; Aya, F.; Arance, A.; Malvehy, J.; et al. Timeline of Adverse Events during Immune Checkpoint Inhibitors for Advanced Melanoma and Their Impacts on Survival. *Cancers* 2022, 14, 1237. https://doi.org/10.3390/cancers14051237

Academic Editor: David Wong

Received: 14 January 2022
Accepted: 19 February 2022
Published: 27 February 2022

Publisher's Note: MDPI stays neutral with regard to jurisdictional claims in published maps and institutional affiliations.

1. Introduction

Immune checkpoint inhibitors (ICIs) have transformed cancer treatment by providing clear survival benefits in a wide range of cancers, with an acceptable safety profile [1–5].

However, ICIs can cause excessive immune "invigoration" and a diverse spectrum of immune-related adverse events (irAEs) [6–8].

There have, in fact, been irAEs that have led to fatal events (mainly colitis, myocarditis, pneumonitis) or to the development of lifelong secondary immune disorders (diabetes, hypothyroidism) [9–11]. Recently, management algorithms have improved approaches to the most common irAEs, resulting in a reduction of serious toxicities and related deaths [9,12,13]. Currently, however, there are no biomarkers that are able to predict the occurrence of an irAE, and only a few studies have described intrinsic factors related to the patient or to the primary tumor that could lead to an increased risk of adverse events [7,14].

Knowing the pattern of appearance of irAEs is useful to improve the safety of the treatment and to set up early management of any complications [13,15–17].

Dermatological irAEs are considered some of the most frequent reactions during immunotherapy, but data regarding the temporality and influence on therapy management are scarce. Moreover, several studies found a possible role of irAEs as good response markers [18–28], where the presence of an irAE has been linked with multiple favorable cancer outcomes, including ORR, PFS, and OS [19,21,22]. The objective of this study was to characterize the timeline and clinical presentation of dermatologic irAEs and identify the impact of irAEs on survival.

2. Materials and Methods

A cohort study was conducted at the Hospital Clinic of Barcelona, Spain, from January 2011 to May 2019. Patients eligible for inclusion were those diagnosed with melanoma, classified as stages IIIC and IV, according to the 7th edition of the American Joint Committee on Cancer (AJCC) staging system and treated with immunotherapy as first line therapy. All included patients received at least one dose of immunotherapy and were followed up for at least three months. Patients were evaluated according to the results of clinical surveillance of advanced melanoma patients at least every three months with body scan CT imaging and laboratory tests, and additionally in cases where adverse events occurred.

Exclusion criteria were other lines of treatment—previous targeted treatment or chemotherapy—to prevent a delayed effect of the medication from being attributed to the immunotherapy. All patients gave their informed consent to be part of an online up-to-date safe database (Xarxa Catalano Balear de Centres de Melanoma). This study used an approved protocol in accordance with the provisions of regulation (EU) 2016/679 by the Board of Research Ethics of the hospital's CEIM. This study was performed following the 2015 STROBE guidelines [29].

2.1. Adverse Events Registration Protocol

All irAEs were recorded and classified based on the affected organ. In addition, subtypes of adverse reactions were classified according to the organ/system and the degree of affectation according to the Common Terminology Criteria for Adverse Events of the National Cancer Institute Version CTCAE 5.01(2018). An investigative dermatologist was responsible for checking medical records and the consistency of the information provided, with respect to the characteristics of the irAEs. Since adverse events were evaluated retrospectively, in order to homogeneously collect the events, all were classified according to the CTCAE scale v5.01 (2018). The times at which skin reactions appeared were evaluated with respect to any other reactions, the drug associated with this reaction, and the patient's condition at that time.

2.2. Statistical Analysis

Pearson's X^2 test and a trend test for ordinal variables were used to compare categorical and ordinal variables, respectively. For continuous variables, the Wilcoxon test was used for comparison between two groups of samples and the Kruskal–Wallis test for comparing multiple groups [30]. The baseline characteristics of the patients were summarized using the number and percentage for categorical variables, and the median and range for continuous variables, and we plotted the analysis using boxplot and density plots [31].

Survival curves based on Kaplan–Meier methods and a log-rank test were used to investigate differences in post treatment OS with respect to global and skin toxicities. Curves were calculated using the 'survfit' function in the 'survival' package and plotted with the 'survminer" package in R [32–34].

Multivariate survival analyses were performed using Cox's proportional hazards model. Models were fitted using the 'coxph' function in the 'survival' package in R [33,34]. Hazard ratio estimates were calculated for the effect of skin toxicities on OS adjusted for AJCC stages and the duration of treatment (as categorical variables stratified by quartiles).

All statistical tests were two-sided and p values ≤ 0.05 were considered statistically significant. All tests were performed using the computing environment R [35].

3. Results

Initially, 341 patients were identified with advanced melanoma, who had undergone systemic therapy. After applying the inclusion and exclusion criteria, a final cohort of 153 patients was included for analysis. The reasons for exclusion were: treatment other than immunotherapy initiated as the first line ($n = 149$), patients started treatment before 2010 ($n = 9$), and no follow-up performed ($n = 30$).

Patients had a median age of 60 years (interquartile range (IQR) 45–69) at the time of initiating immunotherapy. Globally, (66) 45.2% were women and, according to the AJCC 7TH version, were stratified as 28 Stage IIIC (18.9%), 22 Stage IVA (14.9%), 28 Stage IVB (18.9%), and 70 Stage IVC (48.2%). Most tumors (75.8%) were *BRAF* wild type. The baseline characteristics of the patients, stratified by the presence and type of irAE, are shown in Table 1.

Table 1. Baseline characteristics of the cohort.

		All Toxicities			**Skin Toxicities**		
	Total	**No toxicity**	**Toxicity**	***p* Value**	**No Toxicity**	**Toxicity**	***p* Value**
	N = 153	*N = 37*	*N = 116*		*N = 77*	*N = 76*	
Age, Median *	60.0 [45.0;69.0]	60.0 [43.0;69.0]	61.0 [46.0;68.2]	0.309	62.0 [48.0;71.0]	56.5 [44.8;66.0]	0.171
Gender				0.348			0.575
Female	66 (43.1%)	13 (35.1%)	53 (45.7%)		31 (40.3%)	35 (46.1%)	
Male	87 (56.9%)	24 (64.9%)	63 (54.3%)		46 (59.7%)	41 (53.9%)	
AJCC stage				0.709			0.014
Stage IIIC	28 (18.9%)	5 (14.3%)	23 (20.4%)		9 (12.0%)	19 (26.0%)	
Stage IVA	22 (14.9%)	7 (20.0%)	15 (13.3%)		9 (12.0%)	13 (17.8%)	
Stage IVB	28 (18.9%)	7 (20.0%)	21 (18.6%)		12 (16.0%)	16 (21.9%)	
Stage IVC	70 (47.3%)	16 (45.7%)	54 (47.8%)		45 (60.0%)	25 (34.2%)	
LDH				0.427			0.079
Abnormal	22 (14.9%)	7 (20.6%)	15 (13.2%)		15 (20.8%)	7 (9.2%)	
Normal	126 (85.1%)	27 (79.4%)	99 (86.8%)		57 (79.2%)	69 (90.8%)	
Karnofsky score				0.063			0.055
>80	56 (56.0%)	9 (37.5%)	47 (61.8%)		25 (46.3%)	31 (67.4%)	
≤80	44 (44.0%)	15 (62.5%)	29 (38.2%)		29 (53.7%)	15 (32.6%)	
BRAF status:				0.844			0.053
BRAF mutated	37 (24.2%)	8 (21.6%)	29 (25.0%)		13 (16.9%)	24 (31.6%)	

Table 1. *Cont.*

		All Toxicities			**Skin Toxicities**		
	Total	**No toxicity**	**Toxicity**	*p* **Value**	**No Toxicity**	**Toxicity**	*p* **Value**
Wild type	116 (75.8%)	29 (78.4%)	87 (75.0%)		64 (83.1%)	52 (68.4%)	
Melanoma location				0.598			0.277
Trunk	57 (37.5%)	13 (35.1%)	44 (38.3%)		23 (30.3%)	34 (44.7%)	
Lower limbs	25 (16.4%)	9 (24.3%)	16 (13.9%)		7 (9.2%)	7 (9.2%)	
Head and neck	19 (12.5%)	3 (8.1%)	16 (13.9%)		9 (11.8%)	10 (13.2%)	
Unknown	14 (9.2%)	3 (8.1%)	11 (9.6%)		17 (22.4%)	8 (10.5%)	
Acral	13 (8.6%)	2 (5.4%)	11 (9.6%)		6 (7.9%)	7 (9.2%)	
Upper limbs	13 (8.6%)	5 (13.5%)	8 (7.0%)		5 (6.6%)	6 (7.9%)	
Mucosa	11 (7.2%)	2 (5.4%)	9 (7.8%)		9 (11.8%)	4 (5.3%)	
Breslow index *	3.5 [2.0;6.0]	3.3 [1.9;6.0]	3.5 [2.2;6.0]	0.935	3.7 [2.0;6.6]	3.4 [2.0;4.8]	0.608
Ulceration				0.863			0.224
Absent	48 (38.4%)	11 (35.5%)	37 (39.4%)		20 (32.3%)	28 (44.4%)	
Present	77 (61.6%)	20 (64.5%)	57 (60.6%)		42 (67.7%)	35 (55.6%)	
Mitotic index *	5.0 [3.0;10.0]	6.0 [2.50;9.5]	5.0 [3.00;9.8]	0.974	5.0 [3.0;10.0]	5.5 [4.0;9.2]	0.726
Treatment duration *	126.0 [59;351]	69.0 [40;233]	164 [63;360]	0.012	66.0 [42;181]	312.0 [83;479]	<0.001

* Continuous variables expressed as median (IQR). Abbreviations: AJCC, American Joint Committee on Cancer; IQR, interquartile range; LDH lactate dehydrogenase.

3.1. Adverse Events Profile

3.1.1. Drug-Related AEs

Of the 153 patients included in the study, 116 (75.8%) developed an irAE during or within 3 months of finishing immunotherapy. Regarding the immune checkpoint drugs: an irAE occurred in 64.3% of patients receiving ipilimumab, 79% of pembrolizumab patients, 72.7% of the nivolumab group, and in 77.7% of those receiving the combination regimen ipilimumab plus nivolumab.

3.1.2. Target Organ and Timeline of IrAEs

The analysis by target organ-adverse events shows that the most frequent involvement was the skin (50% of cases), followed by asthenia (38%), gastrointestinal (29%), and hepatobiliary (21%) (Figure 1A). Most irAEs appeared within the first 90 days of treatment and asthenia was the earliest event recorded (median time 42.5 days, IQR 21-104). Density and box plots (Figure 2A,B) show the differences according to the target organ: the onset time of asthenia, skin, gastrointestinal, and endocrine toxicities occurred mainly at the beginning of the treatment, while musculoskeletal and connective tissue toxicities could occur at any time and remained constant over time. The median time to the emergence of dermatological irAEs was 65.5 days (IQR 26-139), whereas for musculoskeletal and connective tissue irAEs, 187 days (IQR 70-316), $p < 0.0001$.

3.1.3. Severity of IrAEs

A total of 9 patients (5.9%) presented 10 moderate/severe irAEs (grades 3–4 according to CTCA v.5). There were no significant differences between the occurrence of grade 3–4 irAEs and the culprit drug. The analysis of target organ irAEs showed that grade 3–4 toxicities manifested more frequently as hepatobiliary (10% of patients), gastrointestinal (7%), and cutaneous (6%) (Figure 1A). Moreover, density and box plots (Figure 2C,D) showed differences related to the median appearance time for grade 3–4 irAEs; median time to dermatological grade 3–4 irAEs was 114 days (IQR 69-218), while median time for gastrointestinal irAEs was 50 days (IQR 44-81), (*p* = 0.05). Importantly, dermatological and hepatobiliary irAEs could be detected at any time during treatment, in contrast to gastrointestinal irAEs, which tended to develop earlier than the other two.

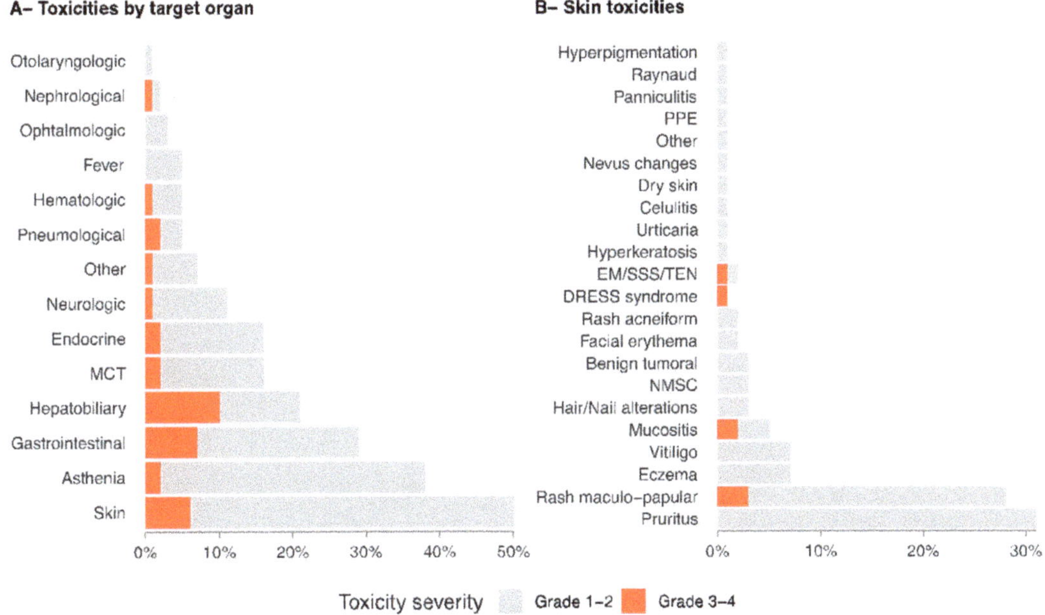

Figure 1. Details of toxicities box plots showing the percentage of patients who presented specific toxicities. Panel (**A**) shows the frequency of immune-related adverse events (irAEs) stratified by target organs. Fever and asthenia were considered as target symptoms due to the frequency of these side effects. Panel (**B**) shows dermatological irAEs in detail. Abbreviations: EM, erythema multiforme; MCT, musculoskeletal and connective tissue; PPE, palmar-plantar erythrodysesthesia; NMSC, non-melanoma skin cancer; SSS, Stevens-Johnson syndrome; TEN, toxic epidermal necrolysis.

3.2. Dermatological Adverse Events (Derm IrAEs)

Of the 153 patients included in the study, 76 (50%) developed dermatological irAEs during the treatment with immunotherapy. Derm irAEs presented a wide spectrum of disorders (Figure 1B). Overall, the most frequent derm irAEs seen were pruritus in 47 patients (31%), maculopapular rash in 42 (28%), eczematous dermatitis in 11 (7%), vitiligo-like depigmentation in 11 (7%), and lichenoid mucositis in 7 patients (5%).

Moderate or severe toxicities (grade 3–4) were observed in 9 patients, and 1 presented two severe derm irAEs simultaneously (Figure 1B shows all derm irAEs observed). A total of 11.8% of derm irAEs were considered grades 3 or 4.

The most severe derm irAEs consisted of generalized maculopapular rash (3%), erosive lichenoid mucositis (2%), DRESS syndrome (1%), and erythema multiforme (1%).

Mild grade 1–2 derm irAEs tended to appear within the first 2 months of therapy with a median time of 65.5 days (IQR 26-139.25), while grade 3–4 derm irAEs appeared later (median 114 days; IQR 69-218) and could be detected at any time during therapy. Importantly, most of the irAEs that appeared after discontinuing therapy were dermatological (85% of delayed onset irAEs).

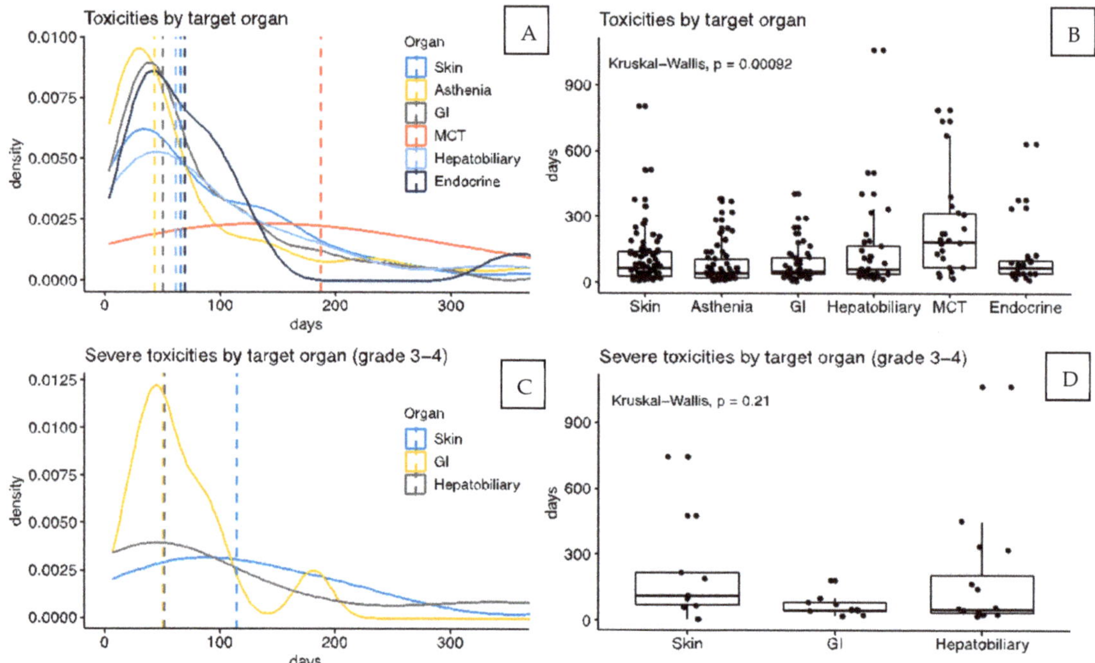

Figure 2. Density plot and boxplot toxicities. Box plots showing the percentages of patients who presented specific toxicities. Panel (**A,B**) show the frequency of immune-related adverse events (irAEs) stratified by target organs. Fever and asthenia were considered as target symptoms due to the frequency of these side effects. Panel (**C,D**) show the frequency of severe (grade 3–4) irAEs stratified by target organs. Abbreviations: EM, erythema multiforme; MCT, musculoskeletal and connective tissue; PPE, palmar-plantar erythrodysesthesia; NMSC, non-melanoma skin cancer; SSS, Stevens-Johnson syndrome; TEN, toxic epidermal necrolysis.

3.3. Survival Analysis

Survival analysis with Kaplan–Meier plots showed no significant improvement in overall survival of those patients who contracted an irAE during treatment. However, the stratified analysis by the target organ of toxicity showed that patients who presented derm irAEs presented a statistically significant survival advantage compared to those who did not ($p < 0.001$). Patients who developed derm irAEs with immune checkpoint therapy presented a five-year overall survival of 53.1% (95% CI, 37.7–74.8%) versus 24.9% (95% CI, 13.5–46.2%) of those who did not present derm irAEs (Figure 3).

A univariate Cox regression analysis showed a hazard ratio of 0.40 (0.24–0.66, $p < 0.001$) in the group of patients who presented derm irAEs. A univariate analysis demonstrated that the baseline clinical–pathological status (AJCC staging) and the duration of immunotherapy were also related to improved survival (HR 6.46 (2.57–16.23, $p < 0.001$) for stage IVC and HR 0.53 (0.32–0.88, $p = 0.014$ for those treated longer than 65 days)).

A multivariate analysis of the benefit of experiencing derm irAEs in overall survival, after adjusting the model for the AJCC stages and for the duration of the therapy, demonstrated that this survival advantage was no longer statistically significant (HR 0.74, 0.44–1.25, $p = 0.263$) (Table 2).

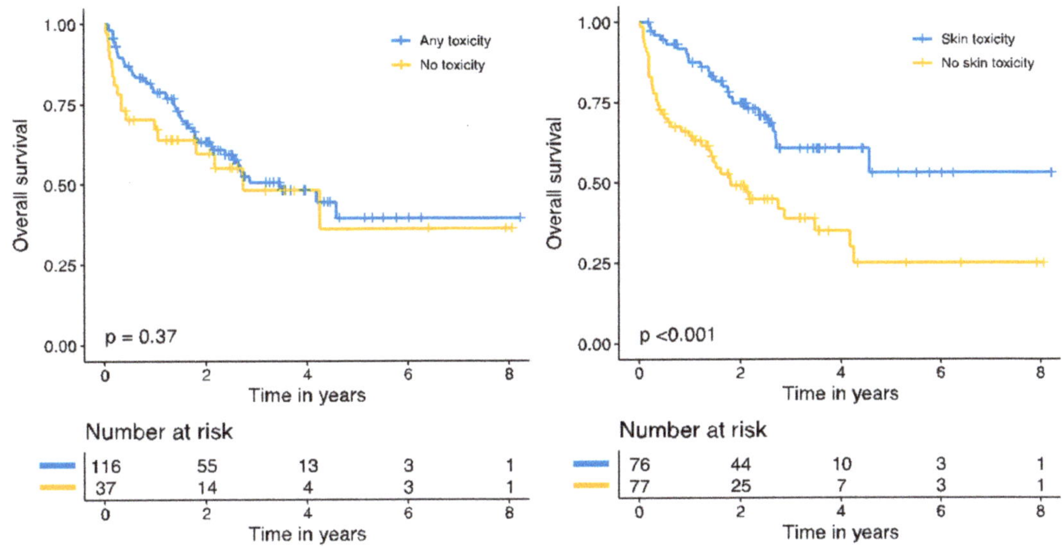

Figure 3. Kaplan–Meier plots for overall survival.

Table 2. Univariate and multivariate analyses of overall survival.

Variables		HR (Univariable)	HR (Multivariable)
irAE	No skin toxicities	-	-
	Skin toxicities	0.40 (0.24–0.66, $p < 0.001$)	0.74 (0.44–1.25, $p = 0.263$)
AJCC Stage	IIIC	-	-
	IVA	1.12 (0.30–4.16, $p = 0.870$)	0.79 (0.21–2.97, $p = 0.726$)
	IVB	1.56 (0.49–4.92, $p = 0.449$)	1.02 (0.31–3.27, $p = 0.980$)
	IVC	6.46 (2.57–16.23, $p < 0.001$)	3.44 (1.32–8.99, $p = 0.012$)
Duration of treatment	<65 days	-	-
	65–331 days	0.53 (0.32–0.88, $p = 0.014$)	0.62 (0.37–1.06, $p = 0.079$)
	>331 days	0.11 (0.05–0.24, $p < 0.001$)	0.20 (0.08–0.48, $p < 0.001$)

Abbreviations: AJCC, American Joint Committee on Cancer; irAE, immune related adverse effect.

4. Discussion

This study into the development of immune-related adverse events (irAEs) during melanoma treatment with ICI demonstrates a specific timeline of presentation, depending on the organ involved and the severity of symptoms. In line with the literature [36,37], derm irAEs are the most frequent and can be detected in about 50% of patients, during treatment or even 3 months after ICI therapy is discontinued. Most patients will present derm irAEs during the first 3 months, similar to gastrointestinal and endocrine irAEs. It is thought that the majority of irAEs, including grade 3–4 events, are developed within 12 weeks of the initial administration [38]. However, certain irAEs can present a distinctive timeline: the most severe derm irAEs (grades 3–4), despite their infrequency (6% of patients), could appear at any time during follow-up, as with hepatobiliary irAEs, but in contrast to gastrointestinal severe irAEs, which present almost exclusively during the first 3 months. Musculoskeletal irAEs are rare and both mild and severe cases may be detected at any time during the therapy.

The second important result is that patients presenting derm irAEs showed a significantly better outcome compared to those with no derm irAEs (5-year overall survival, 53% vs. 25%). However, this apparent benefit in survival was not found after adjusting for the clinical baseline status and for the duration of therapy. Previous studies suggested that patients developing derm irAEs may experience longer intervals without disease progression, and a meta-analysis found that vitiligo-like depigmentation could be a biomarker of good response [1,18,20,21,23–28,39–41]. However, in view of our results, this apparent benefit could also be caused by a longer time under treatment: the longer the treatment, the better response the patient presents, and the more likely the irAEs will occur. In our opinion, the association between derm irAEs and outcome should be studied, and other prognostic markers that could influence this association should always be adjusted for.

On the one hand, it is reasonable to expect a more intense immune "invigoration" in such responses and, consequently, a higher likelihood of irAEs. On the other hand, it is well known that the lower the disease burden at baseline, the better the response to immunotherapy [42]. Moreover, this can be a long-term response leading to improved survival, and indeed, the time to withdraw ICI drugs is still under debate [43]. In the same line, it is reasonable to expect that those patients with better baseline statuses and who respond to treatment will receive long-term immunotherapy. The question is whether this population is the most likely to present irAEs as well. The specific timeline of derm irAEs could explain why this is not the case for gastrointestinal or endocrine toxicities, as they tend to appear within the first 3 months.

Derm irAEs offer an opportunity to elucidate some of the major unresolved questions concerning ICI cancer therapy. In general, derm irAEs do not lead to a discontinuation of ICI therapy: in our experience, only 6% of patients presented more severe grade 3 or 4 derm AEs. Derm irAEs are normally reversible and respond well to low doses of systemic steroids, provided there is accurate and early detection, and in exception for life-threatening toxic epidermal necrolysis or DRESS syndrome [44,45]. This finding could be related to a longer duration of treatment despite the appearance of derm irAEs.

Finally, we could not identify any independent marker associated with the emergence of irAEs, in the primary tumor or in the individual demographics, apart from the duration of treatment and baseline status, as discussed above. Another unresolved question involves the development of the late onset irAEs. Notably, in our series, most late onset irAEs were dermatological (11 out of 13 late onset events). In clinical practices, patients receive second line treatments after disease progression and, therefore, the principal causative drug is sometimes difficult to identify. Determining which patients are at higher risk of developing irAEs, even after therapy has been withdrawn, remains elusive.

The main limitations of this study include the retrospective methodology based on a single-center clinical setting allowing the inclusion of a limited sample of patients. The adverse events were all recategorized according to the CTCAE v.5.1, to have a homogeneous collection. The roles played by the specific drugs and therapeutic regimens used were not analyzed; moreover, the suspension of medication was not analyzed in terms of partial or definitive interruption, and the late onset irAEs could appear after the 3-month post-therapy follow-up. However, one strength of the present cohort is that only naïve patients were analyzed, avoiding the role of other therapies in the appearance of adverse events. In the same line, we did not assess the merging role of combined therapies during progression of the disease, such as immunotherapy plus radiotherapy or immunotherapy plus target therapy in the case of *BRAF* mutated tumors.

5. Conclusions

Dermatological immune-related adverse events (derm irAEs) during immunotherapy for advanced melanomas are the most frequent irAEs and tend to appear within the first 3 months of therapy. In view of our results, it is important to keep in mind that (1) any kind of toxicity could occur in the same patient at any time, during or after therapy. (2) The most severe derm irAEs and hepatobiliary irAEs could occur at any time, but tend to appear later than milder events, while gastrointestinal irAEs are more frequent during the first 2 months. (3) Patients who present derm irAEs can expect significantly better outcomes; however, this tendency is not statistically significant after adjusting for the clinical baseline status and the duration of the therapy.

Prospective large-scale studies, along with careful dermatological evaluation of derm irAEs, will help to elucidate which irAEs play relevant roles in predicting a better response.

Author Contributions: Conceptualization, L.V.-C., S.P. (Sebastian Podlipnik), S.P. (Sebastian Podlipnik) and C.C.; methodology, S.P. (Sebastian Podlipnik) and C.C.; software, S.P. (Sebastian Podlipnik); validation, S.P. (Sebastian Podlipnik) and C.C.; formal analysis, S.P. (Sebastian Podlipnik); investigation, L.V.-C., S.P. (Sebastian Podlipnik) and C.C.; resources, S.P. (Sebastian Podlipnik) and C.C.; data curation, L.V.-C., S.P. (Sebastian Podlipnik), N.A., C.I., P.G., P.I., M.D., F.A., A.A., J.M. and C.C.; writing—original draft preparation, L.V.-C., S.P. (Sebastian Podlipnik) and C.C.; writing—review and editing, S.P. (Sebastian Podlipnik) and C.C.; visualization, S.P. (Sebastian Podlipnik) and C.C.; supervision, C.C.; project administration, S.P. (Susana Puig) and C.C.; funding acquisition, none. All authors have read and agreed to the published version of the manuscript.

Funding: The study in the Melanoma Unit, Hospital Clínic, Barcelona, was supported in part by grants from Fondo de Investigaciones Sanitarias PI 12/00840, PI15/00956, PI15/00716, PI18/0959 and Spain; by the CIBER de Enfermedades Raras of the Instituto de Salud Carlos III, Spain, co-funded by "Fondo Europeo de Desarrollo Regional (FEDER); by the AGAUR 2014_SGR_603 and 2017_SGR_1134 of the Catalan Government, Spain"; by a grant from "Fundació La Marató de TV3, 201331-30", Catalonia, Spain; by the European Commission under the 6th Framework Programme, Contract ref. LSHC-CT-2006-018702 (GenoMEL); by CERCA Programme/Generalitat de Catalunya and by a Research Grant from "Fundación Científica de la Asociación Española Contra el Cáncer" GCB15152978SOEN, Spain. Dr. Cristina Carrera had a personal grant "Pla estratègic de recerca i innovació en salut 2018-2020 (ref. BDNS 357800) PERIS" from the Health Department of the Catalan Government (SLT006/17/00296).

Institutional Review Board Statement: The study was conducted in accordance with the Declaration of Helsinki, and approved by the Institutional Review Board (or Ethics Committee) of Hospital Clínic Barcelona (protocol code IRB HCB/2015/0298 and HCB/2021/0496 date of final approval 29/APR/21).

Informed Consent Statement: Informed consent to be included in the "Xarxa de Melanoma Catalano-Balear" was obtained from all subjects involved in the study, as they signed it upon the first consultation at the Melanoma Unit in Dermatology Department.

Data Availability Statement: The data presented in this study are available in this article.

Acknowledgments: We thank all members of the melanoma unit at the Hospital Clinic of Barcelona. We thank our patients and their families, who are the main reason for our studies; the nurses (especially to M Eugenia Moliner, Maria Soledad Escudero, Carmen López Bermudo) from the Melanoma Unit at Hospital Clínic of Barcelona, and all melanoma scholarship fellows who play a fundamental role in our unit. We appreciate the English edition and critical review made by Paul Hetherington.

Conflicts of Interest: The authors have no conflict of interest to declare.

References

1. Freeman-Keller, M.; Kim, Y.; Cronin, H.; Richards, A.; Gibney, G.; Weber, J.S. Nivolumab in resected and unresectable metastatic melanoma: Characteristics of immune-related adverse events and association with outcomes. *Clin. Cancer Res.* **2016**, *22*, 886–894. [CrossRef]
2. Hao, C.; Tian, J.; Tian, J.; Liu, H.; Li, F.; Niu, H.; Zhu, B. Efficacy and safety of anti-PD-1 and anti-PD-1 combined with anti-CTLA-4 immunotherapy to advanced melanoma: A systematic review and meta-analysis of randomized controlled trials. *Medicine* **2017**, *96*, e7325. [CrossRef] [PubMed]
3. Kruger, S.; Ilmer, M.; Kobold, S.; Cadilha, B.L.; Endres, S.; Ormanns, S.; Schuebbe, G.; Renz, B.W.; D'Haese, J.G.; Schloesser, H.; et al. Advances in cancer immunotherapy 2019—Latest trends. *J. Exp. Clin. Cancer Res.* **2019**, *38*, 1–11. [CrossRef] [PubMed]

4.   Schachter, J.; Ribas, A.; Long, G.V.; Arance, A.; Grob, J.J.; Mortier, L.; Daud, A.; Carlino, M.S.; McNeil, C.; Lotem, M.; et al. Pembrolizumab versus ipilimumab for advanced melanoma: Final overall survival results of a multicentre, randomised, open-label phase 3 study (KEYNOTE-006). *Lancet* **2017**, *390*, 1853–1862. [CrossRef]
5.   Duan, J.; Cui, L.; Zhao, X.; Bai, H.; Cai, S.; Wang, G.; Zhao, Z.; Zhao, J.; Chen, S.; Song, J.; et al. Use of Immunotherapy with Programmed Cell Death 1 vs Programmed Cell Death Ligand 1 Inhibitors in Patients with Cancer: A Systematic Review and Meta-analysis. *JAMA Oncol.* **2020**, *6*, 375–384. [CrossRef]
6.   Mavropoulos, J.C.; Wang, T.S. Managing the skin toxicities from new melanoma drugs. *Curr. Treat. Options Oncol.* **2014**, *15*, 281–301. [CrossRef]
7.   Simeone, E.; Grimaldi, A.M.; Festino, L.; Trojaniello, C.; Vitale, M.G.; Vanella, V.; Palla, M.; Ascierto, P.A. Immunotherapy in metastatic melanoma: A novel scenario of new toxicities and their management. *Melanoma Manag.* **2019**, *6*, MMT30. [CrossRef]
8.   Dolladille, C.; Ederhy, S.; Sassier, M.; Cautela, J.; Thuny, F.; Cohen, A.A.; Fedrizzi, S.; Chrétien, B.; Da-Silva, A.; Plane, A.F.; et al. Immune Checkpoint Inhibitor Rechallenge after Immune-Related Adverse Events in Patients with Cancer. *JAMA Oncol.* **2020**, *6*, 865–871. [CrossRef]
9.   Abdel-Wahab, N.; Alshawa, A.; Suarez-Almazor, M.E. Adverse events in cancer immunotherapy. In *Advances in Experimental Medicine and Biology*; Springer: New York, NY, USA, 2017; Volume 995, pp. 155–174. [CrossRef]
10. Michot, J.M.; Bigenwald, C.; Champiat, S.; Collins, M.; Carbonnel, F.; Postel-Vinay, S.; Berdelou, A.; Varga, A.; Bahleda, R.; Hollebecque, A.; et al. Immune-related adverse events with immune checkpoint blockade: A comprehensive review. *Eur. J. Cancer* **2016**, *54*, 139–148. [CrossRef]
11. Alhatem, A.; Patel, K.; Eriksen, B.; Bukhari, S.; Liu, C. Nivolumab-Induced Concomitant Severe Upper and Lower Gastrointestinal Immune-Related Adverse Effects. *ACG Case Rep. J.* **2019**, *6*, e00249. [CrossRef]
12. Gordon, R.; Kasler, M.K.; Stasi, K.; Shames, Y.; Errante, M.; Ciccolini, K.; Skripnik Lucas, A.; Raasch, P.; Fischer-Cartlidge, E. Checkpoint inhibitors: Common immune-related adverse events and their management. *Clin. J. Oncol. Nurs.* **2017**, *21*, 45–52. [CrossRef] [PubMed]
13. Weber, J.S.; Postow, M.; Lao, C.D.; Schadendorf, D. Management of Adverse Events Following Treatment With Anti-Programmed Death-1 Agents. *Oncologist* **2016**, *21*, 1230–1240. [CrossRef] [PubMed]
14. Postow, M.A.; Sidlow, R.; Hellmann, M.D. Immune-related adverse events associated with immune checkpoint blockade. *N. Engl. J. Med.* **2018**, *378*, 158–168. [CrossRef]
15. Ciccarese, C.; Alfieri, S.; Santoni, M.; Santini, D.; Brunelli, M.; Bergamini, C.; Licitra, L.; Montironi, R.; Tortora, G.; Massari, F.; et al. New toxicity profile for novel immunotherapy agents: Focus on immune-checkpoint inhibitors. *Expert Opin. Drug Metab. Toxicol.* **2016**, *12*, 57–74. [CrossRef]
16. Zhao, B.; Zhao, H.; Zhao, J. Fatal adverse events associated with programmed cell death protein 1 or programmed cell death-ligand 1 monotherapy in cancer. *Ther. Adv. Med. Oncol.* **2020**, *12*, 1758835919895753. [CrossRef]
17. Wang, Y.; Zhou, S.; Yang, F.; Qi, X.; Wang, X.; Guan, X.; Shen, C.; Duma, N.; Vera Aguilera, J.; Chintakuntlawar, A.; et al. Treatment-Related Adverse Events of PD-1 and PD-L1 Inhibitors in Clinical Trials: A Systematic Review and Meta-analysis. *JAMA Oncol.* **2019**, *5*, 1008–1019. [CrossRef]
18. Hua, C.; Boussemart, L.; Mateus, C.; Routier, E.; Boutros, C.; Cazenave, H.; Viollet, R.; Thomas, M.; Roy, S.; Benannoune, N.; et al. Association of Vitiligo With Tumor Response in Patients With Metastatic Melanoma Treated With Pembrolizumab. *JAMA Dermatol.* **2016**, *152*, 45–51. [CrossRef] [PubMed]
19. Ascierto, P.A.; Long, G.V.; Robert, C.; Brady, B.; Dutriaux, C.; Di Giacomo, A.M.; Mortier, L.; Hassel, J.C.; Rutkowski, P.; McNeil, C.; et al. Survival Outcomes in Patients with Previously Untreated BRAF Wild-Type Advanced Melanoma Treated with Nivolumab Therapy: Three-Year Follow-up of a Randomized Phase 3 Trial. *JAMA Oncol.* **2019**, *5*, 187–194. [CrossRef]
20. Chan, L.; Hwang, S.J.E.; Byth, K.; Kyaw, M.; Carlino, M.S.; Chou, S.; Fernandez-Penas, P. Survival and prognosis of individuals receiving programmed cell death 1 inhibitor with and without immunologic cutaneous adverse events. *J. Am. Acad. Dermatol.* **2020**, *82*, 311–316. [CrossRef]
21. Min Lee, C.K.; Li, S.; Tran, D.C.; Zhu, G.A.; Kim, J.; Kwong, B.Y.; Chang, A.L.S. Characterization of dermatitis after PD-1/PD-L1 inhibitor therapy and association with multiple oncologic outcomes: A retrospective case-control study. *J. Am. Acad. Dermatol.* **2018**, *79*, 1047–1052. [CrossRef]
22. Suo, A.; Chan, Y.; Beaulieu, C.; Kong, S.; Cheung, W.Y.; Monzon, J.G.; Smylie, M.; Walker, J.; Morris, D.; Cheng, T.; et al. Anti-PD1-Induced Immune-Related Adverse Events and Survival Outcomes in Advanced Melanoma. *Oncologist* **2020**, *25*, 438–446. [CrossRef] [PubMed]
23. Sanlorenzo, M.; Vujic, I.; Daud, A.; Algazi, A.; Gubens, M.; Luna, S.A.; Lin, K.; Quaglino, P.; Rappersberger, K.; Ortiz-Urda, S. Pembrolizumab cutaneous adverse events and their association with disease progression. *JAMA Dermatol.* **2015**, *151*, 1206–1212. [CrossRef] [PubMed]
24. Brunot, A.; Grob, J.J.; Jeudy, G.; Grange, F.; Guillot, B.; Kramkimel, N.; Mortier, L.; Le Corre, Y.; Aubin, F.F.; Mansard, S.; et al. Association of Anti–Programmed Cell Death 1 Antibody Treatment With Risk of Recurrence of Toxic Effects After Immune-Related Adverse Events of Ipilimumab in Patients With Metastatic Melanoma. *JAMA Dermatol.* **2020**, *156*, 982–986. [CrossRef]
25. Bottlaender, L.; Amini-Adle, M.; Maucort-Boulch, D.; Robinson, P.; Thomas, L.; Dalle, S. Cutaneous adverse events: A predictor of tumour response under anti-PD-1 therapy for metastatic melanoma, a cohort analysis of 189 patients. *J. Eur. Acad. Dermatol. Venereol.* **2020**, *34*, 2096–2105. [CrossRef]

26. Indini, A.; Di Guardo, L.; Cimminiello, C.; Prisciandaro, M.; Randon, G.; De Braud, F.; Del Vecchio, M. Immune-related adverse events correlate with improved survival in patients undergoing anti-PD1 immunotherapy for metastatic melanoma. *J. Cancer Res. Clin. Oncol.* **2019**, *145*, 511–521. [CrossRef]

27. Zhao, J.J.; Wen, X.Z.; Ding, Y.; Li, D.D.; Zhu, B.Y.; Li, J.J.; Weng, D.S.; Zhang, X.; Zhang, X.S. Association between immune-related adverse events and efficacy of PD-1 inhibitors in Chinese patients with advanced melanoma. *Aging* **2020**, *12*, 10663–10675. [CrossRef]

28. Swami, U.; Monga, V.; Bossler, A.D.; Zakharia, Y.; Milhem, M. Durable Clinical Benefit in Patients with Advanced Cutaneous Melanoma after Discontinuation of Anti-PD-1 Therapies Due to Immune-Related Adverse Events. *J. Oncol.* **2019**, *2019*, 1856594. [CrossRef]

29. Vandenbroucke, J.P.; von Elm, E.; Altman, D.G.; Gøtzsche, P.C.; Mulrow, C.D.; Pocock, S.J.; Poole, C.; Schlesselman, J.J.; Egger, M.; STROBE initiative. Strengthening the Reporting of Observational Studies in Epidemiology (STROBE): Explanation and elaboration. *Ann. Intern. Med.* **2007**, *147*, W-163–W-194. [CrossRef]

30. Subirana, I.; Sanz, H.; Vila, J. Building Bivariate Tables: The compareGroups Package for R. *J. Stat. Softw.* **2014**, *57*, 1–16. [CrossRef]

31. Kassambara, A. ggpubr: "ggplot2" Based Publication Ready Plots. 2019. Available online: https://cran.r-project.org/package=ggpubr (accessed on 13 January 2022).

32. Kassambara, A.; Kosinski, M.; Biecek, P. survminer: Drawing Survival Curves Using "ggplot2. 2019. Available online: https://cran.r-project.org/package=survminer (accessed on 13 January 2022).

33. Therneau, T.M. A Package for Survival Analysis in S. 2015. Available online: https://cran.r-project.org/package=survival (accessed on 13 January 2022).

34. Therneau, T.M.; Grambsch, P.M. *Modeling Survival Data: Extending the Cox Model*; Springer: New York, NY, USA, 2000.

35. R Core Team. R: A Language and Environment for Statistical Computing. 2019. Available online: https://www.r-project.org/ (accessed on 13 January 2022).

36. Naidoo, J.; Page, D.B.; Li, B.T.; Connell, L.C.; Schindler, K.; Lacouture, M.E.; Postow, M.A.; Wolchok, J.D. Toxicities of the anti-PD-1 and anti-PD-L1 immune checkpoint antibodies. *Ann. Oncol. Off. J. Eur. Soc. Med. Oncol.* **2016**, *27*, 1362. [CrossRef] [PubMed]

37. Marin-Acevedo, J.A.; Chirila, R.M.; Dronca, R.S. Immune Checkpoint Inhibitor Toxicities. *Mayo Clin. Proc.* **2019**, *94*, 1321–1329. [CrossRef] [PubMed]

38. Cappelli, L.C.; Gutierrez, A.K.; Bingham, C.O.; Shah, A.A. Rheumatic and Musculoskeletal Immune-Related Adverse Events Due to Immune Checkpoint Inhibitors: A Systematic Review of the Literature. *Arthritis Care Res.* **2017**, *69*, 1751–1763. [CrossRef] [PubMed]

39. Lau, K.S.; Liu, R.; Wong, C.C.; Siu, W.K.S.; Yuen, K.K. Clinical outcome and toxicity for immunotherapy treatment in metastatic cancer patients. *Ann. Palliat. Med.* **2019**, *8*, 1003. [CrossRef] [PubMed]

40. Teulings, H.E.; Limpens, J.; Jansen, S.N.; Zwinderman, A.H.; Reitsma, J.B.; Spuls, P.I.; Luiten, R.M. Vitiligo-like depigmentation in patients with stage III–IV melanoma receiving immunotherapy and its association with survival: A systematic review and meta-analysis. *J. Clin. Oncol.* **2015**, *33*, 773–781. [CrossRef]

41. Eggermont, A.M.M.; Kicinski, M.; Suciu, S. Management of Immune-Related Adverse Events Affecting Outcome in Patients Treated with Checkpoint Inhibitors-Reply. *JAMA Oncol.* **2020**, *6*, 1301. [CrossRef] [PubMed]

42. Huang, A.C.; Postow, M.A.; Orlowski, R.J.; Mick, R.; Bengsch, B.; Manne, S.; Xu, W.; Harmon, S.; Giles, J.R.; Wenz, B.; et al. T-cell invigoration to tumour burden ratio associated with anti-PD-1 response. *Nature* **2017**, *545*, 60–65. [CrossRef]

43. Robert, C.; Ribas, A.; Hamid, O.; Daud, A.; Wolchok, J.D.; Joshua, A.M.; Hwu, W.J.; Weber, J.S.; Gangadhar, T.C.; Joseph, R.W.; et al. Durable complete response after discontinuation of pembrolizumab in patients with metastatic melanoma. *J. Clin. Oncol.* **2018**, *36*, 1668–1674. [CrossRef]

44. Sibaud, V. Dermatologic Reactions to Immune Checkpoint Inhibitors: Skin Toxicities and Immunotherapy. *Am. J. Clin. Dermatol.* **2018**, *19*, 345–361. [CrossRef]

45. Boada, A.; Carrera, C.; Segura, S.; Collgros, H.; Pasquali, P.; Bodet, D.; Puig, S.; Malvehy, J. Cutaneous toxicities of new treatments for melanoma. *Clin. Transl. Oncol.* **2018**, *20*, 1373–1384. [CrossRef]

Article

Characterization and Clinical Utility of $BRAF^{V600}$ Mutation Detection Using Cell-Free DNA in Patients with Advanced Melanoma

Piotr Rutkowski [1], Patrick Pauwels [2,3], Joseph Kerger [4], Bart Jacobs [5], Geert Maertens [5], Valerie Gadeyne [6], Anne Thielemans [6], Katrien de Backer [6] and Bart Neyns [7,*]

[1] Maria Sklodowska-Curie National Research Institute of Oncology, 00-001 Warsaw, Poland; piotr.rutkowski@pib-nio.pl
[2] Department of Pathology, Antwerp University Hospital, 2650 Edegem, Belgium; patrick.pauwels@uza.be
[3] Centre for Oncological Research (CORE), Antwerp University, 2610 Wilrijk, Belgium
[4] Institut Jules Bordet, Université Libre de Bruxelles (ULB), 1000 Brussels, Belgium; joseph.kerger@bordet.be
[5] Biocartis, 2800 Mechelen, Belgium; bart.jacobs@novartis.com (B.J.); gmaertens@biocartis.com (G.M.)
[6] F. Hoffmann–La Roche, 4070 Basel, Switzerland; valerie.gadeyne@roche.com (V.G.); anne.thielemans@contractors.roche.com (A.T.); katrien.de_backer@contractors.roche.com (K.d.B.)
[7] Department of Medical Oncology, Vrije Universiteit Brussel (VUB), Universitair Ziekenhuis Brussel (UZ Brussel), 1090 Brussels, Belgium
* Correspondence: bart.neyns@uzbrussel.be; Tel.: +32-2-4775-694

Citation: Rutkowski, P.; Pauwels, P.; Kerger, J.; Jacobs, B.; Maertens, G.; Gadeyne, V.; Thielemans, A.; de Backer, K.; Neyns, B. Characterization and Clinical Utility of $BRAF^{V600}$ Mutation Detection Using Cell-Free DNA in Patients with Advanced Melanoma. *Cancers* **2021**, *13*, 3591. https://doi.org/10.3390/cancers13143591

Academic Editor: Eduardo Nagore

Received: 19 May 2021
Accepted: 14 July 2021
Published: 17 July 2021

Publisher's Note: MDPI stays neutral with regard to jurisdictional claims in published maps and institutional affiliations.

Simple Summary: The choice of cancer drug(s) for the treatment of advanced melanoma is based on the types of gene alterations that are present in the patient's tumor(s). Sometimes, the tumor sample that is obtained from surgery may be degraded, and the test does not provide a reliable result, leading to the selection of the wrong treatment, and, consequently, poor outcomes for the patient. Surgery to obtain fresh tumor samples is inconvenient. In recent years, scientists have learned that fragments of genes from dying cells, including tumors, are constantly being released into the blood. This study shows that the presence of altered genes can be reliably determined using easy-to-obtain blood samples. The study also shows that, while there is a small rate of error with the commonly used tests based on the tumor tissue sample, retests using blood samples may be a less invasive and rapid alternative for identifying the *BRAF* mutation status and selecting the right treatment for these patients.

Abstract: Tissue-based tests for $BRAF^{V600}$ mutation-positive melanoma involve invasive biopsy procedures, and can lead to an erroneous diagnosis when the tumor samples degrade. Herein, we explored a minimally invasive, cell-free deoxyribonucleic acid (cfDNA)-based platform, to retest patients for $BRAF^{V600}$ mutations. This phase 2 study enrolled adult patients with unresectable/metastatic melanoma. A prescreening testing phase evaluated the concordance between a prior tissue-based $BRAF^{V600}$ mutation test result and a subsequent plasma cfDNA-based test result. A treatment phase evaluated the patients who were confirmed as $BRAF^{V600}$ mutation-positive, and were treated with cobimetinib plus vemurafenib. It was found that 35/54 patients (64.8%) with a mutant *BRAF* status by prior tissue test had a positive $BRAF^{V600}$ mutation with the cfDNA test. Further, 7/118 patients (5.9%) with a wild-type *BRAF* status had a positive $BRAF^{V600}$ mutation cfDNA test; tissue retests on archival samples confirmed $BRAF^{V600}$ mutation positivity in 5/7 patients (71.4%). One of these patients received BRAF pathway-targeted therapy (cobimetinib plus vemurafenib), and had progression-free survival commensurate with previous experience. In the overall cobimetinib plus vemurafenib-treated population, 29/36 patients (80.6%) had an objective response. The median progression-free survival was 13.6 months (95% confidence interval, 9.5–16.5). Cell-free DNA–based tests may be a fast and convenient option to identify *BRAF* mutation status in melanoma patients, and help inform treatment decisions.

Keywords: $BRAF^{V600}$ mutations; melanoma; cell-free DNA

1. Introduction

Approximately half of all patients with cutaneous melanoma harbor *BRAF* mutations [1–4]. *BRAF* pathway-targeted therapies are an optimal treatment option among patients with advanced disease [5,6], and also as an adjuvant treatment for high-risk patients [7–9]. The eligibility for BRAF pathway-targeted therapy is predicated on the detection of a $BRAF^{V600}$ mutation in the tumor [10–12]. Currently, multiple diagnostic methodologies are used to detect the presence of *BRAF* mutations in tumor tissue, including high-resolution melt polymerase chain reaction (PCR); the Cobas 4800 BRAF V600 mutation test (Roche Diagnostics, Indianapolis, IN, USA); real-time allele-specific amplification PCR; next-generation sequencing; digital droplet PCR; immunohistochemistry, which is restricted to the $BRAF^{V600E}$ mutation; and Idylla, an automated, PCR-based, molecular platform (Biocartis, Jersey City, NJ, USA) [1,11,13–15]. All of these techniques have varying degrees of inherent sensitivity and specificity [1,2,16].

Sample quality is a key factor underlying the reliability of the *BRAF* mutation tests [17]. Traditionally, the samples used for the detection of *BRAF* mutations in patients with melanoma, are formalin-fixed paraffin-embedded tissue biopsies [17,18]. The inherent limitations of formalin-fixed paraffin-embedded tissue samples are at least two-fold. First, the presence of formaldehyde degrades nucleic acids and denatures protein epitopes, which reduces the detectable signal in the sample [19]. Second, discrepancies have been reported between the analysis of primary versus metastatic lesions, especially when there has been a long interval between both the diagnoses [20]. Both the preceding situations can lead to false-negative results that are potentially deleterious for the patient, because they would lead to the selection of the wrong treatment. The other obvious challenge with tumor tissue biopsies is that they constitute an invasive procedure that cannot always be safely repeated. One approach to overcome the limitation of a tissue biopsy-based approach is to use circulating plasma DNA as the test sample [21]. Small fragments of DNA from tumor cells are constantly being released into the lymph and blood. Therefore, the circulating cell-free DNA (cfDNA) has tumor DNA that is representative of all the lesions present in an individual patient. Further, blood sampling is far less invasive procedure than tumor biopsy, and can be easily repeated in an outpatient setting [21]. Thus, plasma cfDNA-based tests could allow for easy rechecking of the *BRAF* mutation status in patients with advanced melanoma and correct diagnoses.

Accordingly, we designed a multicenter phase 2 study to assess the ability of the Idylla platform to detect the presence of $BRAF^{V600}$ mutations in cfDNA in the plasma of patients with advanced melanoma, and to determine the clinical benefit derived from the subsequent test-informed treatment decision.

2. Materials and Methods

2.1. Study Design

This single-arm, open-label, multicenter, phase 2 study enrolled adult patients with unresectable or metastatic melanoma. The study comprised two sequential phases. The first, the prescreening testing phase, provided a preliminary estimate of the concordance between a prior tissue-based $BRAF^{V600}$ mutation test result and a subsequent plasma cfDNA-based test result. The second, the treatment phase, evaluated clinical outcomes in all patients confirmed to have $BRAF^{V600}$ mutation-positive tumors (either by a prior tissue test, or confirmed by tissue retest for patients who had a positive plasma cfDNA test result following a prior negative tissue test). Within the treatment phase, a substudy ($n = 8$), to evaluate the correlation between the duration of response (DOR), progression-free survival (PFS), and concentration of the $BRAF^{V600}$ mutation in patient plasma, was also conducted. To accomplish this substudy evaluation, additional plasma cfDNA tests were performed on plasma samples that were collected prior to the first treatment dose of cobimetinib and vemurafenib, on day 15 of the first cycle, prior to the first dose of each subsequent treatment cycle, and upon disease progression.

2.2. Patients

Adult patients with a diagnosis of stage IIIC unresectable locally advanced or stage IV metastatic cutaneous melanoma (as per American Joint Committee on Cancer 7th Edition TNM Staging System) [22], and a tissue test result for the $BRAF^{V600}$ mutation, were eligible to enroll in the prescreening phase. The patients entering the treatment phase could have received prior systemic treatment for metastatic melanoma—with the exception of prior BRAF/MEK pathway-inhibitor treatment. Any adverse events (AEs) from prior therapy should have been resolved or be of ≤grade 1 severity per the common terminology criteria for adverse events, version 4.03 [23]. Additional eligibility criteria for the treatment phase included an Eastern Cooperative Oncology Group performance status of 0–2 and adequate renal, hepatic, and end-organ function (see Supplementary Materials Appendix for full eligibility criteria).

2.3. Samples and Mutation Analysis

Prior to entering the prescreening phase of the study, the patients had documented $BRAF^{V600}$ mutation testing performed by Cobas® or routine PCR testing. Once enrolled, patient plasma was prepared from 10 mL of venous blood collected in an EDTA (ethylenediaminetetraacetic acid) tube for cfDNA analysis using the Idylla ctBRAF mutation assay on the Idylla platform (see Supplementary Materials for additional details) [15,20].

For patients with a prior *BRAF* wild-type tissue test result and for whom a $BRAF^{V600}$ mutation was detected using plasma cfDNA, tissue samples were obtained to perform a new tissue analysis. If tissue material from the prior test was still available, it was retested using the Idylla platform, by means of the Idylla BRAF mutation assay. If tissue was not available, but it was possible to obtain a new tissue sample from a recent metastasis, this was tested using the Idylla platform. If it was not possible to obtain a new tissue sample from a recent metastasis, the patient was discontinued from the study. To assess the concentration of $BRAF^{V600}$ and *NRAS* mutant cfDNA in samples collected during treatment, an Idylla ctNRAS-BRAF mutation assay was performed retrospectively once all samples for each individual patient included in the substudy were collected (Figure 1).

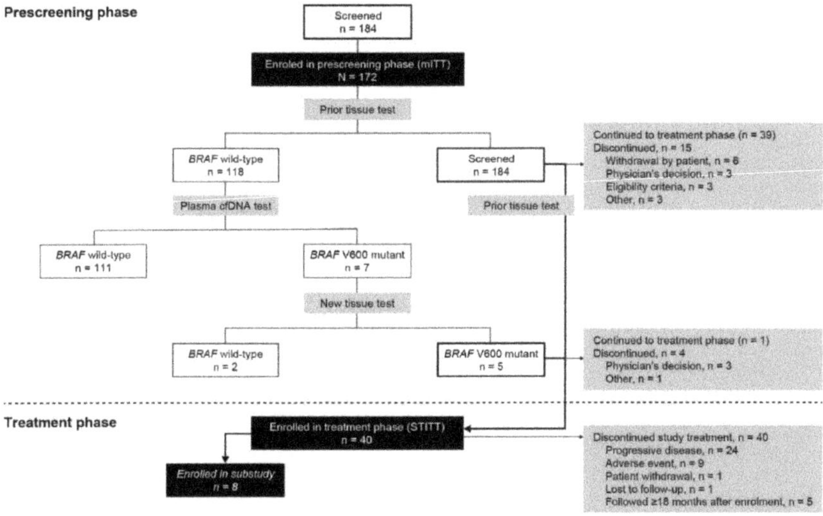

Figure 1. Patient disposition. mITT, modified intention-to-treat; STITT, study treatment intention-to-treat.

All the patients who satisfied all the inclusion criteria for the prescreening testing phase of the study were also tested for the presence of the *NRAS* mutation using plasma cfDNA. This *NRAS/BRAF* mutation testing on serial liquid biopsies was performed as an exploratory post hoc analysis; the results of these analyses did not influence treatment decisions.

2.4. Treatments

The patients with tumors that were $BRAF^{V600}$ mutation-positive, based on a tumor tissue sample, received oral cobimetinib 60 mg once daily for 21 days, followed by a 7-day rest period in each 28-day cycle and oral vemurafenib 960 mg twice daily. The eligible patients were treated until disease progression as per investigator assessment, unacceptable toxicity, or consent withdrawal—whichever occurred first.

2.5. Outcomes and Assessments

Medical history obtained within 28 days of cycle 1 day 1, included confirmation of melanoma, clinically significant diseases within the previous 3 years, major surgeries, and cancer history. The progression of disease during screening was documented.

All measurable and non-measurable lesions were documented at screening, within 28 days prior to cycle 1 day 1. On-treatment tumor assessments were performed every 8 weeks until investigator-determined progression or death, and evaluation of tumor response was conducted—conforming to response evaluation criteria in solid tumors, version 1.1. All patients had a brain screening by computed tomography or magnetic resonance imaging within 28 days prior to cycle 1 day 1, to assess for brain metastases, and subsequently as per local standard of care and as clinically indicated.

Efficacy assessments included objective response rates as per RECIST version 1.1 and Kaplan–Meier estimates for DOR, PFS, and overall survival. For the substudy, the association of these efficacy endpoints with the level of *BRAF* mutation was analyzed.

Safety assessments consisted of monitoring and recording AEs, including serious AEs and nonserious AEs of special interest, performing protocol-specified safety laboratory assessments, measuring protocol-specified vital signs, and conducting other protocol-specified tests that were deemed critical to the safety evaluation of the study. Adverse event severity was assessed using common terminology criteria for adverse events (CTCAE), version 4.03.

2.6. Statistical Considerations

Testing of the $BRAF^{V600}$ mutation using plasma cfDNA was considered to be of clinical interest if $\geq$10% of the patient tumors designated as *BRAF* wild-type based on a prior tissue test were subsequently shown to be $BRAF^{V600}$ mutation-positive based on cfDNA and confirmed in a new tissue sample or retest of the archival tissue using the Idylla platform. The plasma test was considered not to be of clinical value if $\leq$3% of *BRAF* wild-type tumors were reclassified as $BRAF^{V600}$ mutation-positive tumors.

Based on a Fleming one-stage design with a one-sided α of 0.05 and a power of 0.90, a sample size of $\geq$104 patients for whom the prior tissue test indicates wild-type status should be identified. Assuming an equal number of patients with $BRAF^{V600}$ mutation and wild-type *BRAF* mutation based on the prior tissue test, a total of 208 patients were to be enrolled in the prescreening phase of the study. For 104 patients, the power to show that the frequency of reclassifying tumors as $BRAF^{V600}$ mutation-positive exceeds 3% is 84%, 75%, and 63% for frequencies of interest of 9%, 8%, and 7%, respectively.

The modified intention-to-treat (mITT) population comprised all enrolled patients with a documented $BRAF^{V600}$ tissue test result on melanoma tumor tissue at study entry and an available result of the plasma cfDNA test. The study treatment intention-to-treat (STITT) population comprised all enrolled patients with a plasma cfDNA test who received $\geq$1 dose of cobimetinib or vemurafenib. The analysis of the primary objective was performed on all patients in the mITT population with $BRAF^{V600}$ wild-type status based on a prior tissue test.

The analysis of the secondary and the exploratory objectives was performed on all patients in the mITT population. The analysis of the secondary objectives related to clinical outcome (tumor response) was performed on the STITT population. Patients in the substudy were a subset of patients from the STITT population.

3. Results

3.1. Patient Disposition

A total of 184 patients consented, and were evaluated for eligibility for the prescreening phase. Of these patients, 10 did not have documented $BRAF^{V600}$ mutation test results on melanoma tissue upon study entry, and two had no plasma cfDNA test results (Figure 1). The remaining 172 patients represented the mITT population. Of these, 118 (68.6%) had a wild-type *BRAF* status and 54 (31.4%) had a mutant $BRAF^{V600}$ status. Among the 118 patients with an initial wild-type *BRAF* status, the tumors in five of these patients were confirmed to be $BRAF^{V600}$ mutant in the tissue retests, and one patient continued into the treatment phase. Of the 54 patients with a mutant $BRAF^{V600}$ status, based on the prior tissue test, 39 continued into the treatment phase. The baseline characteristics of the patients are shown in Table 1.

Table 1. Baseline demographic, disease, biopsy, and mutation characteristics.

	mITT	STITT
	$n = 172$	$n = 40$
Age, median years (range)	62.5 (20–93)	56.5 (26–82)
Sex, n (%)		
Male	84 (48.8)	18 (45.0)
Female	88 (51.2)	22 (55.0)
ECOG score, n (%)		
0	NE	21 (52.5)
1	NE	16 (40.0)
2	NE	3 (7.5)
Time since diagnosis of metastases, median months (range)	19.3 (0.1–260.7)	12.4 (0.8–260.7)
Age at diagnosis of metastases, median years (range)	59.1 (18–91)	53.5 (26–83)
Disease stage at study entry, n (%)		
Unresectable stage IIIC	16 (9.3)	1 (2.5)
Stage IV	156 (90.7)	29 (97.5)
Measurable disease at study entry, n (%)	147 (85.5)	37 (92.5)
Number of target lesions, n (%)		
0–3	NE	22 (55.0)
>3	NE	18 (45.0)
Type of tissue material, n (%)		
Archival	140 (81.4)	30 (75.0)
Recent	32 (18.6)	10 (25.0)
Prior tissue *BRAF* mutation test result, n (%)		
BRAF wild-type	118 (68.6)	1 (2.5)
$BRAF^{V600}$ mutation	54 (31.4)	39 (97.5)
Prior therapy, n (%)		
Immunotherapy	–	9 (33)
Targeted therapy	–	0
Other systemic therapy	–	7 (17.5)
Investigational treatment	–	1 (2.5)
Radiotherapy	–	8 (20.0)

ECOG, Eastern Cooperative Oncology Group; mITT, modified intention-to-treat; NE, not evaluated; STITT, study treatment intention-to-treat.

3.2. Prescreening Phase

3.2.1. Primary Endpoint

Among the 118 patients with a wild-type *BRAF* status, seven (5.9%) had a positive $BRAF^{V600}$ mutation test result, based on the plasma cfDNA test. Tissue retests on archival samples confirmed the $BRAF^{V600}$ mutation in five (71.4%) of the seven patients. Thus, for 4.2% of the patients with a wild-type *BRAF* status, the presence of a $BRAF^{V600}$ mutation was confirmed in a tissue retest; the one-sided test, comparing this frequency with the a priori hypothesis of 3%, was not statistically significant ($p = 0.215$).

3.2.2. Secondary Endpoint

The plasma cfDNA test was concordant with the prior tissue test in 35 of 54 of the patients who were classified as $BRAF^{V600}$ mutation-positive, and in 111 of 118 of the patients with a wild-type status (Table 2). Relative to tissue testing, the plasma test demonstrated 64.8% sensitivity and 94.1% specificity. Of the 42 patients with a $BRAF^{V600}$ mutant plasma cfDNA test result, 35 had a mutant status, based on the prior tissue test. The positive predictive value (i.e., the likelihood that the mutation detected with the plasma test corresponds to the tissue test) was 83.3%. Of the 130 patients with a wild-type plasma test result, 111 had a wild-type status, based on the prior tissue test. The negative predictive value (i.e., the likelihood that the wild-type status detected with the plasma test corresponds to the tissue test) was 85.4%.

Table 2. Comparison of tissue-based and cfDNA-based mutation tests. (**A**) Comparison of plasma and tissue $BRAF^{V600}$ mutation test results, and (**B**) comparison of $BRAF$ and $NRAS$ plasma mutation tests results in the mITT population. Grey-shaded cells designate the discordance between the respective tests. cfDNA, cell-free DNA; mITT, modified intention-to-treat.

A. BRAF		**Plasma cfDNA Test Result**		
		Wild-Type	**Mutant**	
Tissue test result	Wild-type	111	7	Patients with wild-type tissue test = 118
	Mutant	19	35	Patients with mutant tissue test = 54
		Patients with wild-type plasma test = 130	Patients with mutant plasma test = 42	
B. Plasma		**BRAF cfDNA test result**		
		Wild-type	**Mutant**	
NRAS cfDNA test result	Wild-type	107	42	Patients with wild-type NRAS = 149
	Mutant	22	0	Patients with mutant NRAS = 22
		Patients with wild-type plasma test = 129	Patients with mutant plasma test = 42	

3.2.3. Exploratory Endpoint

Of the 172 patients in the mITT population, 171 (99.4%) had an available $NRAS$ mutation test result. Of these patients, 149 (87.1%) had an $NRAS$ wild-type status and 22 (12.9%) had a mutant status. A summary of the relationship of $NRAS$ mutant status and $BRAF^{V600}$ mutant plasma test results is presented in Table 2. The mutations in $BRAF$ and $NRAS$ were mutually exclusive.

3.3. Treatment Phase

3.3.1. Efficacy Outcomes

In the STITT population, 29 of the 36 patients (80.6%) with target lesions and measurable disease, had either a complete response (three patients) or a partial response (26 patients) (Table 3). One patient (2.8%) had progressive disease, and two patients (5.6%) had stable disease. The median PFS (95% confidence interval (CI)) was 13.6 months (9.5–16.5). Of the 29 patients who had a complete or partial response, the median DOR (95% CI) was 11.0 months (9.2—not estimable). The median overall survival time was not estimable at the time of the analysis.

Table 3. Objective response rate and PFS (STITT).

	STITT $n = 40$	Archival Tissue * $n = 29$	Recent Tissue [†] $n = 10$
Objective response rate, ‡ n (%)			
CR	3 (8.3)	3 (11.5)	–
PR	26 (72.2)	20 (76.9)	6 (60.0)
SD	2 (5.6)	–	2 (20.0)
PD	1 (2.8)	–	1 (10.0)
Other nonresponders	4 (11.1)	3 (11.5)	1 (10.0)
Responders (CR + PR), n/n' (%)	29/36 (80.6)	23/26 (88.5)	6/10 (60.0)
95% CI	64.0–91.8	69.8–97.6	26.2–87.8
Duration of Response, n' ‡	29	23	$n' = 6$
Median, months (95% CI)	11.0 (9.2–NE)	11.0 (9.2–NE)	8.8 (3.6–NE)
Progression-free survival			
Median months (95% CI)	13.6 (9.5–16.5)	14.5 (10.8–NE)	6.2 (3.6–NE)

CI, confidence interval; CR, complete response; n', number of patients considered in the analysis; percentages calculated based on n'; NE, not estimable; PD, progressive disease; PR, partial response; SD, stable disease; STITT, study treatment intention-to-treat. * Patients for whom treatment was based on the mutation identified by prior tissue test and for whom archival tissue was used. [†] Patients for whom treatment was based on the mutation identified by prior tissue test and for whom recent tissue was used. ‡ For patients with measurable disease at screening having responded during the study (CR or PR).

3.3.2. Correlation of Mutation Testing and Clinical Outcome

The $BRAF^{V600}$ mutation was detected in the plasma at prescreening for 26 of the 40 treated patients. The median quantitation cycle (Cq) for the $BRAF$ mutation was 40.25. The median PFS was 12.8 months and 10.8 months for the patients below and above the median mutation Cq, respectively; however, the Cq was not significantly associated with PFS duration ($p = 0.653$). Of these 26 patients, 22 had a complete or partial response. No significant relationship was found between mutation Cq and the DOR ($p = 0.409$). The median DOR was similar for these patients, regardless of whether the mutation Cq was above or below the median (10.7 vs. 10.9 months, respectively).

The clinical outcome was analyzed separately for patients with a $BRAF^{V600}$ mutation, based on detection of the mutation in the plasma test at the prescreening. The median PFS was 14.8 months for the patients with a $BRAF$ wild-type plasma test result, and 12.8 months for the patients with a $BRAF^{V600}$ mutation plasma test result ($p = 0.286$). The median DOR was not evaluable for the patients with a $BRAF$ wild-type plasma test result, and was 10.9 months for the patients with a $BRAF^{V600}$ mutation plasma test result ($p = 0.220$).

3.3.3. Substudy

The concentration of the $BRAF^{V600}$ mutation before the first dose was evaluated as a prognostic factor/metric for PFS duration in the eight patients who were participating in the substudy. The $BRAF^{V600}$ mutation was observed at the start of cycle 1 for five of the eight patients (62.5%). The frequency of detection of the $BRAF^{V600}$ mutation then decreased to 28.6% at the last assessment. All of these mutations were $BRAF^{V600E/E2/D}$, except for one instance of $BRAF^{V600K/R/W}$. No correlation between the plasma concentration of the mutation (average Cq for mutation) at the start of cycle 1, and the PFS duration, was observed (Spearman's rank correlation coefficient: -0.100; $p = 0.873$).

3.3.4. Exposure and Safety

The mean duration of the treatment with both cobimetinib and vemurafenib was 11.2 ± 9.6 months (median, 7.9 months; range, 0.4–36.8 months). The median number of cycles was 9.0 for both cobimetinib and vemurafenib. The most frequently reported AEs were rash (47.5%), blood creatinine phosphokinase level increase (32.5%), diarrhea (32.5%), photosensitivity reaction (27.5%), pyrexia (27.5%), arthralgia (27.5%), and maculopapular rash (22.5%) (Table 4).

Table 4. Common treatment-emergent adverse events (incidence $\geq$10% by preferred term).

System Organ Class/Preferred Term	*n* = 40
Any treatment-emergent adverse event, *n* (%)	39 (97.5)
Skin and subcutaneous tissue disorders, *n* (%)	33 (82.5)
Rash	19 (47.5)
Photosensitivity reaction	11 (27.5)
Maculopapular rash	9 (22.5)
Alopecia	4 (10.0)
Investigations, *n* (%)	22 (55.0)
Blood creatinine phosphokinase increased	13 (32.5)
Gamma-glutamyltransferase increased	5 (12.5)
C-reactive protein increased	4 (10.0)
General disorders and administration site conditions, *n* (%)	21 (52.5)
Pyrexia	11 (27.5)
Fatigue	8 (20.0)
Oedema peripheral	5 (12.5)
Infections and infestations, *n* (%)	21 (52.5)
Upper respiratory tract infection	8 (20.0)
Conjunctivitis	6 (15.0)
Urinary tract infection	4 (10.0)
Gastrointestinal disorders, *n* (%)	20 (50.0)
Diarrhea	13 (32.5)
Vomiting	6 (15.0)
Nausea	5 (12.5)
Musculoskeletal and connective tissue disorders, *n* (%)	14 (35.0)
Arthralgia	11 (27.5)
Musculoskeletal pain	4 (10.0)
Myalgia	4 (10.0)
Pain in extremity	4 (10.0)
Eye disorders, *n* (%)	12 (30.0)
Vision blurred	6 (15.0)
Chorioretinopathy	4 (10.0)
Nervous system disorders, *n* (%)	12 (30.0)
Headache	6 (15.0)
Metabolism and nutrition disorders, *n* (%)	7 (17.5)
Decreased appetite	4 (10.0)
Hypokalemia	4 (10.0)
Hypertension	5 (12.5)

4. Discussion

The current study aimed to evaluate the clinical utility of a plasma-derived cfDNA test for the detection of $BRAF^{V600}$ mutations in patients with metastatic melanoma, in whom the original testing using tumor biopsy falsely diagnosed the $BRAF^{V600}$ wild-type status. It was observed that ~6% of the patients who were originally classified as having *BRAF* wild-type tumors tested positive for *BRAF* mutations using the cfDNA test. The sensitivity of the plasma cfDNA testing was 64.8%, and its positive and negative predictive values were 83.3% and 85.4%, respectively. No significant difference was observed between the median PFS and the median DOR for the patients with a *BRAF* wild-type versus $BRAF^{V600}$ mutation-positive test result using cfDNA testing.

Previous studies have shown that cfDNA testing may be a surrogate for determining *BRAF* mutations in patients with melanoma [24–28]. In our study, the presence of *BRAF* mutations was confirmed by a subsequent tissue-based retest in five of the seven patients who showed *BRAF* mutations using cfDNA testing. The remaining two patients showed *BRAF* wild-type tumors upon the retest, which could potentially be a result of the archival tissue sample being degraded. Although the study did not meet its predefined criteria for the clinical significance of cfDNA testing ($\geq$10% of patients designated as *BRAF* wild-type based on a prior tissue test to subsequently show $BRAF^{V600}$ mutations on cfDNA testing), ~6% of *BRAF* wild-type patients were subsequently shown to be $BRAF^{V600}$-mutation-positive, based on cfDNA tests, suggesting that cfDNA testing may have a place in the clinical setting. Furthermore, one of the five *BRAF* mutation-positive patients went on to receive *BRAF*-targeted therapy and responded to treatment with a PFS that was in a range that was observed in previous experience [29]. These results suggest that cfDNA testing may be a quick and easy alternative to identify *BRAF* mutations in patients for whom an archival tumor tissue sample may not be available in sufficient quality to allow for a retesting population, so they can benefit from the early initiation of *BRAF*-targeted therapy. However, it should be noted that cfDNA testing demonstrated a sensitivity of 64.8% in the present study. Therefore, although the cfDNA technique may be beneficial in the initial screening for the *BRAF* mutation status (owing to the ease of obtaining a blood sample versus a tumor re-biopsy, and the possibility of loss of the archival sample due to degradation, as seen in two patients in this study), retesting of the archival tumor tissue, if available, should be the primary choice for confirmation of the *BRAF* mutation status.

This study also evaluated any potential correlation with the cell-free *BRAF* mutation load of the plasma and treatment outcomes, as suggested by previous reports [29–31]; however, no appreciable differences were observed in either the median PFS or DOR among patients in whom the Cq for the *BRAF* mutation was above or below the median Cq of 40.25. The substudy, which tested *BRAF* mutation load as a metric of treatment response, showed that there was a decrease in the frequency of detection of the $BRAF^{V600}$ mutation during the treatment, and the assay failed to detect the $BRAF^{V600}$ mutation in a number of patients as the treatment progressed. Overall, the treatment outcomes and the observed safety profile with combination treatment with cobimetinib plus vemurafenib, were consistent with the pivotal phase 3 data in a population of patients with advanced melanoma and *BRAF* mutations [29,32].

A key limitation of the study was the relatively small number of patients who were included. In addition, only one patient whose tumor was originally classified as *BRAF* wild-type, and subsequently flagged as *BRAF* mutation-positive by the cell-free assay, went on to receive and benefit from targeted therapy with cobimetinib plus vemurafenib. The strengths of the study include its prospective nature, and that consecutive eligible and consenting patients at the study centers were enrolled (i.e., there was no selection bias). Furthermore, the study confirmed that a high percentage of patients that are designated as having *BRAF*-mutated tumors, using tissue-based tests, have positive results in cfDNA testing.

5. Conclusions

The study shows that plasma testing using cfDNA is a less invasive and rapid alternative to confirm the presence of *BRAF* mutations versus tumor biopsies, and may help in the early identification of patients who are more likely to benefit from treatment with *BRAF* inhibitors. However, cfDNA testing should be used with caution in patients undergoing *BRAF* inhibitor therapy, owing to its decreased sensitivity with treatment in these patients. No significant correlation between the use of cfDNA testing, and PFS and DOR could be determined among the patients receiving cobimetinib plus vemurafenib treatment. Additional studies in a larger patient population are warranted to confirm these results.

Supplementary Materials: The following is available online at https://www.mdpi.com/article/10.3390/cancers13143591/s1, Supplementary material appendix: patient eligibility criteria.

Author Contributions: Conceptualization, P.R., P.P., J.K., G.M., V.G. and B.N.; data curation, V.G., A.T. and K.d.B.; formal analysis, B.J., V.G., A.T. and K.d.B.; investigation, P.R. and J.K.; methodology, P.R., P.P., J.K., G.M., V.G. and B.N.; project administration, V.G., A.T. and K.d.B.; resources, B.J. and G.M.; supervision, V.G., A.T. and K.d.B.; validation, V.G., A.T. and K.d.B.; visualization, V.G., A.T. and K.d.B.; writing—original draft, P.R. and B.N.; writing—review and editing, P.R., P.P., B.J., V.G., A.T., K.d.B. and B.N. All authors have read and agreed to the published version of the manuscript.

Funding: This study was sponsored by F. Hoffmann–La Roche Ltd.; Genentech, Inc., a member of the Roche Group; and Biocartis Group NV.

Institutional Review Board Statement: The study was registered with ClinicalTrials.gov. (NCT02768207) and EudraCT (2015-001731-20). The study was conducted in accordance with the good clinical practice guidelines, and the principles of the Declaration of Helsinki. The protocol was approved by an independent ethics committee at each site.

Informed Consent Statement: Written informed consent was obtained for all the patients before performing any study-specific prescreening test or evaluation. A second written informed consent form was obtained prior to the conduct of any study-related procedures in relation to the treatment phase. The patients who participated in the translational substudy signed a separate (third) informed consent form.

Data Availability Statement: Qualified researchers may request access to individual patient-level data through the clinical study data request platform (https://vivli.org/, accessed on 19 January 2021). Further details on Roche's criteria for eligible studies are available here (https://vivli.org/members/ourmembers/, accessed on 19 January 2021). For further details on Roche's global policy on the sharing of clinical information and how to request access to related clinical study documents, see here (https://www.roche.com/research_and_development/who_we_are_how_we_work/clinical_trials/our_commitment_to_data_sharing.htm, accessed on 19 January 2021).

Acknowledgments: We thank the patients and investigators who participated in the study, and Jerome Sah (ApotheCom, San Francisco, CA, USA) for providing medical writing and editorial support, which was funded by F. Hoffmann—La Roche Ltd.

Conflicts of Interest: Piotr Rutkowski reports consultancy fees from Novartis, Bristol Myers Squibb, Merck Sharp & Dohme Pierre Fabre, and Blueprint Medicines; research grants to his institution and/or himself, from Pfizer and Bristol Myers Squibb; and fees for speaking engagements from Novartis, Roche, Pierre Fabre, Bristol Myers Squibb, Merck Sharp & Dohme, Sanofi, Merck, and Pfizer. Patrick Pauwels reports consultancy fees from AstraZeneca, Pfizer, Roche, Merck Sharp & Dohme, Bristol Myers Squibb, Takeda, Biocartis, and Novartis; and grants from AstraZeneca, Roche, and Biocartis. Joseph Kerger has no conflicts to disclose. Bart Jacobs reports employment with Biocartis and Novartis. Geert Maertens reports employment with, patents with, and royalties from, Biocartis. Valerie Gadeyne, Anne Thielemans, and Katrien de Backer report employment with F. Hoffmann—La Roche. Bart Neyns reports the following paid monies to his institution: consultancy fees from Novartis, and research grants and payments for speaking engagements from Novartis, Pfizer, Bristol Myers Squibb, and Merck Sharp & Dohme.

References

1. Bisschop, C.; Ter, E.A.; Bosman, L.J.; Platteel, I.; Jalving, M.; van den Berg, A.; Diepstra, A.; van Hemel, B.; Diercks, G.F.H.; Hospers, G.A.P.; et al. Rapid *BRAF* mutation tests in patients with advanced melanoma: Comparison of immunohistochemistry, Droplet Digital PCR, and the Idylla Mutation Platform. *Melanoma Res.* **2018**, *28*, 96–104. [CrossRef]
2. Bruno, W.; Martinuzzi, C.; Andreotti, V.; Pastorino, L.; Spagnolo, F.; Dalmasso, B.; Cabiddu, F.; Gualco, M.; Ballestrero, A.; Bianchi-Scarrà, G.; et al. Heterogeneity and frequency of *BRAF* mutations in primary melanoma: Comparison between molecular methods and immunohistochemistry. *Oncotarget* **2017**, *8*, 8069–8082. [CrossRef]
3. Davies, H.; Bignell, G.R.; Cox, C.; Stephens, P.; Edkins, S.; Clegg, S.; Teague, J.; Woffendin, H.; Garnett, M.J.; Bottomley, W.; et al. Mutations of the *BRAF* gene in human cancer. *Nature* **2002**, *417*, 949–954. [CrossRef]
4. Li, Y.; Umbach, D.M.; Li, L. Putative genomic characteristics of *BRAF* ^{V600K} versus ^{V600E} cutaneous melanoma. *Melanoma Res.* **2017**, *27*, 527–535. [CrossRef]
5. Keilholz, U.; Ascierto, P.A.; Dummer, R.; Robert, C.; Lorigan, P.; van Akkooi, A.; Arance, A.; Blank, C.U.; Sileni, V.C.; Donia, M.; et al. ESMO consensus conference recommendations on the management of metastatic melanoma: Under the auspices of the ESMO Guidelines Committee. *Ann. Oncol.* **2020**, *31*, 1435–1448. [CrossRef] [PubMed]
6. Swetter, S.M.; Thompson, J.A.; Albertini, M.R.; Barker, C.A.; Boland, G.; Carson, W.E. National Comprehensive Cancer Network® (NCCN®) Clinical Practice Guidelines in Oncology—Cutaneous Melanoma Version 4.2020. 2020. Available online: https://www.nccn.org/professionals/physician_gls/pdf/cutaneous_melanoma.pdf (accessed on 16 September 2020).
7. Cohen, J.V.; Buchbinder, E.I. The evolution of adjuvant therapy for melanoma. *Curr. Oncol. Rep.* **2019**, *21*, 106. [CrossRef] [PubMed]
8. Dimitriou, F.; Long, G.V.; Menzies, A.M. Novel adjuvant options for cutaneous melanoma. *Ann. Oncol.* **2021**, *32*, 854–865. [CrossRef] [PubMed]

9. Blankenstein, S.A.; van Akkooi, A.C.J. Adjuvant systemic therapy in high-risk melanoma. *Melanoma Res.* **2019**, *29*, 358–364. [CrossRef]

10. Noor, R.; Trinh, V.; Kim, K.; Hwu, W.J. BRAF-targeted therapy for metastatic melanoma: Rationale, clinical activity and safety. *Clin. Investig.* **2011**, *1*, 1127–1139. [CrossRef]

11. Cheng, L.; Lopez-Beltran, A.; Massari, F.; MacLennan, G.T.; Montironi, R. Molecular testing for *BRAF* mutations to inform melanoma treatment decisions: A move toward precision medicine. *Mod. Pathol.* **2018**, *31*, 24–38. [CrossRef]

12. Cheng, L.Y.; Haydu, L.E.; Song, P.; Nie, J.; Tetzlaff, M.T.; Kwong, L.N.; Gershenwald, J.E.; Davies, M.A.; Zhang, D.Y. High sensitivity sanger sequencing detection of *BRAF* mutations in metastatic melanoma FFPE tissue specimens. *Sci. Rep.* **2021**, *11*, 9043. [CrossRef]

13. Harle, A.; Salleron, J.; Franczak, C.; Dubois, C.; Filhine-Tressarieu, P.; Leroux, A.; Merlin, J.-L. Detection of *BRAF* mutations using a fully automated platform and comparison with high resolution melting, real-time allele specific amplification, immunohistochemistry and next generation sequencing assays, for patients with metastatic melanoma. *PLoS ONE* **2016**, *11*, e0153576. [CrossRef]

14. Huang, W.; Kuo, T.T.; Wu, C.E.; Cheng, H.-Y.; Hsieh, C.-H.; Hsieh, J.-J.; Shen, Y.-C.; Hou, M.-M.; Hsu, T.; Chang, J.W.-C. A comparison of immunohistochemical and molecular methods used for analyzing the $BRAF^{V600E}$ gene mutation in malignant melanoma in Taiwan. *Asia Pac. J. Clin. Oncol.* **2016**, *12*, 403–408. [CrossRef] [PubMed]

15. Janku, F.; Claes, B.; Huang, H.J.; Falchook, G.S.; Devogelaere, B.; Kockx, M.; Bempt, I.V.; Reijans, M.; Naing, A.; Fu, S.; et al. *BRAF* mutation testing with a rapid, fully integrated molecular diagnostics system. *Oncotarget* **2015**, *6*, 26886–26894. [CrossRef]

16. Alrabadi, N.; Gibson, N.; Curless, K.; Cheng, L.; Kuhar, M.; Chen, S.; Warren, S.J.P.; Alomari, A.K. Detection of driver mutations in *BRAF* can aid in diagnosis and early treatment of dedifferentiated metastatic melanoma. *Mod. Pathol.* **2019**, *32*, 330–337. [CrossRef]

17. Kong, B.Y.; Carlino, M.S.; Menzies, A.M. Biology and treatment of *BRAF* mutant metastatic melanoma. *Melanoma Manag.* **2016**, *3*, 33–45. [CrossRef]

18. Vanni, I.; Tanda, E.T.; Spagnolo, F.; Andreotti, V.; Bruno, W.; Ghiorzo, P. The current state of molecular testing in the *BRAF*-mutated melanoma landscape. *Front. Mol. Biosci.* **2020**, *7*, 113. [CrossRef] [PubMed]

19. Donadio, E.; Giusti, L.; Cetani, F.; Da Valle, Y.; Ciregia, F.; Giannaccini, G.; Pardi, E.; Saponaro, F.; Torregrossa, L.; Basolo, F.; et al. Evaluation of formalin-fixed paraffin-embedded tissues in the proteomic analysis of parathyroid glands. *Proteome Sci.* **2011**, *9*, 29. [CrossRef] [PubMed]

20. Janku, F.; Huang, H.J.; Claes, B.; Falchook, G.S.; Fu, S.; Hong, D.; Ramzanali, N.M.; Nitti, G.; Cabrilo, G.; Tsimberidou, A.M.; et al. *BRAF* mutation testing in cell-free DNA from the plasma of patients with advanced cancers using a rapid, automated molecular diagnostics system. *Mol. Cancer Ther.* **2016**, *15*, 1397–1404. [CrossRef]

21. Calapre, L.; Warburton, L.; Millward, M.; Ziman, M.; Gray, E.S. Circulating tumour DNA (ctDNA) as a liquid biopsy for melanoma. *Cancer Lett.* **2017**, *404*, 62–69. [CrossRef]

22. Balch, C.M.; Gershenwald, J.E.; Soong, S.J.; Thompson, J.F.; Atkins, M.B.; Byrd, D.R.; Buzaid, A.C.; Cochran, A.J.; Coit, D.G.; Ding, S.; et al. Final version of 2009 AJCC melanoma staging and classification. *J. Clin. Oncol.* **2009**, *27*, 6199–6206. [CrossRef]

23. National Institutes of Health. Common Terminology Criteria for Adverse Events (CTCAE) v4.03. 2009. Available online: https://www.eortc.be/services/doc/ctc/CTCAE_4.03_2010-06-14_QuickReference_5x7.pdf (accessed on 16 September 2020).

24. Long-Mira, E.; Ilie, M.; Chamorey, E.; Leduff-Blanc, F.; Montaudié, H.; Tanga, V.; Allégra, M.; Lespinet-Fabre, V.; Bordone, O.; Bonnetaud, C.; et al. Monitoring *BRAF* and *NRAS* mutations with cell-free circulating tumor DNA from metastatic melanoma patients. *Oncotarget* **2018**, *9*, 36238–36249. [CrossRef]

25. Molina-Vila, M.A.; de-Las-Casas, C.M.; Bertran-Alamillo, J.; Jordana-Ariza, N.; González-Cao, M.; Rosell, R. cfDNA analysis from blood in melanoma. *Ann. Transl. Med.* **2015**, *3*, 309. [CrossRef]

26. Burjanivova, T.; Malicherova, B.; Grendar, M.; Minarikova, E.; Dusenka, R.; Vanova, B.; Bobrovska, M.; Pecova, T.; Homola, I.; Lasabova, Z.; et al. Detection of $BRAF^{V600E}$ mutation in melanoma patients by digital PCR of circulating DNA. *Genet. Test. Mol. Biomark.* **2019**, *23*, 241–245. [CrossRef]

27. Knol, A.C.; Vallée, A.; Herbreteau, G.; Nguyen, J.M.; Varey, E.; Gaultier, A.; Théoleyre, S.; Saint-Jean, M.; Peuvrel, L.; Brocard, A.; et al. Clinical significance of *BRAF* mutation status in circulating tumor DNA of metastatic melanoma patients at baseline. *Exp. Dermatol.* **2016**, *25*, 783–788. [CrossRef]

28. Kozak, K.; Kowalik, A.; Gos, A.; Wasag, B.; Lugowska, I.; Jurkowska, M.; Krawczynska, N.; Kosela-Paterczyk, H.; Switaj, T.; Teterycz, P.; et al. Cell-free DNA *BRAF* ^{V600E} measurements during *BRAF* inhibitor therapy of metastatic melanoma: Long-term analysis. *Tumori* **2020**. [CrossRef]

29. Larkin, J.; Ascierto, P.A.; Dreno, B.; Atkinson, V.; Liszkay, G.; Maio, M.; Mandalà, M.; Demidov, L.; Stroyakovskiy, D.; Thomas, L.; et al. Combined vemurafenib and cobimetinib in *BRAF*-mutated melanoma. *N. Engl. J. Med.* **2014**, *371*, 1867–1876. [CrossRef] [PubMed]

30. Schreuer, M.; Jansen, Y.; Planken, S.; Chevolet, I.; Seremet, T.; Kruse, V.; Neyns, B. Combination of dabrafenib plus trametinib for *BRAF* and *MEK* inhibitor pretreated patients with advanced *BRAF*(V600)-mutant melanoma: An open-label, single arm, dual-centre, phase 2 clinical trial. *Lancet Oncol.* **2017**, *18*, 464–472. [CrossRef]

31. Schreuer, M.; Meersseman, G.; Van Den Herrewegen, S.; Jansen, Y.; Chevolet, I.; Bott, A.; Wilgenhof, S.; Seremet, T.; Jacobs, B.; Buyl, R.; et al. Quantitative assessment of $BRAF^{V600}$ mutant circulating cell-free tumor DNA as a tool for therapeutic monitoring in metastatic melanoma patients treated with BRAF/MEK inhibitors. *J. Transl. Med.* **2016**, *14*, 95. [CrossRef] [PubMed]
32. Ascierto, P.A.; McArthur, G.A.; Dréno, B.; Atkinson, V.; Liszkay, G.; Di Giacomo, A.M.; Mandalà, M.; Demidov, L.; Stroyakovskiy, D.; Thomas, L.; et al. Cobimetinib combined with vemurafenib in advanced BRAF(V600)-mutant melanoma (coBRIM): Updated efficacy results from a randomised, double-blind, phase 3 trial. *Lancet Oncol.* **2016**, *17*, 1248–1260. [CrossRef]

Article

Psychological Distress of Metastatic Melanoma Patients during Treatment with Immune Checkpoint Inhibitors: Results of a Prospective Study

Lisa Wiens [1], Norbert Schäffeler [2], Thomas Eigentler [1], Claus Garbe [1] and Andrea Forschner [1,*]

[1] Department of Dermatology, University Hospital Tübingen, 72076 Tübingen, Germany; wiens_lisa@web.de (L.W.); thomas.eigentler@med.uni-tuebingen.de (T.E.); claus.garbe@med.uni-tuebingen.de (C.G.)

[2] Department of Psychosomatic Medicine and Psychotherapy, University Hospital Tübingen, 72076 Tübingen, Germany; Norbert.Schaeffeler@med.uni-tuebingen.de

* Correspondence: andrea.forschner@med.uni-tuebingen.de; Tel.: +49-7071-29-84555

Simple Summary: A large proportion of patients with metastatic melanoma suffer from psychological distress. Early identification of these patients is important to be able to offer them adequate support. This longitudinal study aimed to investigate the extent to which the psychological distress of patients with malignant melanoma might change during their first three months of treatment with immune checkpoint inhibitors. We found a high proportion of distressed patients in a cohort of 113 patients at the beginning of immunotherapy, which decreased during therapy. A binary logistic regression analysis provided additional factors indicating an increased risk of developing distress— female gender and occurrence of adverse events correlated significantly with distress values above the threshold. The strongest factor was patients' self-assessment. When initiating immunotherapy, it is also important to consider the needs of patients and offer them psycho-oncological support.

Abstract: Background: Immune checkpoint inhibitors (ICI) provide effective treatment options for advanced melanoma patients. However, they are associated with high rates of immune-related side effects. There are no data on the distress of melanoma patients during their ICI treatment. We, therefore, conducted a prospective longitudinal study to assess distress and the need for psychooncological support in these patients. Methods: Questionnaires were completed before initiation of ICI (T0), after 6–8 weeks (T1), and after 12–14 weeks (T2). We furthermore included the Hornheide Screening Instrument (HSI), distress thermometer (DT), and patients' self-assessment. Binary logistic regression was performed to identify factors indicating a need for psychooncological support. Results: 36.3%/55.8% (HSI / DT) of the patients were above the threshold, indicating a need for psychooncological support at T0, and 7.8% of the patients reported practical problems. In contrast, at T2, the distress values had decreased to 29.0%/40.2% (HSI/DT), respectively. Female gender and occurrence of side effects significantly correlated to values above the threshold. The strongest factor was the patient's self-assessment. Conclusion: With the beginning of ICI, psychooncological support should be offered. Furthermore, practical problems should be considered, e.g., transport to therapy. Female patients and patients with side effects should be given special attention, as well as the patient self-assessment.

Keywords: metastatic melanoma; checkpoint inhibitors; immune therapy; psychooncology; distress; his; dt; problems

Citation: Wiens, L.; Schäffeler, N.; Eigentler, T.; Garbe, C.; Forschner, A. Psychological Distress of Metastatic Melanoma Patients during Treatment with Immune Checkpoint Inhibitors: Results of a Prospective Study. *Cancers* **2021**, *13*, 2642. https://doi.org/10.3390/cancers13112642

Academic Editor: Salvatore Grisanti

Received: 27 April 2021
Accepted: 25 May 2021
Published: 27 May 2021

Publisher's Note: MDPI stays neutral with regard to jurisdictional claims in published maps and institutional affiliations.

1. Introduction

Melanoma is one of the most aggressive cancer types with an increasing incidence worldwide [1]. Receiving a diagnosis of melanoma is a deep cut in the lives of the person affected and can lead to numerous stress reactions [2]. Despite the high incidence of melanoma and an increasing number of new treatment options, only limited data is available on psychological distress in melanoma patients [3,4]. Some retrospective research indicated that patients on systemic therapy appear to be at lower risk for psychological distress, possibly due to the closer physician contact during therapy [5]. However, this was a retrospective survey that did not refer to the time of therapy initiation.

The first-line treatment for patients suffering from melanoma is surgical excision of the primary tumor and, if present, of metastases [6]. For melanoma patients with stage III or IV and unresectable metastases, immune checkpoint inhibitors such as pembrolizumab or nivolumab (PD-1 Inhibitors) as monotherapy and nivolumab in combination with ipilimumab (CTLA-4 Inhibitor) are available and improved prognosis of metastasized melanoma patients considerably [7–10]. Pembrolizumab and nivolumab have also been approved in the adjuvant setting [6,11].

We aimed to investigate possible changes in psychological distress of melanoma patients during their first three months of treatment with immune checkpoint inhibitors.

The study was designed to provide a comprehensive overview of the patient cohort, including tumor-specific, social and psychological aspects, to better understand the individual psychooncological needs of the patients during their course of immune checkpoint inhibitor therapy.

2. Materials and Methods

We conducted a prospective, longitudinal study on melanoma patients from the Center for Dermatooncology at the University of Tuebingen.

Inclusion criteria were German-speaking women and men of full age who had been scheduled for treatment with immune checkpoint inhibitors (ICI) for metastatic melanoma (AJCC-Stage III/IV). Exclusion criteria were severe problems with the German language, severe psychological or neurological symptoms (e.g., psychosis, dementia), or lack of consent. The study was performed between December 2017 and February 2019. A total of 160 patients were informed about the study. Of these, 39 patients declined to participate. Thus, 121 patients decided to participate and provided written informed consent. Eight patients did not complete the first or second follow-up questionnaire, completed it only partially or too late, or decided against further participation in the study. The basis of the statistical evaluation was, therefore, 113 patients at T0.

After detailed information about the possibility of participating in this study and the patient's consent, the questionnaire was handed out before the start of ICI. The patient filled out the baseline questionnaire before the initiation of ICI, directly on site. It was directly transferred to the electronic Psychooncological Screening System (ePOS) (Supplementary File S1).

The two follow-up questionnaires were conducted within 6–8 and 12–14 weeks after the initiation of ICI (Supplementary Files S2 and S3). Moreover, patients completed additional questionnaires at T1 and T2. In these questionnaires, the patients were asked to state whether they had had contact with a member of the psychooncological support team to that time point and whether the ICI therapy had been carried out as planned or not. Furthermore, they were asked to indicate whether side effects had occurred. During the survey, six patients died before completing the follow-up questionnaires. Nonetheless, they remained included. This is the reason for the statistical basis at time T1: 109 patients and T2: 107 patients.

Written consent was obtained from all patients included in this study, and ethical approval had been obtained from the local ethics committee of the Medical Faculty of the University of Tuebingen (file number 689/2017BO2).

2.1. Screening Instruments

2.1.1. Distress Thermometer (DT)

The DT consists of a scale from 0–10. The patient is asked to indicate on the thermometer how much stress he/she has felt during the last week, including the day of survey. The value 0 indicates "not stressed at all", the value 10 "extremely stressed" [12]. The DT is supplemented by an additional "problem list", consisting of 36 items concerning family, emotional, spiritual, and physical aspects [13]. Following the recommendations of Mehnert et al. [13], we considered patients with DT values $\geq$5 as over-threshold, and thus, were in need of psychooncological support.

2.1.2. Hornheide Screening Instrument (HSI)

The HSI represents the shortened form of the Hornheide Questionnaire (HF). The HSI is used to identify oncological patients in need of psychooncological support [14]. The seven items used in the HSI cover the following areas: 1—body condition, 2—mental condition, 3—disease-independent stress, 4—availability of a trusted person, 5—stress in the family due to the hospital stay, 6—the presence of inner peace, and 7—in what way the patient feels informed about the disease and treatment. For items 1, 2, and 7 the answer categories are 0 = rather good, 1 = medium and 2 = rather bad. The answer categories for items 3 and 5 are: 2 = Yes and 0 = No, for items 4 and 6: 0 = Yes and 2 = No. In computer-assisted screening, the analysis is carried out within seconds. The discriminant function on which the HSI is based identifies the patient as "in need of care" or "not in need of care". The need for care is present with values >0.30 [15] using the discriminant function [16].

2.1.3. Self-Assessment for the Need for Psychooncological Support

The subjective need for psychooncological support was assessed by the following question: "Do you currently need support in coping with the disease or psychooncological counselling?" [17].

2.2. Central Malignant Melanoma Registry

Tumor-specific data were obtained from the central registry for malignant melanoma. In this registry, information on the age and gender of the patients, date of the first diagnosis, metastasis, and further course of the disease are available. All patients included in this registry had provided written consent to this documentation.

2.3. Statistical Analyses

Demographic and clinical data were presented descriptively. The categorical variables were presented as absolute and relative frequencies, the continuous variables as mean values. A Mann–Whitney U test or Wilcoxon test was performed to analyze differences between patient groups (over-threshold values in DT present or absent). For small sample sizes, the exact Fischer test was used.

Furthermore, a binary logistic regression analysis was calculated to check for potential risk factors indicating the need for psychooncological support. Statistical analysis was performed using the statistical program for social sciences SPSS Version 25 (IBM, New York, NY, USA).

3. Results

One hundred and thirteen patients were included in the study, 54 women (47.8%) and 59 men (52.2%). The mean age of the women at the initial survey was 61.4 years (30–90 years, SD = 14.81), that of the men 62.4 years (30–89 years, SD = 13.80).

Three patients were on daily antidepressants with citalopram or mirtazapine and one patient had a daily medication of lorazepam. Beyond that, none of the patients had a permanent medication with antidepressants or other psychopharmaceuticals. Of the 113 patients included at baseline (T0), 109 were alive at the first follow-up (T1) and 107 at T2 (Figure 1).

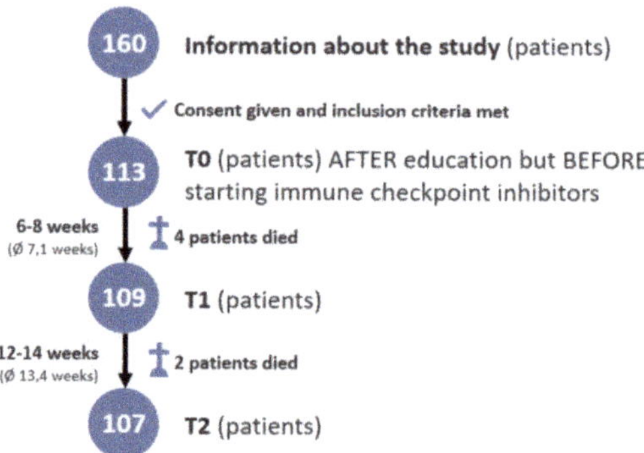

Figure 1. Flow chart.

3.1. Clinical Data

Cutaneous melanoma was the most common type (66.4%, *n* = 75) (Figure 2).

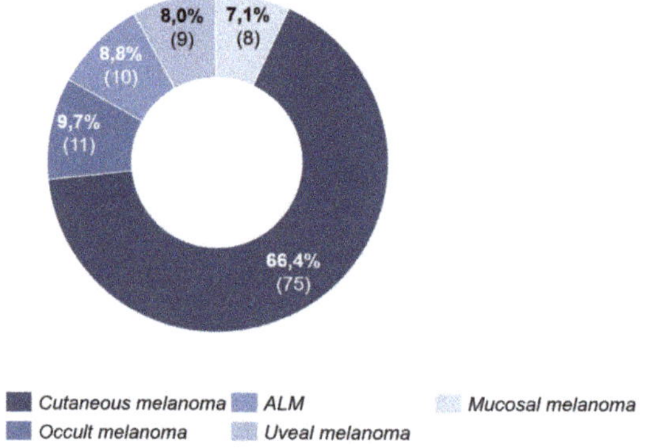

Figure 2. Melanoma type.

At the time of the initial survey (T0), 68 patients (60.2%) were classified as tumor stage IV and 45 patients (39.8%) as stage III (Figure 3).

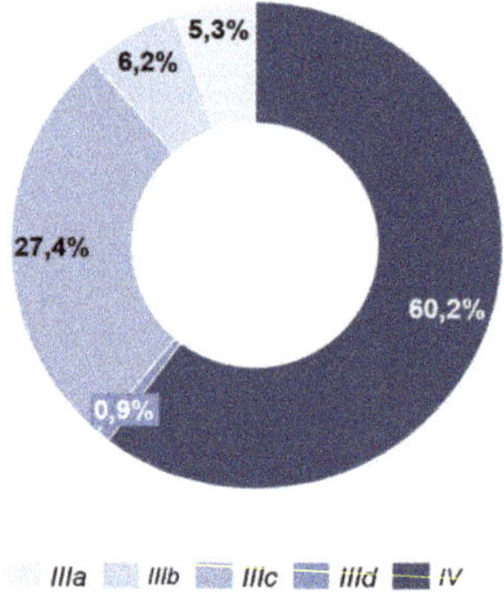

Figure 3. Tumor stage.

Combined systemic therapy with ipilimumab and nivolumab received 48 patients (42.5%). Ten patients (8.8%) started monotherapy with PD-1 antibody nivolumab, nine patients (8.0%) with PD-1 antibody pembrolizumab. Among the total cohort of 113 patients, 44 patients (38.9%) had nivolumab and two patients (1.8%) pembrolizumab in an adjuvant setting (Figure 4).

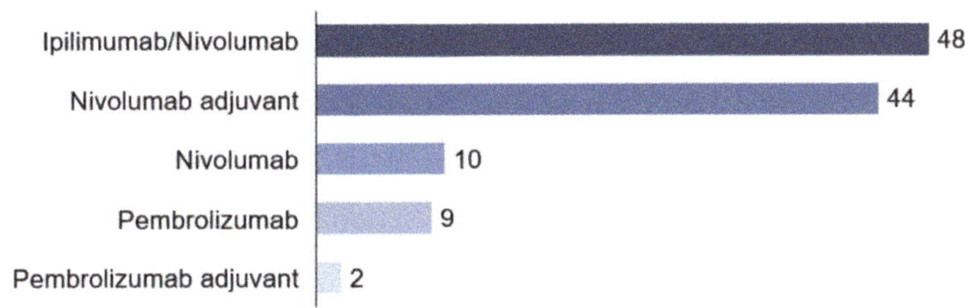

Figure 4. Therapy.

For evaluating the documented adverse events, patients were classified by treatment regime. On the one hand, patients with combined ICI (ipilimumab and nivolumab ($n = 48$)) and, on the other hand, patients who received a PD-1 antibody monotherapy with nivolumab or pembrolizumab ($n = 65$).

In 65.2% ($n = 30$) of the patients with combined ICI, side effects occurred up to time T2. These appeared on average 33 days (4–63 days, SD = 18.14) after the start of systemic therapy. Most frequently (29.7%), fatigue was documented. The second most frequently documented side effect was colitis with 17.2% (Figure 5).

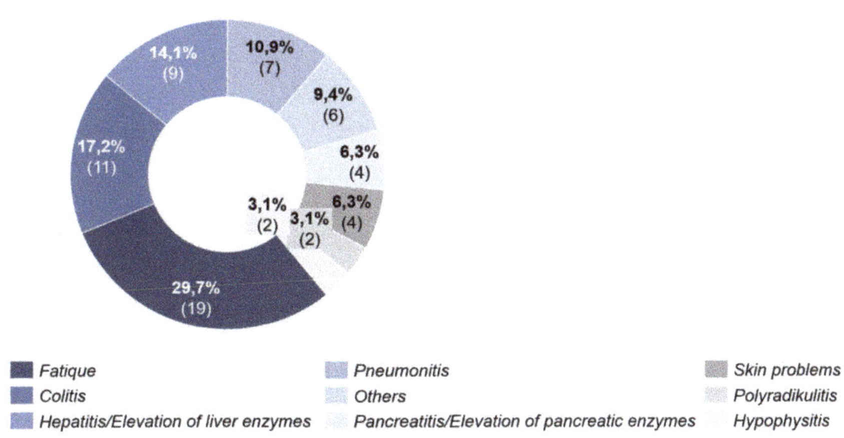

Figure 5. Side effects under combined immunotherapy.

Under PD-1 antibody therapy, 27.0% ($n = 17$) of patients experienced side effects. These were seen after an average of 39 days (20–89 days, SD = 20.34) after initiation of ICI. Here, fatigue was also the most frequent cause with 32.5%. Colitis was in second place with 20.0% (Figure 6).

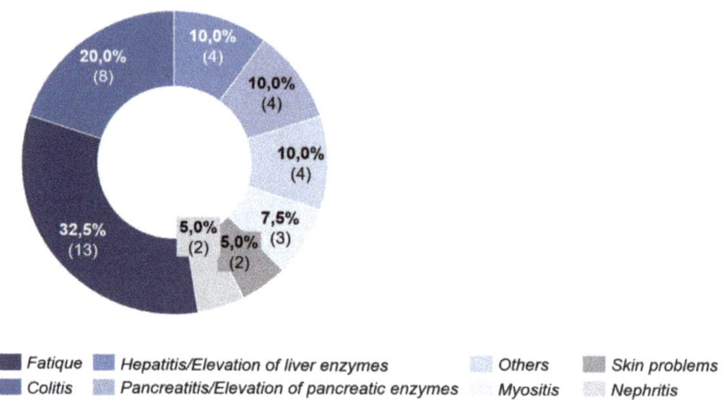

Figure 6. Side effects under PD-1 antibody therapy.

3.2. Psychological Distress

We found 55.8% of the patients were screened as significantly distressed by the Distress thermometer (cutoff $\geq$ 5). At T1, the percentage was 54.1%. At T2, the percentage dropped down to 40.2%.

Concerning the related problem list, physical problems were the priority at all measurement points. They reached their maximum value at time T1 (T0: 54.3%; T1: 67.0%; T2: 63.9%). Exhaustion, skin problems, sleep disorders, and pain were most frequently mentioned. Emotional problems came second to problem areas and were highest at the beginning of immunotherapy (T0: 35.9%; T1: 27.0%; T2: 29.2%). Above all, worries and fears occurred in half of the patients interviewed. Practical problems took third place. Family problems and religious issues were extremely rare (Figure 7 and Table 1).

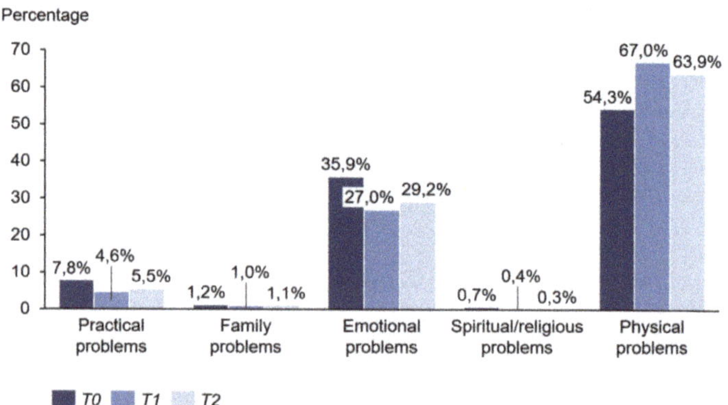

Figure 7. Problem areas.

Measured by HSI, 36.3% of patients showed over-threshold values (cutoff > 0.3) before starting therapy. At T1, the percentage was 36.7%. At time T2, the average exposure dropped to 29.0%.

Thirteen patients (11.5%) indicated a subjective need for psychooncological support at T0, nine patients (8.3%) at T1, and four patients (3.7%) at the last survey time T2. Within the survey period, twelve women and three men (13.8%; $n = 15$) had contact with a member of the psychooncological support team and had made an appointment for an interview. Of these 15 patients, seven had previously indicated a subjective need, another six patients showed over threshold values in HSI or DT, and one patient neither indicated a subjective need nor achieved values above the cutoff in the survey. At time T2, 21 patients (22.8%) who had not had contact with the psychooncological support team indicated that, in retrospective view, a conversation with the psychooncological team would probably have helped them to better cope with the disease.

Table 1. Problem list at T0, T1, T2.

Problem List	T0 ($n = 113$)	T1 ($n = 109$)	T2 ($n = 107$)
Practical problems			
Housing	4.4% ($n = 5$)	3.7% ($n = 4$)	3.7% ($n = 4$)
Insurance	3.5% ($n = 4$)	3.7% ($n = 4$)	5.6% ($n = 6$)
Work/school	13.3% ($n = 15$)	10.1% ($n = 11$)	13.1% ($n = 14$)
Transportation	16.8% ($n = 19$)	11.9% ($n = 13$)	10.3% ($n = 11$)
Childcare	1.8% ($n = 2$)	0.0% ($n = 0$)	0.0% ($n = 0$)
Family problems			
Dealing with partner	6.2% ($n = 7$)	4.6% ($n = 5$)	4.7% ($n = 5$)
Dealing with children	0.0% ($n = 0$)	1.8% ($n = 2$)	0.9% ($n = 1$)
Emotional problems			
Worry	51.3% ($n = 58$)	56.9% ($n = 62$)	62.6% ($n = 67$)
Fears	49.6% ($n = 56$)	47.7% ($n = 52$)	48.6% ($n = 52$)
Sadness	24.8% ($n = 28$)	23.9% ($n = 26$)	21.5% ($n = 23$)
Depression	10.6% ($n = 12$)	8.3% ($n = 9$)	9.3% ($n = 10$)
Nervousness	29.2% ($n = 33$)	19.3% ($n = 21$)	17.8% ($n = 19$)
Loss of interest in usual activities	14.2% ($n = 16$)	12.8% ($n = 14$)	5.6% ($n = 6$)
Spiritual/religious issues			
In relation to God	3.5% ($n = 4$)	2.8% ($n = 3$)	1.9% ($n = 2$)
Loss of faith	0.0% ($n = 0$)	0.0% ($n = 0$)	0.0% ($n = 0$)
Physical problems			
Pain	24.8% ($n = 28$)	40.4% ($n = 44$)	36.4% ($n = 39$)
Nausea	12.4% ($n = 14$)	22.9% ($n = 25$)	16.8% ($n = 18$)
Fatigue	40.7% ($n = 46$)	62.4% ($n = 68$)	64.5% ($n = 69$)
Sleep	31.9% ($n = 36$)	42.2% ($n = 46$)	43.9% ($n = 47$)
Getting around	25.7% ($n = 29$)	30.3% ($n = 33$)	20.6% ($n = 22$)
Bathing. dressing	3.5% ($n = 4$)	5.5% ($n = 6$)	4.7% ($n = 5$)
Appearance	3.5% ($n = 4$)	9.2% ($n = 10$)	4.7% ($n = 5$)
Breathing	11.5% ($n = 13$)	16.5% ($n = 18$)	10.3% ($n = 11$)
Mouth sores	1.8% ($n = 2$)	7.3% ($n = 8$)	1.9% ($n = 2$)
Eating	10.6% ($n = 12$)	13.8% ($n = 15$)	11.2% ($n = 12$)
Indigestion	12.4% ($n = 14$)	31.2% ($n = 34$)	33.6% ($n = 36$)
Constipation	9.7% ($n = 11$)	11.0% ($n = 12$)	5.6% ($n = 6$)
Diarrhea	9.7% ($n = 11$)	17.4% ($n = 19$)	12.1% ($n = 13$)
Changes in urination	9.7% ($n = 11$)	3.7% ($n = 4$)	3.7% ($n = 4$)
Fevers	0.0% ($n = 0$)	8.3% ($n = 9$)	6.5% ($n = 7$)
Skin dry/itchy	21.2% ($n = 24$)	45.0% ($n = 49$)	43.9% ($n = 47$)
Nose dry/congested	3.5% ($n = 4$)	11.0% ($n = 12$)	12.1% ($n = 13$)
Tingling in hands/feet	13.3% ($n = 15$)	11.9% ($n = 13$)	15.0% ($n = 16$)
Feeling swollen	7.1% ($n = 8$)	12.8% ($n = 14$)	12.1% ($n = 13$)
Memory/concentration	15.0% ($n = 17$)	15.6% ($n = 17$)	6.5% ($n = 7$)
Sexual	7.1% ($n = 8$)	5.5% ($n = 6$)	4.7% ($n = 5$)

The following table (Table 2) shows the individual values of the distress thermometer at all three survey times. In total, 63 patients marked a value of five or higher at T0 and were above the cutoff. At T1, there were 59 patients, and at T2, 43 patients with suprathreshold values.

Table 2. Distress thermometer at T0, T1, and T2.

DT Values	T0 (n = 113)	T1 (n = 109)	T2 (n = 107)
Value 0	2.7% (n = 3)	1.8% (n = 2)	1.9% (n = 2)
Value 1	6.2% (n = 7)	5.5% (n = 6)	10.3% (n = 11)
Value 2	8.8% (n = 10)	11.0% (n = 12)	15.0% (n = 16)
Value 3	13.3% (n = 15)	12.8% (n = 14)	14.0% (n = 15)
Value 4	13.3% (n = 15)	14.7% (n = 16)	18.7% (n = 20)
Value 5	17.7% (n = 20)	16.5% (n = 18)	20.6% (n = 22)
Value 6	15.9% (n = 18)	19.3% (n = 21)	13.1% (n = 14)
Value 7	9.7% (n = 11)	7.3% (n = 8)	3.7% (n = 4)
Value 8	4.4% (n = 5)	9.2% (n = 10)	2.8% (n = 3)
Value 9	4.4% (n = 5)	1.8% (n = 2)	0.0% (n = 0)
Value 10	3.5% (n = 4)	0.0% (n = 0)	0.0% (n = 0)

3.3. Comparison of Groups with and without Over-Threshold Values in DT

Patients with (55.8%, n = 63) and without (44.2%, n = 50) over-threshold values in DT differed significantly in gender before the start of ICI. Women more often showed a value above the cutoff of $\geq$5 (Table 3). Significant differences were found in gender, tumor stage, type of therapy, and side effects at time T1 (Table 4). There were significant differences between the two groups in the occurrence of side effects and the outcome in staging at time T2 (Table 5).

Table 3. Comparison of the groups at T0.

Characteristic	Overall n = 113 (100%)	DT < 5 n = 50 (44.2%)	DT $\geq$ 5 n = 63 (55.8%)	p-Value
Demographic characteristics				
Gender (%)				p = 0.041 [C] V = 0.210 [1]
Female	n = 54 (100%)	n = 18 (33.3%)	n = 36 (66.7%)	
Male	n = 59 (100%)	n = 32 (54.2%)	n = 27 (45.8%)	
Age (in Years)	61.9 (SD = 14.24)	61.6 (SD = 12.64)	62.2 (SD = 15.48)	p = 0.840 [M]
Clinical characteristics				
Disease duration (in months)	38.2 (SD = 54.7)	35.8 (SD = 40.05)	40.1 (SD = 64.29)	p = 0.833 [M]
AJCC-Stage				p = 0.317 [C]
III	n = 45 (100%)	n = 23 (51.1%)	n = 22 (48.9%)	
IV	n = 68 (100%)	n = 27 (39.7%)	n = 41 (60.3%)	
Histological subtype				p = 0.415 [C]
Cutaneous melanomas	n = 75 (100%)	n = 35 (46.7%)	n = 40 (53.3%)	
ALM	n = 10 (100%)	n = 2 (20%)	n = 8 (80%)	
Mucous membrane	n = 8 (100%)	n = 5 (62.5%)	n = 3 (37.5%)	
Uveal melanoma	n = 9 (100%)	n = 3 (33.3%)	n = 6 (66.7%)	
Occult melanoma	n = 11 (100%)	n = 5 (45.5%)	n = 6 (54.5%)	
Localisation				p = 0.192 [C]
Head/neck	n = 21 (100%)	n = 8 (38.1%)	n = 13 (61.9%)	
Trunk	n = 31 (100%)	n = 18 (58.1%)	n = 13 (41.9%)	
Upper extremity	n = 10 (100%)	n = 5 (50%)	n = 5 (50%)	
Lower extremity	n = 32 (100%)	n = 9 (28.1%)	n = 23 (71.9%)	
Mucosa	n = 8 (100%)	n = 5 (62.5%)	n = 3 (37.5%)	
Occult melanoma	n = 11 (100%)	n = 5 (45.5%)	n = 6 (54.5%)	
Therapy				p = 0.294 [C]
Combined immunotherapy	n = 48 (100%)	n = 18 (37.5%)	n = 30 (62.5%)	
PD-1-antibody monotherapy	n = 65 (100%)	n = 32 (49.2%)	n = 33 (50.8%)	

[C] = Chi-Quadrat-Test. [M] = Mann-Whitney U Test. [1] V = Cramers V (Effect power: 0.10 small, 0.30 medium, 0.50 large).

3.4. Indicators for the Need of Psychooncological Support According to HSI

A binary logistic regression was performed to check for the potential influence of patients' subjective assessment and other objective factors on the need for psychooncological support. Since HSI is a widely used tool for assessing psychooncological support, we defined the HSI result as a response variable (HSI > 0.3: the need for care, HSI ≤ 0.3: no need for care).

3.4.1. Parameters at the Beginning of Immunotherapy

The model included the following variables: positive subjective assessment, female gender, stage IV, and melanoma in the visible body parts. Other metric variables included age and duration of disease in months. The complete model, which included all parameters, was statistically significant, indicating that the model could distinguish between patients needing support and the patients not in need of psychooncological support. The model explained between 16.9% (Cox and Snell R-squared) and 23.2% (Nagelkerkes R-squared) of the variance and correctly classified 69.9% of cases. As shown in Table 6, 3 variables made a unique, statistically significant contribution to the model (positive subjective assessment, age at start of therapy, and female gender). According to HSI, the strongest predictor of psychooncological need was the variable "positive subjective assessment," with an odds ratio of 5.76. This means that patients who reported a need for psychooncological support were more than five times more likely to achieve values above the threshold at HSI. Other significant parameters for values above the threshold in HSI indicating a need for psychooncological support were female gender and higher age. Increasing age increased the percentage of patients needing psychooncological support (Table 6).

Table 4. Comparison of the groups at T1.

Characteristic	Overall *n* = 109 (100%)	DT < 5 *n* = 50 (45.9%)	DT ≥ 5 *n* = 59 (54.1%)	*p*-Value
Demographic characteristics				
Gender (%)				$p = 0.041$ [C] $V = 0.210$ [1]
Female	*n* = 50 (100%)	*n* = 16 (32.0%)	*n* = 34 (68.0%)	
Male	*n* = 59 (100%)	*n* = 34 (57.6%)	*n* = 25 (42.4%)	
Age (in Years)	62.45 (SD = 14.10)	61.3 (SD = 12.81)	63.4 (SD = 15.12)	$p = 0.314$ [M]
Clinical characteristics				
AJCC-Stage				$p = 0.004$ [C] $V = 0.293$ [1]
III	*n* = 44 (100%)	*n* = 28 (63.6%)	*n* = 16 (36.4%)	
IV	*n* = 65 (100%)	*n* = 22 (33.8%)	*n* = 43 (66.2%)	
Therapy				$p = 0.003$ [C] $V = 0.302$ [1]
Combined immunotherapy	*n* = 46 (100%)	*n* = 13 (28.3%)	*n* = 33 (71.7%)	
PD-1-antibody monotherapy	*n* = 63 (100%)	*n* = 37 (58.7%)	*n* = 26 (41.3%)	
Side effects				$p = 0.005$ [C] $V = 0.287$ [1]
No	*n* = 71 (100%)	*n* = 40 (56.3%)	*n* = 31 (43.7%)	
≥1	*n* = 38 (100%)	*n* = 10 (26.3%)	*n* = 28 (73.7%)	
Staging				$p = 0.270$ [C]
No	*n* = 83 (100%)	*n* = 35 (42.2%)	*n* = 48 (57.8%)	
Stable Disease	*n* = 19 (100%)	*n* = 12 (63.2%)	*n* = 7 (36.8%)	
Progress	*n* = 7 (100%)	*n* = 3 (42.9%)	*n* = 4 (57.1%)	

[C] = Chi-Quadrat-Test. [M] = Mann-Whitney U Test. [1] V = Cramers V (Effect power: 0.10 small, 0.30 medium, 0.50 large).

Table 5. Comparison of the groups at T2.

Characteristic	Overall $n = 107$ (100%)	DT < 5 $n = 64$ (59.8%)	DT $\geq$ 5 $n = 43$ (40.2%)	*p*-Value
Demographic characteristics				
Gender (%)				$p = 0.057$ C
Female	$n = 49$ (%)	$n = 24$ (49.0%)	$n = 25$ (51.0%)	
Male	$n = 58$ (%)	$n = 40$ (69.0%)	$n = 18$ (31.0%)	
Age (in Years)	62.2 (SD = 14.08)	61.3 (SD = 14.15)	63.5 (SD = 14.02)	$p = 0.429$ M
Clinical characteristics				
AJCC-Stage				$p = 0.382$ C
III	$n = 44$ (100%)	$n = 29$ (65.9%)	$n = 15$ (34.1%)	
IV	$n = 63$ (100%)	$n = 35$ (55.6%)	$n = 28$ (44.4%)	
Therapy				$p = 0.078$ C
Combined immunotherapy	$n = 45$ (100%)	$n = 22$ (48.9%)	$n = 23$ (51.1%)	
PD-1-antibody monotherapy	$n = 62$ (100%)	$n = 42$ (67.7%)	$n = 20$ (32.3%)	
Side effects				$p = 0.000$ C $V = 0.428$ 1
No	$n = 83$ (100%)	$n = 59$ (71.1%)	$n = 24$ (28.9%)	
$\geq$1	$n = 24$ (100%)	$n = 5$ (20.8%)	$n = 19$ (79.2%)	
Staging				$p = 0.001$ C $V = 0.349$ 1
No	$n = 88$ (100%)	$n = 54$ (61.4%)	$n = 34$ (38.6%)	
Stable Disease	$n = 9$ (100%)	$n = 8$ (88.9%)	$n = 1$ (11.1%)	
Progress	$n = 10$ (100%)	$n = 2$ (20.0%)	$n = 8$ (80.0%)	

C = Chi-Quadrat-Test. M = Mann-Whitney U Test. 1 V = Cramers V (Effect power: 0.10 small, 0.30 medium, 0.50 large.).

Table 6. Odds ratio for the need for psychooncological support according to the HSI.

Characteristic	Odds Ratio	Standard Error	*p*-Value	95% CI	Odds Ratio
Positiv subjective need	**5.76**	**0.76**	**0.020**	**1.31**	**25.36**
Female Gender	**2.93**	**0.47**	**0.022**	**1.17**	**7.33**
Higher Age before start therapy	**1.04**	**0.02**	**0.014**	**1.01**	**1.08**
AJCC-Stage IV	1.16	0.47	0.747	0.46	2.92
Longer duration of illness in month	1.00	0.00	0.678	0.99	1.01
Melanoma on visible parts of the body	0.45	0.60	0.179	0.14	1.45

The significant results are displayed first and marked in bold; sorted by odds ratio, starting with the highest value.

3.4.2. Parameters during Immunotherapy

The following variables were included in the model: positive subjective need, female gender, stage IV, side effects, and the type of therapy. The metric variable included age. The overall model was statistically significant, indicating that the model could distinguish between patients requiring care and those not requiring care. The model explained between 20.0% (Cox and Snell R-squared) and 27.3% (Nagelkerkes R-squared) of the variance and correctly classified 76.1% of cases. As shown in Table 7, 3 variables also made a statistically significant contribution to the model (positive subjective assessment, age at start of therapy, and the occurrence of side effects). The strongest predictor of psychooncological need after HSI was also the variable "positive subjective assessment" with an odds ratio of 8.21 (Table 7).

Table 7. Odds ratio for the need for psychooncological support according to the HSI during immunotherapy.

Characteristic	Odds Ratio	Standard Error	*p*-Value	95% CI	Odds Ratio
Positiv subjective need	**8.21**	**0.91**	**0.021**	**1.37**	**49.27**
Occurrence of side effects	**3.00**	**0.47**	**0.019**	**1.20**	**7.48**
Higher Age before start therapy	**1.04**	**0.02**	**0.037**	**1.00**	**1.08**
Combined immunotherapy	1.49	0.69	0.566	0.39	05.72
AJCC-Stage IV	1.01	0.70	0.985	0.26	4.02
Female gender	0.93	0.48	0.875	0.36	2.39
Staging	0.56	0.61	0.339	0.17	1.84

The significant results are displayed first and marked in bold, sorted by odds ratio, starting with the highest value.

4. Discussion

In this prospective study on distress in metastatic melanoma patients during ICI, we identified more than 50% of distressed patients at baseline before initiating ICI. This percentage is slightly higher than that of other studies on the psycho-oncological burden of melanoma patients [2,5,18]. It can certainly be explained by the fact that our cohort included only advanced melanoma patients who were to receive systemic ICI. During ICI, the proportion of patients with values above the threshold decreased, whereas the physical problems increased to a clear peak at time T1.

It could be assumed that uncertainties about the detailed course and tolerability at the beginning of ICI (T0) led to a higher percentage of distressed patients.

When comparing the patient groups with and without DT values above the threshold, we found that female patients more often suffered distress than male patients at times T0 and T1. Consistent with our results, others have shown that women are distressed to a higher percentage than men during their disease [19–21]. One explanation for the remarkably higher percentage of female patients with DT values above the threshold could be that women might be more concerned about their families and children. Additionally, it has been shown that women are more restricted in their daily activities during their illness than men, which could explain an increased percentage of distressed patients [22].

Furthermore, patients with a higher AJCC stage (stage IV compared to stage III), combined immunotherapy (compared to PD-1 antibodies as monotherapy) and patients who developed side effects compared to those without, showed more distress at T1. We found a significant correlation between the type of immunotherapy and the occurrence of side effects. Thus, significantly more patients suffered from side effects when receiving combined immunotherapy. When comparing combined immunotherapy with monotherapy, it is noticeable that immune-related side effects occurred in 63% of patients with combined immunotherapy and 32% of patients with PD-1 antibody monotherapy. Similarly, several studies revealed that severe or life-threatening adverse events (CTCAE grade 3–4) occur in 17–21% with PD-1 antibody monotherapy and 59% with combined immunotherapy [23,24].

The immune-related side effects suffered by our patients occurred on average after 38 days. This timing of onset fits well with what has been reported in other studies, and it also correlates very well with the time-point of the second survey (T1). This certainly explains the peak of physical problems at time T1. In particular, colitis, and pneumonitis may entail prolonged adverse event management under inpatient conditions [25].

Interestingly, the proportion of patients indicating emotional problems decreases at time T1. In particular, the items "anxiety" and "depression" dropped at time T1 and rose again slightly at time T2, perhaps due to the approaching moment of staging.

One explanation could be that the patients focused on their physical, immune-related problems at T1 so that all other items of the problem list lost their importance beside it. Another explanation could be that patients associated and, to some extent, equated the occurrence of immune-mediated side effects with a good response to therapy.

In both combined immunotherapy and PD1-AK monotherapy, "fatigue" has been documented as the most common side effect. The same in other countries, regardless of the type of cancer, fatigue was the most often mentioned physical problem in cancer patients [26,27]. Nonetheless, fatigue seems to be still underdiagnosed, and support for these patients has not been adequately ensured [28]. Fatigue, in particular, is often stated by cancer patients as the main cause of impaired quality of life and increased distress [29].

The proportion of patients with progressive disease in staging showed an increase in DT at T2. In contrast, patients who showed a decrease in the tumor or metastases or who did not experience worsening the situation were significantly less distressed. This highlights the importance of empathic communication of the staging results in a face-to-face conversation and, if needed, prompt psycho-oncological support, which will primarily be the case with progressive findings.

Some factors can help to identify affected patients and thus to provide psychooncological support. The strongest predictor for patients above threshold by HSI, immediately before starting the systemic therapy (T0) and during its course, was the variable "subjective need." For melanoma patients, who consider themselves to need psychooncological support, it can be assumed with high probability that they also have a psychometrically determined need, measured by HSI. In this context, a subjective need is of enormous importance and should be considered an independent, most probably even the simplest and most intuitive screening tool. This is also recommended in the guideline in the S3 guideline ("Psychooncological Diagnosis, Counseling and Treatment of Adult Cancer Patients"). Patients should be asked for their personal estimation on the need for psychooncological support [12]. However, this question alone should not replace the screenings as they contain several questions helping the patient be self-reflective. Likely, some patients would not have noticed that they had not been feeling well in the last few days, or they might have realized that they do NOT have anyone to talk to about their difficult situation. The screenings can help to induce a deeper self-reflection.

Therefore, at the beginning and during therapy, considering the patients' subjective perception, psychooncological support should be offered. In this way, psychological comorbidities, such as anxiety or depression might be assessed early on. In this way moreover, compliance could probably be ensured. DiMatteo and colleagues found that patients with comorbid depression were three times more likely not to follow treatment recommendations or even discontinue therapy [30]. Through the availability of psychooncological support, patient care could be ensured, which could be lifesaving for cancer patients.

However, in contrast to the high proportion of patients with over threshold values in HSI and DT, only 11.5% indicated a subjective need for psychooncological support at the beginning of immunotherapy. This subjective need decreased even further during the treatment (T1: 8.3%; T2: 3.7%).

Possible reasons for patients not considering themselves to need support despite values above the threshold could be an insufficient knowledge about what the psychooncological support looks like in detail, a kind of fear of stigmatization, inability to cope with the disease and the resulting misjudgment, or other support they already receive [31,32].

The individual ability of the patients to cope with stress and use resources and possible intrinsic predispositions to feel anxious or worried may have biased the results of our study. Intrapsychic personality characteristics were not assessed. A follow-up study could focus on these possibly influencing factors in combination with further personality characteristics.

During the survey period, more women than men made use of psychooncological support. This is in line with the results of the study by Zwaan et al., in which significantly more women than men had contact with the psychooncological service [33]. This gender difference is probably due to the traditional distribution of roles. For men, it might still be an obstacle to show weakness and ask for help. Based on these findings, patients should not only be relied on to seek psychooncological support themselves. They should be actively informed about possible supports and offers of help. Special care should be taken to reduce possible prejudices against "psychooncology" and to encourage men to seek external help.

5. Conclusions

Melanoma patients who are to be treated by ICI should be screened by DT or HSI for psychological distress, complemented by patients' self-assessment. In our study, the percentage of values above the threshold decreased during ICI. At the beginning of ICI therapy, practical problems should be given special consideration, e.g., transport to therapy. It should be considered that there are special situations that might require additional support. Patients with immune-related adverse events had more often physical problems, and patients suffering progressive disease were significantly more often above-threshold, thus in need of psychooncological support.

A problem list, as in DT, can rapidly help the treating physician get an overview of the patient's current problems. For the future, studies would be desirable to investigate the extent to which psycho-oncological support can reduce the proportion of over-threshold patients.

Supplementary Materials: The following are available online at https://www.mdpi.com/article/10.3390/cancers13112642/s1, Supplementary File S1: Hornheide Screening Instrument, Distress Thermometer, Supplementary File S2: first follow-up questionnaire, Supplementary File S3: second follow-up questionnaire.

Author Contributions: Conceptualization, L.W., A.F., and C.G.; methodology, A.F., N.S.; software, L.W., N.S., A.F.; formal analysis, L.W., T.E., N.S., A.F; investigation, L.W.; resources, L.W; data curation, L.W.; writing—original draft preparation, L.W., A.F.; writing—review and editing, all authors; visualization, L.W., C.G., A.F. supervision, C.G., A.F.; project administration, A.F.; All authors have read and agreed to the published version of the manuscript.

Funding: This research received no external funding.

Institutional Review Board Statement: The study was conducted according to the guidelines of the Declaration of Helsinki and approved by the Ethics Committee of the Medical Faculty of the University of Tuebingen (file number 689/2017BO2).

Informed Consent Statement: Informed consent was obtained from all subjects involved in the study.

Data Availability Statement: The datasets used and/or analyzed during the current study are available from the corresponding author on reasonable request. The data are not publicly available due to patient confidentiality.

Acknowledgments: The authors thank the whole team of the melanoma outpatient department for their care for our melanoma patients.

Conflicts of Interest: A.F. served as a consultant to Roche, Novartis, MSD, BMS, Pierre-Fabre; received travel support from Roche, Novartis, BMS, Pierre-Fabre, received speaker fees from Roche, Novartis, BMS, MSD, and CeGaT, outside the submitted work. She reports institutional research grants from BMS Stiftung Immunonkologie outside the submitted work. T.E. reports personal fees from Amgen, grants and personal fees from Novartis, personal fees from Philogen, grants, and personal fees from Roche, grants and personal fees from Sanofi, personal fees from BMS, personal fees from MSD, outside the submitted work. C.G. reports personal fees from Amgen, grants and personal fees from NeraCare, grants and personal fees from Novartis, personal fees from Philogen, grants and personal fees from Roche, grants and personal fees from Sanofi, personal fees from BMS, personal fees from MSD, outside the submitted work. The other authors state no conflict of interest.

References

1. Leiter, U.; Keim, U.; Garbe, C. Epidemiology of Skin Cancer: Update 2019. In *Sunlight, Vitamin D and Skin Cancer*, 3rd ed.; Advances in Experimental Medicine and Biology; Reichrath, J., Ed.; Springer: Berlin/Heidelberg, Germany, 2020; Volume 1268, pp. 123–139. [CrossRef]
2. Loquai, C.; Scheurich, V.; Syring, N.; Schmidtmann, I.; Rietz, S.; Werner, A.; Grabbe, S.; Beutel, M.E. Screening for Distress in Routine Oncological Care—A Survey in 520 Melanoma Patients. *PLoS ONE* **2013**, *8*, e66800. [CrossRef]
3. Taube, K.-M. Psychoonkologie in der Dermatologie. *Hautarzt* **2013**, *64*, 424–428. [CrossRef]
4. Dunn, J.; Watson, M.; Aitken, J.F.; Hyde, M.K. Systematic review of psychosocial outcomes for patients with advanced melanoma. *Psycho-Oncology* **2017**, *26*, 1722–1731. [CrossRef]

5. Forschner, A.; Riedel, P.; Gassenmaier, M.; Scheu, A.; Kofler, L.; Garbe, C.; Schäffeler, N.; Mayer, S.; Eigentler, T.K. The demand for psycho-oncological support in 820 melanoma patients: What are the determinants for the development of distress? *JCO* **2017**, *35* (Suppl. 15), 9514. [CrossRef]

6. Michielin, O.; van Akkooi, A.C.J.; Ascierto, P.A.; Dummer, R.; Keilholz, U. Cutaneous melanoma: ESMO Clinical Practice Guidelines for diagnosis, treatment and follow-up. *Ann. Oncol.* **2019**, *30*, 1884–1901. [CrossRef] [PubMed]

7. Steininger, J.; Gellrich, F.F.; Schulz, A.; Westphal, D.; Beissert, S.; Meier, F. Systemic Therapy of Metastatic Melanoma: On the Road to Cure. *Cancers* **2021**, *13*, 1430. [CrossRef] [PubMed]

8. Robert, C.; Ribas, A.; Schachter, J.; Arance, A.; Grob, J.-J.; Mortier, L.; Daud, A.; Carlino, M.S.; McNeil, C.M.; Lotem, M.; et al. Pembrolizumab versus ipilimumab in advanced melanoma (KEYNOTE-006): Post-hoc 5-year results from an open-label, multicentre, randomised, controlled, phase 3 study. *Lancet Oncol.* **2019**, *20*, 1239–1251. [CrossRef]

9. Robert, C.; Long, G.V.; Brady, B.; Dutriaux, C.; Maio, M.; Mortier, L.; Hassel, J.C.; Rutkowski, P.; McNeil, C.; Kalinka-Warzocha, E.; et al. Nivolumab in previously untreated melanoma without BRAF mutation. *N. Engl. J. Med.* **2015**, *372*, 320–330. [CrossRef] [PubMed]

10. Larkin, J.; Chiarion-Sileni, V.; Gonzalez, R.; Grob, J.-J.; Rutkowski, P.; Lao, C.D.; Cowey, C.L.; Schadendorf, D.; Wagstaff, J.; Dummer, R.; et al. Five-Year Survival with Combined Nivolumab and Ipilimumab in Advanced Melanoma. *N. Engl. J. Med.* **2019**, *381*, 1535–1546. [CrossRef] [PubMed]

11. Terheyden, P.; Krackhardt, A.; Eigentler, T. The Systemic Treatment of Melanoma. *Dtsch. Arztebl. Int.* **2019**, *116*, 497–504. [CrossRef] [PubMed]

12. Pflugfelder, A.; Kochs, C.; Blum, A.; Capellaro, M.; Czeschik, C.; Dettenborn, T.; Dill, D.; Dippel, E.; Eigentler, T.; Feyer, P.; et al. Malignant melanoma S3-guideline "diagnosis, therapy and follow-up of melanoma". *J.Ger. Soc. Dermatol.* **2013**, *11* (Suppl. 6), 116–126. [CrossRef]

13. Mehnert, A.; Müller, D.; Lehmann, C.; Koch, U. Die deutsche Version des NCCN Distress-Thermometers. *Zeitschrift Psychiatrie Psychol. Psychother.* **2006**, *54*, 213–223. [CrossRef]

14. Rumpold, G.; Augustin, M.; Zschocke, I.; Strittmatter, G.; Söllner, W. Die Validität des Hornheider Fragebogens zur psychosozialen Unterstützung bei Tumorpatienten—Eine Untersuchung an zwei repräsentativen ambulanten Stichproben von Melanompatienten. *Psychother. Psychosom. Med. Psychol.* **2001**, *51*, 25–33. [CrossRef]

15. Strittmatter, G.; Tilkorn, M.; Mawick, R. How to Identify Patients in Need of Psychological Intervention. In *Cancers of the Skin: Proceedings of the 8th World Congress*; Recent Results in Cancer Research; Dummer, R., Nestle, F.O., Burg, G., Eds.; Springer: Berlin/Heidelberg, Germany, 2002; Volume 160, pp. 353–361. [CrossRef]

16. Herschbach, P.; Weis, J. Screeningverfahren in der Psychoonkologie:Testinstrumente zur Identifikation betreuungsbedürftiger Krebspatienten. Eine Empfehlung der PSO für die Psychoonkologische Behandlungspraxis. Available online: https://www.dapo-ev.de/wp-content/uploads/2017/04/pso_broschuere2.pdf (accessed on 3 February 2021).

17. Schaeffeler, N.; Pfeiffer, K.; Ringwald, J.; Brucker, S.; Wallwiener, M.; Zipfel, S.; Teufel, M. Assessing the need for psychooncological support: Screening instruments in combination with patients' subjective evaluation may define psychooncological pathways. *Psycho-Oncology* **2015**, *24*, 1784–1791. [CrossRef] [PubMed]

18. Mehnert, A.; Brähler, E.; Faller, H.; Härter, M.; Keller, M.; Schulz, H.; Wegscheider, K.; Weis, J.; Boehncke, A.; Hund, B.; et al. Four-week prevalence of mental disorders in patients with cancer across major tumor entities. *J. Clin. Oncol.* **2014**, *32*, 3540–3546. [CrossRef] [PubMed]

19. Blum, A.; Blum, D.; Stroebel, W.; Rassner, G.; Garbe, C.; Hautzinger, M. Psychosoziale Belastung und subjektives Erleben von Melanompatienten in der ambulanten Nachsorge. *Psychotherapie Psychosomatik Medizinische Psychologie* **2003**, *53*, 258–266. [CrossRef]

20. Tang, M.H.; Castle, D.J.; Choong, P.F.M. Identifying the prevalence, trajectory, and determinants of psychological distress in extremity sarcoma. *Sarcoma* **2015**, *2015*, 745163. [CrossRef] [PubMed]

21. Carlson, L.E.; Angen, M.; Cullum, J.; Goodey, E.; Koopmans, J.; Lamont, L.; MacRae, J.H.; Martin, M.; Pelletier, G.; Robinson, J.; et al. High levels of untreated distress and fatigue in cancer patients. *Br. J. Cancer* **2004**, *90*, 2297–2304. [CrossRef] [PubMed]

22. Johnson, R.L.; Gold, M.A.; Wyche, K.F. Distress in women with gynecologic cancer. *Psycho-Oncology* **2010**, *19*, 665–668. [CrossRef]

23. Wolchok, J.D.; Chiarion-Sileni, V.; Gonzalez, R.; Rutkowski, P.; Grob, J.-J.; Cowey, C.L.; Lao, C.D.; Wagstaff, J.; Schadendorf, D.; Ferrucci, P.F.; et al. Overall Survival with Combined Nivolumab and Ipilimumab in Advanced Melanoma. *N. Engl. J. Med.* **2017**, *377*, 1345–1356. [CrossRef]

24. Schachter, J.; Ribas, A.; Long, G.V.; Arance, A.; Grob, J.-J.; Mortier, L.; Daud, A.; Carlino, M.S.; McNeil, C.; Lotem, M.; et al. Pembrolizumab versus ipilimumab for advanced melanoma: Final overall survival results of a multicentre, randomised, open-label phase 3 study (KEYNOTE-006). *Lancet* **2017**, *390*, 1853–1862. [CrossRef]

25. Postow, M.A.; Sidlow, R.; Hellmann, M.D. Immune-Related Adverse Events Associated with Immune Checkpoint Blockade. *N. Engl. J. Med.* **2018**, *378*, 158–168. [CrossRef] [PubMed]

26. Koornstra, R.H.T.; Peters, M.; Donofrio, S.; van den Borne, B.; de Jong, F.A. Management of fatigue in patients with cancer—A practical overview. *Cancer Treat. Rev.* **2014**, *40*, 791–799. [CrossRef] [PubMed]

27. Oh, H.S.; Seo, W.S. Systematic Review and Meta-Analysis of the Correlates of Cancer-Related Fatigue. *Worldviews Evid. Based Nurs.* **2011**, *8*, 191–201. [CrossRef]

28. Mitchell, S.A.; Hoffman, A.J.; Clark, J.C.; DeGennaro, R.M.; Poirier, P.; Robinson, C.B.; Weisbrod, B.L. Putting Evidence Into Practice: An Update of Evidence-Based Interventions for Cancer-Related Fatigue During and Following Treatment. *Clin. J. Oncol. Nurs.* **2014**, *18*, 38–58. [CrossRef]

29. Tibubos, A.N.; Ernst, M.; Brähler, E.; Fischbeck, S.; Hinz, A.; Blettner, M.; Zeissig, S.R.; Weyer, V.; Imruck, B.H.; Binder, H.; et al. Fatigue in survivors of malignant melanoma and its determinants: A register-based cohort study. *Support. Care Cancer* **2019**, *27*, 2809–2818. [CrossRef] [PubMed]

30. DiMatteo, M.R.; Lepper, H.S.; Croghan, T.W. Depression is a risk factor for noncompliance with medical treatment: Meta-analysis of the effects of anxiety and depression on patient adherence. *Arch. Intern. Med.* **2000**, *160*, 2101–2107. [CrossRef]

31. Bachthaler, S. Psychoonkologie—Das Erstgespräch und die weitere Begleitung. *Psychiatr. Prax.* **2015**, *42*, 53–54. [CrossRef]

32. Söllner, W.; Mairinger, G.; Zingg-Schir, M.; Fritsch, P. Krankheitsprognose, psychosoziale Belastung und Einstellung von Melanompatienten zu unterstützenden psychotherapeutischen Massnahmen. *Hautarzt* **1996**, *47*, 200–205. [CrossRef] [PubMed]

33. De Zwaan, M.; Mösch, P.; Sinzinger, H.; Stresing, K.; Oberhof, P.; Kohl, C.; Schilke, C.; Müller, A. Der Zusammenhang zwischen psychoonkologischem Betreuungsbedarf, Wunsch nach Unterstützung und tatsächlicher Behandlung bei Krebspatientinnen und-patienten. *Neuropsychiatrie* **2012**, *26*, 152–158. [CrossRef]

Review

Update on Management Recommendations for Advanced Cutaneous Squamous Cell Carcinoma

Jesús García-Foncillas [1,2,*], Antonio Tejera-Vaquerizo [3,4], Onofre Sanmartín [5], Federico Rojo [1], Javier Mestre [6], Salvador Martín [7], Ignacio Azinovic [1] and Ricard Mesía [8]

1   Departamento de Oncología, Hospital Universitario Fundación Jiménez Díaz, 28040 Madrid, Spain; FRojo@fjd.es (F.R.); Ignacio.azinovic@quironsalud.es (I.A.)
2   Department of Medicine, Faculty of Medicine, Universidad Autónoma de Madrid, 28040 Madrid, Spain
3   Instituto Dermatológico GlobalDerm, Palma del Río, 14700 Cordoba, Spain; antoniotejera@aedv.es
4   Unidad de Oncología Cutánea, Hospital San Juan de Dios, 14012 Cordoba, Spain
5   Instituto Valenciano de Oncología, 46009 Valencia, Spain; osanmartinj@gmail.com
6   Hospital Universitario Miguel Servet, 50009 Zaragoza, Spain; javiermestre@yahoo.com
7   Clínica Universitaria de Navarra, 28027 Madrid, Spain; smalgarra@unav.es
8   B-ARGO Group, Medical Oncology Department, Institut Català d'Oncologia (ICO), Badalona, 08908 Barcelona, Spain; rmesia@iconcologia.net
*   Correspondence: jgfoncillas@quironsalud.es; Tel.: +34-900-815-019

Citation: García-Foncillas, J.; Tejera-Vaquerizo, A.; Sanmartín, O.; Rojo, F.; Mestre, J.; Martín, S.; Azinovic, I.; Mesía, R. Update on Management Recommendations for Advanced Cutaneous Squamous Cell Carcinoma. *Cancers* **2022**, *14*, 629. https://doi.org/10.3390/cancers14030629

Academic Editor: Aimilios Lallas

Received: 2 December 2021
Accepted: 21 January 2022
Published: 27 January 2022

Publisher's Note: MDPI stays neutral with regard to jurisdictional claims in published maps and institutional affiliations.

Simple Summary: Cutaneous squamous cell carcinoma (cSCC) is the second most common form of skin cancer, which predominantly occurs on the head and neck. Early detection and treatment of primary tumours is crucial to limit progression and local invasion of deep tissues. While high-risk markers of poor prognosis have been identified, factors predicting regional control or survival remain uncertain. Therefore, diagnosis and management of cSCC should be performed individually, considering patient's clinicopathological profile and the best available treatment options. Surgical excision, radiotherapy, and/or systemic treatments can be selected depending on patient's status and tumour stage. Considering that a more comprehensive assessment will be provided by a multidisciplinary team, we aimed to generate a practical document that may assist oncologists and dermatologists on the prognosis, diagnosis, management, and follow-up of patients with advanced cSCC.

Abstract: Cutaneous squamous cell carcinoma (cSCC) is the second most common form of skin cancer, the incidence of which has risen over the last years. Although cSCC rarely metastasizes, early detection and treatment of primary tumours are critical to limit progression and local invasion. Several prognostic factors related to patients' clinicopathologic profile and tumour features have been identified as high-risk markers and included in the stratification scales, but their association with regional control or survival is uncertain. Therefore, decision-making on the diagnosis and management of cSCC should be made based on each individual patient's characteristics. Recent advances in non-invasive imaging techniques and molecular testing have enhanced clinical diagnostic accuracy. Surgical excision is the mainstay of local treatment, whereas radiotherapy (RT) is recommended for patients with inoperable disease or in specific circumstances. Novel systemic treatments including immunotherapies and targeted therapies have changed the therapeutic landscape for cSCC. The anti-PD-1 agent cemiplimab is currently the only FDA/EMA-approved first-line therapy for patients with locally advanced or metastatic cSCC who are not candidates for curative surgery or RT. Given the likelihood of recurrence and the increased risk of developing multiple cSCC, close follow-up should be performed during the first years of treatment and continued long-term surveillance is warranted.

Keywords: cutaneous squamous cell carcinoma; prognosis; multidisciplinary management; surgery; systemic therapy

1. Introduction

Cutaneous squamous cell carcinoma (cSCC) is one of the most relevant non-melanoma skin cancers (NMSC), along with basal cell carcinoma (BCC). With more than 700,000 cases diagnosed each year in the U.S. [1], the likelihood of patients developing at least a second cSCC is 13% [2]. In Spain, cSCC is the second cause of skin cancer-related mortality, with an estimated incidence of 40 cases per 100,000 person-years [3]. It predominantly occurs on the head and neck (cSCCHN). Sun exposure, particularly chronic UV exposure, is the most relevant risk factor for cSCC [4]. Artificial ultraviolet radiation, including PUVA therapy with a load of more than 350 sessions, has also been associated with a higher incidence of cSCC [5]. Immunosuppression, especially in the context of solid organ transplants and patients with chronic lymphocytic leukaemia, is associated with a higher frequency among younger patients and more aggressive carcinomas [6].

Diagnosis of advanced cSCC should involve experts from different specialties, including dermatologists, anatomical pathologists, surgeons, radiation oncologists, and medical oncologists, thereby enabling a more comprehensive patient assessment. Likewise, decision-making for patient management and follow-up should be guided by this multidisciplinary approach, selecting the best available therapeutic options from an individualized perspective [7,8]. In recent years, increased knowledge of molecular biology and the implementation of novel techniques for surgery and radiotherapy (RT) have led to new treatment lines which have significantly improved the chance of local control and survival. Moreover, a broader therapeutic arsenal beyond surgery and RT is currently available that includes novel systemic treatments, such as new chemotherapeutic agents, targeted therapies, or immunotherapies [8,9].

This review was prompted by several considerations. First, while the recent update of tumour classification system has improved its prognostic value, the most relevant factors predicting patient outcomes remain unclear. Second, standardized protocols are required to ensure the consistency of clinical and pathological reports, thus allowing the appropriate assessment of clinical risk from a multidisciplinary approach. Finally, a practical document that includes the characteristics, usefulness, and role of diagnostic tests and available treatments may assist both oncologists and dermatologists. To further support the evidence herein addressed and its applicability in clinical practice, a questionnaire was conducted among a multidisciplinary panel of specialists with the aim of providing consensus recommendations on the prognosis, diagnosis, treatment, and follow-up of cSCC (References [10–12] are cited in the Supplementary Material).

2. Tumour Staging and Prognostic Factors

Tumour staging of cSCC arising from the head and neck region is based on the American Joint Committee on Cancer (AJCC) classification, 8th ed. [13], which additionally considers tumour thickness [14], perineural infiltration diameter, and tumour invasion depth to classify primary tumours [15] (Table 1). Alternatively, the Brigham and Women's Hospital (BWH) classification has also been proposed as a simpler and more intuitive classification system for localized stage disease. The BWH system may be a better predictor of regional nodal relapse or disease-related mortality than AJCC 8th ed. [16], although no differences have been observed for immunosuppressed patients [17]. In a recent systematic review, the BWH and AJCC systems similarly predicted the presence of metastasis after selective sentinel node biopsy [18].

Table 1. Summary of tumour classification systems AJCC 8th edition and BWH.

	AJCC-8 Classification		BWH Classification
	Primary tumour (T)		
T1	Tumour < 2 cm in greatest dimension	**T1**	0 high-risk factors [b]
T2	Tumour ≥ 2 cm and <4 cm in greatest dimension	**T2a**	1 high-risk factor
T3	Tumour ≥ 4 cm in greatest dimension or minor bone erosion or PNI or deep invasion [a]	**T2b**	2–3 high-risk factors
T4a	Tumour with gross cortical bone/marrow invasion	**T3**	4 high-risk factors or bone invasion
T4b	Tumour with axial skeleton invasion including foraminal involvement and vertebral foramen involvement to the epidural space		
	Regional lymph nodes (N)		
Nx	Regional lymph nodes cannot be assessed		
N1	Metastasis in a single ipsilateral lymph node ≤3 cm in greatest dimension and ENE (−)		
N2a	Metastasis in a single ipsilateral lymph node >3 cm and ≤6 cm in greatest dimension and ENE (−)		
N2b	Metastasis in multiple ipsilateral nodes all ≤6 cm in greatest dimension and ENE (−)		
N2c	Metastasis in bilateral or contralateral lymph node(s), all ≤6 cm in greatest dimension and ENE (−)		
N3a	Metastasis in a lymph node >6 cm in greatest dimension and ENE (−)		
N3b	Metastasis in any lymph node(s) and ENE (+)		
	Distant metastasis (M)		
M0	No distant metastasis		
M1	Distant metastasis		

AJCC, American Joint Committee on Cancer; BWH, Brigham and Women's Hospital; ENE, extranodal extension; PNI, perineural invasion. [a] Deep invasion defined as invasion beyond the subcutaneous fat or >6 mm (as measured from the granular layer of adjacent normal epidermis to the base of the tumour); perineural invasion for T3 classification is defined as tumour cells within the nerve sheath of a nerve lying deeper than the dermis or measuring 0.1 mm or larger in calibre or presenting with clinical or radiographic involvement of tumour named nerves without skull base invasion or transgression [19]. [b] Risk factors include tumour diameter 2 cm or larger, poorly differentiated histology, perineural invasion, and tumour invasion beyond the subcutaneous fat (excluding bone, which automatically upgrades to T3) [20].

Tumour staging according to the TNM classification, which is applicable for cSCC from all regions, is shown in Table 2. The N category referring to regional metastatic affectation is based on a scheme shared with other head and neck tumours. It should be noted that the AJCC classification system is only valid for these tumours. Other prognostic factors not included in this classification, such as degree of differentiation, growth rate [21], and the presence of budding, have also been described [22].

Table 2. Staging based on AJCC TNM classification 8th edition for head and neck cSCC.

T	N	M	Stage
T1	N0	M0	Stage I
T2	N0	M0	Stage II
T3	N0, N1	M0	Stage III
T1	N1	M0	Stage III
T2	N1	M0	Stage III
T1–3	N2	M0	Stage IV
T1–4	N3	M0	Stage IV
T4	N0–3	M0	Stage IV
T1–4	N0–3	M1	Stage IV

AJCC, American Joint Committee on Cancer; M, distant metastasis; N, regional lymph nodes; T, primary tumour.

3. Molecular Pathology and Emerging Biomarkers

cSCC presents a higher number of mutations than other tumours, i.e., up to 5 times that of lung cancer or 4 times that of melanoma [23]. By accumulating these mutations, an area of skin can—usually in response to UV light damage—progress through increased levels of dysplasia to become a cSCC, and in fact, areas of photo-exposed skin share many of the same carcinoma features [24]. Although there is a large range of tumour mutational burden (TMB) in cSCC, the median TMB is the highest compared with other tumour types. Exome sequencing in the cSCC has revealed approximately 1300 nucleotide somatic variations per exome. This observation has led to the hypothesis that constant damage to basal keratinocytes by UV radiation may be responsible for multiple cSCC mutational events [25] (Figure 1).

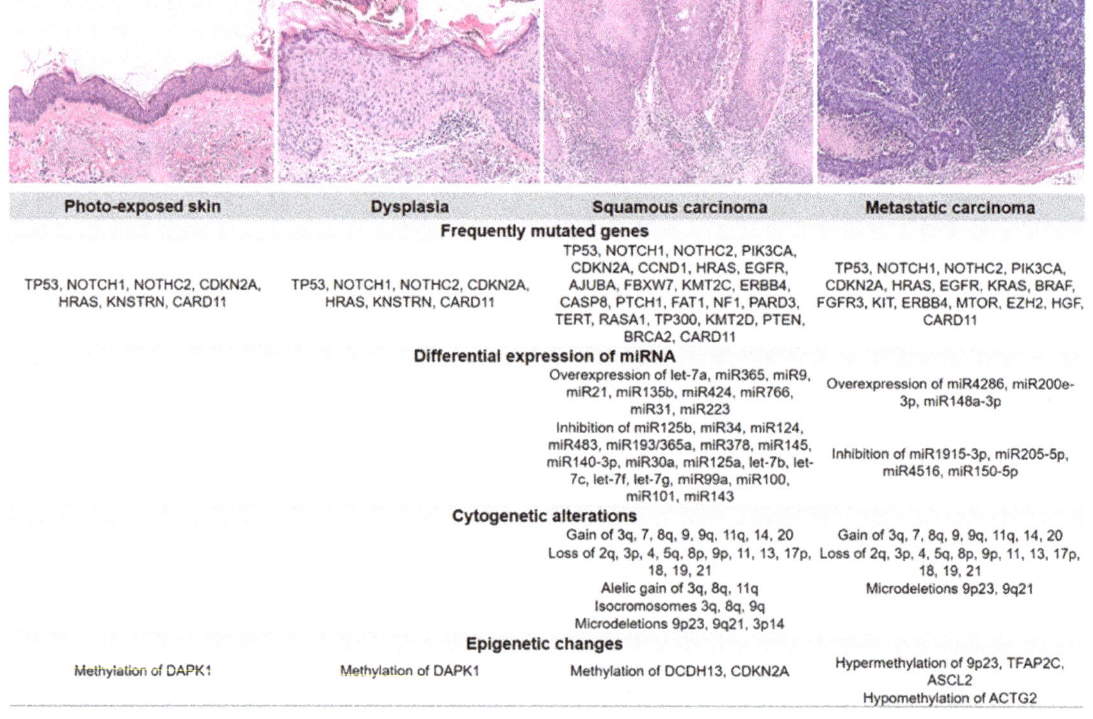

Figure 1. Histology and molecular pathology of photo-exposed skin, dysplasia, squamous carcinoma, and metastatic carcinoma.

The most frequent mutations in cSCC are associated with ultraviolet radiation and affect TP53 and NOTCH signalling pathway genes. While mutations in the TP53 family decrease the expression of other tumour suppressors, such as IFR6 [26], modifications in NOTCH signalling may have an oncogenic or suppressor role depending on the cellular microenvironment [27]. Activating mutations on genes of the RAS family, mainly HRAS, have been identified in up to 9% of SCC [25], with increasing incidence in up to 60% of tumours developed in patients treated with BRAF/MEK inhibitors [28]. CDKN2A, which is mutated in 31% of primary and metastatic SCC, is associated with tumour aggressiveness [29] and epigenetic changes due to hypermethylation of the CpG islands of the FRZB, TFAP2C, ASCL2 [30], and KMT2C [23] genes. Less frequently, alterations in STAT3, PIK3CA, KIT, RIPK4, and RAS1 genes have been also described [23,31]. Evidence also suggests that phenotypic changes caused by tumour–stromal interaction and the secretion of VEGF-C by tumour-associated fibroblasts might be involved in the cSCC metastatic process [32,33].

4. Diagnosis

4.1. Dermatology

In recent years, optical non-invasive diagnostic techniques have been increasingly applied in clinical practice to improve diagnostic accuracy and to characterise the tumour in vivo before surgery or biopsy [9,34]. Dermoscopy can be used to identify the main characteristics of cSCC, such as clustered vascular pattern, glomerular vessels, and hyperkeratosis. Several additional features can be observed using reflectance confocal microscopy, such as atypical honeycomb or disarranged pattern of the spinous-granular layer of the epidermis, round nucleated bright cells in the epidermis, and round vessels in the dermis [35].

While dermoscopy is performed more often in pigmented than in non-pigmented lesions, it is also useful in the differential diagnosis of equivocal cases. For instance, the identification of glomerular-like, clustered or hairpin vessels, and the scale and alignment of dots and vessels can be useful in particular scenarios, such as in minimally invasive cSCC or in pigmented forms [19]. In situ, cSCC frequently presents clinically as an erythematous scaly patch or slightly elevated plaque, whereas invasive cSCC is usually ulcerated and can be patchy, papulonodular, papillomatous, or exophytic [9]. Progression of in situ cSCC to microinvasive cSCC can be detected on dermoscopy examination with the appearance of a thicker lesion, hairpin and/or linear-irregular vessels, and a keratotic centre and/or ulceration (Figure 2). In invasive cSCC, a vertical growth phase reflecting dermal invasion is typically characterized by an increased number of polymorphic vessels, such as linear irregular, hairpin, and grouped glomerular/dotted vessels over a whitish background with a central mass of keratin or ulceration [9,35,36].

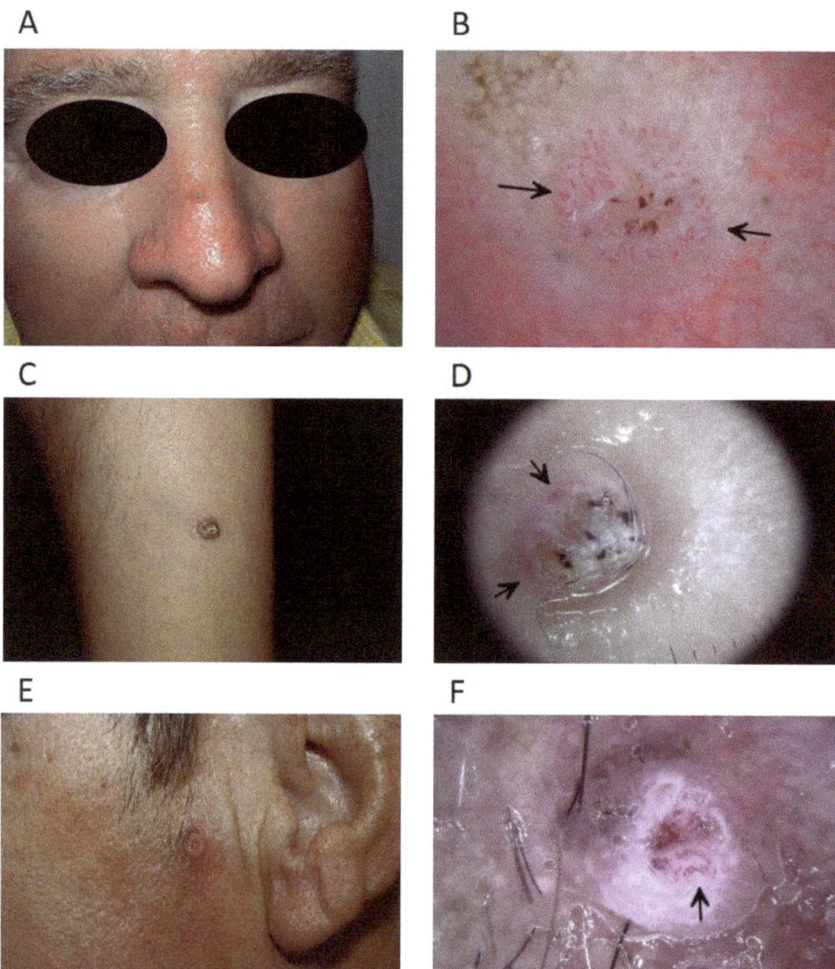

Figure 2. Dermoscopy of cutaneous squamous cell carcinoma. (**A**) Wart-like tumour lesion on the dorsum of the nose; (**B**) dermoscopy with polarised light showing a predominantly vascular pattern with serpentine, hairpin and irregular vessels (arrows), central ulceration and blood staining; (**C**) crateriform keratinising tumour lesion; (**D**) polarised light dermoscopy with central whitish crust and presence of irregular and comma-shaped vessels (arrows) in the periphery; (**E**) crateriform tumour lesion; (**F**) polarised light dermoscopy showing white unstructured areas with irregular groups of white perifollicular circles, central vascular pattern with hairpin and irregular vessels (arrows).

4.2. Histopathology

The histological diagnosis of cSCC is made using haematoxylin–eosin staining, although in cases of uncertain diagnosis, especially in non-keratinizing tumours, additional immunohistochemical studies for cytokeratins or stratified epithelia may be necessary. The morphological image shows bundles of atypical keratinocytes originating in the epidermis and infiltrating the dermis. Morphological features of differentiation can include the formation of corneal pearls, parakeratosis, and dyskeratosis. cSCC ranges from well-differentiated tumours to poorly differentiated neoplasms that show pleomorphic nuclei with a high degree of atypia, frequent mitosis, and very little keratin formation, if any [37].

Well-differentiated histological subtypes with low metastatic potential have been described, including keratoacanthoma and verrucous and fusocellular carcinoma, as well as other aggressive subtypes, such as acantholytic squamous, desmoplastic, and adenosquamous carcinomas, that have a risk of metastasis of 16.1% (95% CI 6.6–39.5) [14]. Other relevant morphological prognostic factors include the degree of dermal infiltration, tumour diameter (greater than 2.0 cm), the presence of perineural (especially in nerves greater than 0.1 mm) and lymphovascular invasion, and resection margins [38].

4.3. Medical Imaging

Radiological investigations are indicated when extensive disease is suspected, to determine bone or soft tissue involvement, invasion of surrounding areas, and the development of metastasis [7]. Main characteristics of imaging modalities used in cSCC are summarized in Table 3.

Table 3. Main characteristics of imaging modalities used in cSCC.

Imaging Modality	Optimal Use in cSCC	Advantages	Disadvantages	Sensitivity/Specificity for H&N Nodal Disease [a]
CT	Bone or lymph node disease	Less expensive, more widely available, and faster image acquisition than MRI	Exposure to contrast dye and ionizing radiation	52%/93%
MRI	Perineural, CNS, deep soft tissue, BM, or lymph node disease	No exposure to ionizing radiation	Less widely available, longer acquisition time, more expensive than CT	65%/81%
US	Superficial lymph node disease and image-guided FNA	Least expensive, no exposure to contrast dye or ionizing radiation, rapid image acquisition, global accessibility	Operator and technique-dependent, limited visualization of deep structures	66%/78%
PET/CT	Distant metastases	Functional and anatomic information, distinguishes postoperative scar tissue from recurrence	Most expensive, lesions less than 10 mm are below resolution for FDG-PET	66%/87%

BM, bone marrow; CNS, central nervous system; cSCC, cutaneous squamous cell carcinoma; CT, computed tomography; FDG, fluorodeoxyglucose; FNA, fine needle aspiration; H&N, head and neck; MRI, magnetic resonance imaging; PET, positron emission tomography; US, ultrasonography. [a] Adapted from Liao et al., 2012 [39].

Computed tomography (CT) scans evaluate cartilage and bone involvement, and three-dimensional imaging is very useful in planning the surgical approach and subsequent reconstruction. Magnetic resonance imaging (MRI) is preferred to assess the involvement of deep soft tissue and structures such as bone, parotid glands, and major nerves [19,40]. Ultrasound (US) is a very sensitive method for identifying lymph node metastases and serves to guide the radiologist during needle biopsy [7,19]. Positron emission tomography (PET) is useful for detecting metastases in sites where other studies are inconclusive, as in the case of fibrosis, necrosis, and previously radiated tissues. However, it should be noted that the presence of infectious and inflammatory processes may lead to false positives [40]. The use of combined PET-CT increases CT sensitivity [40], thus permitting the accurate detection of metastasis in distant organs [19].

4.4. Assessment of Comorbidities

In NMSC patients, three comorbidity assessment tools are frequently used in clinical trials and could be considered in clinical practice: the Charlson Comorbidity Index (CCI), the American Society of Anesthesiologists (ASA) risk classification system, and the Adult Comorbidity Evaluation-27 (ACE-27) [41]. In a systematic review of 22 studies, the CCI was used most often to assess comorbidities (82% of cases), probably because of its extensive use among skin cancer patients and validation in other cancer populations [41]. While higher CCI scores were significantly associated with age > 80 years in patients with head and neck tumours [42], correlation between CCI and life expectancy in nonagenarians undergoing Mohs micrographic surgery is still unclear [43,44].

The ASA risk classification has been traditionally used as a predictor of risk in the preoperative screening of patients undergoing general anaesthesia [45]. Although it was not formally developed as a comorbidity index, its widespread use enables data retrieval from medical records of surgical skin cancer patients [41]. The ACE-27, which is a modification of the Kaplan–Feinstein Index, performs better than a standard medical interview in the identification of comorbidities among skin cancer patients [46]. Since it includes more comorbid conditions and enables further grading than the CCI score, the ACE-27 could potentially predict the prognosis of patients more accurately. However, larger studies are warranted [41].

5. Risk Stratification

Current prognostic factors for recurrence, which help determine the role of definitive or adjuvant treatments, are based on clinicopathologic features described in single-centre or large-scale clinical studies or consensus meetings. These factors have been established taking into account local staging, location, depth, and pathological features. The NCCN guidelines establish low- and high-risk features for cSCC (Table 4) [7]. High-risk features are observed in approximately 5% of all NMSC and include locally advanced disease (stages T3-4), nodal involvement, perineural invasion (PNI), local and regional recurrence, or immunosuppression [47].

Table 4. Clinical and pathological features for risk stratification of cSCC.

	Low-Risk cSCC	High-Risk cSCC
Clinical history and parameters		
• Location/size	Area L < 20 mm Area M < 10 mm	Area L ≥ 20 mm Area M ≥ 10 mm Area H
• Borders	Well-defined	Poorly defined
• Primary vs. recurrent	Primary	Recurrent
• Immunosuppression	-	+
• Prior RT or chronic inflammatory process	-	+
• Rapidly growing tumour	-	+
• Neurologic symptoms	-	+
Pathology		
• Degree of differentiation	Well or moderately defined	Poorly defined
• Subtypes		
• Acantholytic (adenoid), adenosquamous, desmoplastic, or metaplastic (carcinosarcomatous) subtypes	-	+
• Depth: thickness or level of invasion	≤6 mm, no invasion beyond subcutaneous fat	>6 mm or invasion of subcutaneous fat
• Perineural, lymphatic, or vascular involvement	-	+

cSCC, cutaneous squamous cell carcinoma; RT, radiotherapy. Area H: "mask areas" of face (central face, eyelids, eyebrows, periorbital, nose, lips (cutaneous and vermilion), chin, mandible, preauricular and postauricular skin/sulci, temple, and ear), genitalia, hands, and feet. Area M: cheeks, forehead, scalp, neck, and pretibial. Area L: trunk and extremities (excluding hands, nail units, pretibial, ankles, and feet). Adapted from NCCN guidelines for SCC, 2018 [7].

PNI is a well-known high-risk factor for local or distant relapse that affects 5% of patients, mostly in cSCC [48]. In a systematic review, patients with clinical PNI (CPNI) showed a significantly increased risk of local recurrence (37% vs. 17%) and disease-specific death (27% vs. 6%) compared with patients with incidental PNI (IPNI), whereas nodal and distant metastasis were similar regardless of PNI classification. In addition, CPNI had significantly poorer mean 5-year recurrence-free survival (61% vs. 76%) and disease-specific survival (70% vs. 88%) than IPNI patients [49].

6. Treatment

6.1. Role of Surgery in Primary Tumours

Whenever possible, and taking into account the patient's status, surgical excision of the tumour is the first-line treatment for cSCC patients, regardless of age and anatomical location [8]. The choice among available modalities, which include conventional surgery and micrographic-controlled surgery, should be based on the patient's risk factors for poor prognosis (Figure 3).

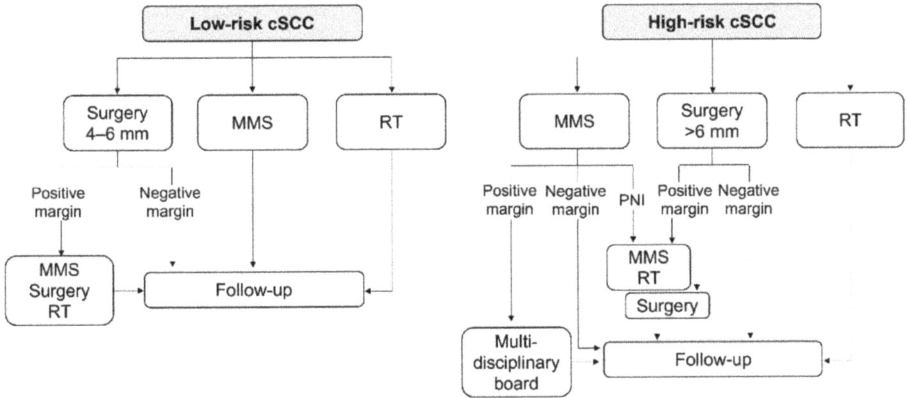

Figure 3. Therapeutic algorithm for low- and high-risk SCC. cSCC, cutaneous squamous cell carcinoma; MMS, Mohs micrographic surgery; PNI, perineural invasion; RT, radiotherapy.

Mohs micrographic surgery (MMS) may be the preferred technique for high-risk patients based on the high cure rates and low recurrence rates reported in retrospective analyses of primary and recurrent cSCC [7]. Alternatively, patients may undergo conventional surgery and intraoperative biopsies. Wide local excision may be more appropriate for large and invasive lesions, as it allows for complete assessment of peripheral and deep margins. To achieve histologically clean margins after surgical excision, the width of the margins should be adapted to the risk of cSCC extension and recurrence. For low-risk lesions, clinical practice guidelines recommend performing an excision with 4–6 mm margins [8]. However, larger margins are needed for larger tumours and when high-risk features, such as perineural or lymphatic invasion, are present. While margins > 6 mm would be required for lesions with a diameter > 1 cm [7], the most appropriate width should be determined by individual risk assessment [8]. From the oncological perspective, a surgical defect should be restored using techniques that do not mobilize surrounding tissues, such as direct closure or grafting [7], while reconstruction with skin flaps may be only advised after ensuring that the tumour has been completely removed.

Particular scenarios have been also described, such as cSCC that develop in association with scars or chronic wounds (Marjolin's ulcer) or previously radiated skin, those that invade deep structures, such as bone, parotid or nerve trunks, and those located in the labial vermilion and the ear [7]. In these cases, amputation of the affected limb, parotidectomy, and prophylactic lymph node drainage may be necessary.

6.2. Locoregional Assessment and Sentinel Node Biopsy

Sentinel lymph node biopsy (SLNB) in patients with cSCC is aimed at the early detection and management of occult nodal metastasis. According to two systematic reviews that included 16 and 23 studies, the positive SLNB rate among cSCC patients ranged between 14% and 8%, respectively [50,51]. A meta-analysis of 19 studies identified microscopic sentinel node involvement in 12.3% of patients who had a tumour diameter greater than 2 cm in all cases. Using the AJCC criteria, a higher risk of positive SLNB was associated with advanced tumour stages, reaching 29.4% in T2b and 50.0% in T3 lesions [52].

The utility of SLNB in detecting cSCC micrometastases not identified by non-invasive examination methods (i.e., instrumental tests) has been discussed. Notably, the rate of micrometastases increased from 3.4% in the overall population to 6.5% in the high-risk group; this rate directly correlated with the depth and diameter of the tumour [53]. In a retrospective analysis of cSCC patients who underwent SLNB and did not present micrometastasis, no relapse event nor local/distant metastases were reported during a mean follow-up of 27.5 months [54].

Nonetheless, given the lack of high-level evidence supporting a real prognostic impact and a well-defined profile of patients who could benefit the most, SLNB is not currently recommended in patients with invasive cSCC outside of a clinical trial setting [19].

6.3. Local Treatment

Surgical excision should be always considered as first-line treatment option for patients with either high- or low-risk cSCC. Nonetheless, based on the risk stratification, local approaches may be considered in patients for whom surgery is not feasible. For low-risk cSCC, several non-surgical treatment modalities are currently available to treat the tumour and field cancerization. Photodynamic therapy consists of a 2-step method that involves the topical application of a photosensitizer, such as 5-aminolevulinic acid or methyl aminolevulinate, followed by an incubation period with light irradiation. Alternatively, a topical therapy with imiquimod up to 5% or 5-fluorouracil may be also applied [9].

6.4. Role of Radiotherapy

RT may be used in first line in patients unable to undergo surgical resection, such as cosmetic or frail patients, to avoid significantly disfiguring surgery or orbital exenteration or in patients too frail to undergo general anaesthesia. A recent meta-analysis involving 21,000 patients showed comparable, extremely low one-year recurrence rates for both surgery and RT: 0.8%, 0.2%, 2%, and 0% for surgical excision, MMS, external beam irradiation, and brachytherapy (BT), respectively, and excellent cosmesis [55]. Different RT techniques, such as external beam RT, BT or electronic low energy sources (ELS), and schedules (i.e., normal fractionation or hypofractionation) can be selected depending on the expertise of the hospital. More recently, newer technologies such as volumetric arc therapy (VMAT) have proven useful in the treatment of scalp lesions or extensive field cancerization in trunk or extremities unamenable to other radiation modalities [56].

The role of RT in NMSC, mainly cSCC, has been defined in a recently published clinical practice guideline, following the ASTRO recommendation grading system [57]. Definitive RT is: (a) advised for patients who cannot undergo or who decline surgery, (b) conditionally recommended in anatomic locations where surgery can compromise function or cosmesis, and (c) not recommended (conditionally) in genetic diseases predisposing to higher radiosensitivity (ataxia-telangiectasia, Gorlin syndrome, Li–Fraumeni syndrome). Definitive RT is only recommended for inoperable patients, whereas elective node irradiation is conditionally recommended in lesions (thickness > 6 mm) in which there is an overlap of the primary tumour with the adjacent nodal basin.

Adjuvant RT is strongly recommended in NMSC patients with gross PNI as well as for cSCC patients with any of the following characteristics: (a) close or positive margins not amenable to re-resection, (b) relapse after prior margin-free surgery, (c) stage T3–T4, (d) desmoplastic reaction or lymphocyte infiltration in chronic immunosuppression. In patients with regional node metastasis, adjuvant RT after therapeutic lymphadenectomy is strongly recommended except for single, <3 cm nodes without extracapsular invasion. The addition of concurrent carboplatin chemotherapy to postoperative RT has not been shown to improve locoregional control in patients with high-risk cSCCHN [47].

For adjuvant RT, conventional doses of 60–66 Gy (2 Gy/day) are the standard, most frequently used fractionation. In case of elective RT, doses of 50–54 Gy are recommended [57]. Hypofractionation (HF) is usually delivered for brachytherapy or ELS, most often at doses of 50 Gy (20 fractions, 4 weeks) or 45 Gy (15 fractions, 3 weeks). Alternatively, doses of 30–40 Gy (5–7 Gy fractions, 1–3 fractions per week) result in excellent local control rates and acceptable toxicity. HF is most advantageous in smaller lesions, frail patients, elderly patients, or anyone with problems coming into the clinic on a daily visit [58,59]. Zaorsky et al. recommend schedules of 50 Gy/15 fractions, 36.75 Gy/7 fractions, or 35 Gy/5 fractions, as cosmesis outcomes are "good" in 80% of patients, depending on frailty [59]. Suggested treatments with RT using external beam irradiation are shown in Table 5.

Table 5. Suggested treatments with RT using external beam irradiation.

Type RT	Observations	Dose (Gy)	Sessions (n)	Times/wk	Fractionation
	Standard, GPS, size > 2 cm	64–66	32–33	5	conventional
	Neck, no surgery	70	35	5	conventional
Definitive	Size < 2 cm	50	20	5	hypofractionation
	Frail patients + inconvenience	50	15	5	hypofractionation
	Frail patients	36,75	7	5	hypofractionation
	Frail patients	35	5	3–5	hypofractionation
	Positive margins	66	33	5	conventional
Adjuvant	Negative margins	60	30	5	conventional
		50	20	5	hypofractionation
	Elective	50–54	25–27	5	conventional

GPS, good performance status; RT, radiotherapy; wk, week.

6.5. Systemic Treatment

6.5.1. Chemotherapy

Systemic treatment is administered to patients with cSCC who have progressed locally and/or with metastases on previous local treatments. Although no chemotherapeutic agents have been specifically approved for cSCCHN, platinum, 5-fluorouracil, anthracycline, or bleomycin alone or in combination are frequently used [60,61]. Moreover, the combined administration of cisplatin, interferon alfa, and 13-cis-retinoic acid has shown to be clinically active in patients with advanced cSCC, resulting in 34% overall and 17% complete response (CR) rates, with a median survival of 14.6 months [62].

6.5.2. Targeted Therapy

Epidermal growth factor receptor (EGFR) expression is present in 90% of cSCC tumours, with overexpression in up to 35% [63]. Four phase II trials tested EGFR tyrosine kinase inhibitors (TKIs), erlotinib [64] and gefitinib [65], and monoclonal antibodies cetuximab [66] or panitumumab [67] in the recurrent/metastatic setting (Table 6). Treatment with TKIs resulted in lower response rates without CR compared with antibodies but with similar disease control and duration of response. Likewise, survival outcomes did not differ, with a median progression-free survival (PFS) of 4–8 months or overall survival (OS) of 8–13 months. The largest cohort reporting clinical outcomes of cetuximab alone in unselected patients with unresectable or metastatic tumours confirmed a high disease control rate of 87% (53% response rate) [68].

Table 6. Trials in loco-regional and/or metastatic unresectable disease with targeted therapies.

Drug	Phase (N)	Patient Characteristics	RR	DC	mDoR	mPFS	mOS	Ref.
Erlotinib	II (39)	PS 0–2 Median age 68 y	10% (no CR)	72%	7.2 mo	4.7 mo	13 mo	[64]
Gefitinib	II (40)	PS 0–2 Median age 67 y	16% (no CR)	51%	31.4 mo	3.8 mo	12.9 mo	[65]
Cetuximab	II (36)	PS 0–2 Strong/moderate EGFR expression Median age 79 y	28% (6% CR)	68%	6.8 mo	4.1 mo	NR	[66]
Panitumumab	II (16)	PS 0–2 Median age 68 y	31% (12% CR)	69%	6 mo	8 mo	11 mo	[67]

CR, complete response; DC, disease control; DoR, duration of response; mo, months; m, median; N, patient number; NR, not reached; OS, overall survival; PFS, progression-free survival; PS, performance status; RR, response rate; y, years.

In contrast with panitumumab or TKIs, cetuximab demonstrated efficacy in cSC-CHN [69,70], so it has been widely used in this setting in monotherapy or combined with chemotherapy [71]. Notably, the administration of cetuximab with chemotherapy and/or anti-EGFR has been associated with significant toxicity, mainly cutaneous. Comorbidities should be evaluated, and geriatric assessment should be performed in elderly patients with cSCC to identify the most effective and tolerable therapy for each patient.

6.5.3. Immunotherapy

Immune checkpoint inhibitors (ICIs) have drastically improved survival outcomes in patients with advanced melanoma, with a considerable proportion of long-term survivors [72]. As tumours with a high mutational burden are known to be more responsive to ICIs and this mutational burden is up to 4 times greater in cSCC than melanoma [33], clinical trials were conducted in patients with cSCC [73]. Currently, anti-programmed cell death-1 (anti-PD-1) antibodies constitute first-line systemic treatment for metastatic or locally advanced cSCC in which curative surgery or RT are not feasible [8]. All systemic treatments are off-label, except for the anti-PD-1 agent cemiplimab, approved by the FDA/EMA in patients with locally advanced or metastatic cSCC who are not candidates for curative surgery or curative radiation [8,74,75].

Cemiplimab is a high-affinity, highly potent human immunoglobulin G4 anti-PD-1 receptor monoclonal antibody [76] that has demonstrated efficacy (overall response rate (ORR) 46.1%) and long-term durable response with very effective disease control (disease control rate: 72.5%) in patients with advanced and metastatic cSCC [77,78]. With a median age of 72 (38–96) years, 67.9% had cSCCHN, 59.6% had metastatic disease, and 40.4% locally advanced cSCC. Up to 66.3% of patients received cemiplimab as first-line treatment, and median observed time to response was 2.1 (1.9–3.7) months. Overall, the CR rate was 16.1% and median time to CR was 11.2 months [78]. Patients had deepening responses over time as evidenced by increasing CR rates [77–79]. Estimated median PFS was 18.4 months (95% CI: 10.3–24.3), while median DOR and OS had not been reached after 15.7 months follow-up. In responding patients, the estimated proportion of patients with ongoing response was 87.8% (95% CI: 78.5–93.3) at 12 months and 69.4% (95% CI: 55.6–79.6) at 24 months [78]. Advanced cSCC patients treated with cemiplimab experienced clinically meaningful improvements in global health status/health-related quality of life, and pain and functional status, and they maintained a low symptom burden [80]. Recently, data from real-world practice have confirmed a similar benefit of cemiplimab to that observed in clinical trials in a cohort of 240 patients with advanced cSCC [81].

Along with cemiplimab, other anti-PD-1 agents are under evaluation in cSCC (Table 7). Pembrolizumab is being tested in a phase II study of patients with recurrent/metastatic or locally advanced unresectable cSCC (NCT03284424). Results of a first interim analysis showed effective antitumour activity with 34.3% ORR and clinically meaningful, durable responses [82]. Pembrolizumab is also being evaluated in patients with locally advanced cSCC after surgery and radiation (NCT03833167). The efficacy of nivolumab monotherapy in patients with locally advanced/metastatic cSCC is being investigated in phase II studies (NCT04204837, NCT03834233).

Table 7. Selected ongoing clinical trials of immunotherapy in cSCC.

Immunotherapy	Treatment	Patients	NCT Code
Cemiplimab	Alone, pre-operative therapy, intralesional	Recurrent cSCC	NCT03889912
	Adjuvant therapy, after surgery and radiotherapy	High-risk cSCC	NCT03969004
	Alone or in combination with RP1	Advanced or metastatic cSCC	NCT04050436
	Alone	Unresectable locally recurrent and/or metastatic cSCC	NCT04242173
	Alone, neoadjuvant therapy	Stage II to IV cSCC	NCT04154943
Pembrolizumab	Alone	Recurrent/metastatic or locally advanced unresectable cSCC	NCT03284424
	Alone	Locally advanced or metastatic cSCC	NCT02964559
	Adjuvant therapy, after surgery and radiotherapy	High risk locally advanced cSCC	NCT03833167
	Combination with cetuximab	Recurrent/metastatic cSCC	NCT03082534
Nivolumab	Alone	Locally advanced/metastatic cSCC	NCT04204837
	Alone	Advanced cSCC	NCT03834233
	Alone or in combination with ipilimumab	Metastatic cSCC in immunosuppressed patients	NCT03816332

7. Immunosuppressed Patients

Immunosuppressed solid organ transplant recipients (SOTRs) present a 65- to 250-fold greater incidence of cSCC than the general population, more aggressive progression, and a higher risk of metastasis and death [83,84]. Moreover, certain transplant recipients are more likely to develop cSCC than others; an increased risk has been reported for patients who undergo heart transplant compared with kidney or liver recipients [85]. In this respect, the type, intensity, and duration of the immunosuppressive regimen seem related to the development of cSCC in SOTRs. A lower incidence of cSCC has been reported in patients who receive sirolimus compared with those treated with calcineurin inhibitors [83].

Immune status has been identified as a strong prognostic factor for disease outcomes in immunosuppressed patients. A multi-institutional study including patients with cSC-CHN reported significantly lower 2-year locoregional recurrence-free survival and PFS in immunocompromised patients compared with immunocompetent patients. Moreover, immunosuppressed status, recurrent disease, poor differentiation, and PNI were significantly associated with locoregional recurrence in this population [86].

A close clinical follow-up, every 3–6 months lifelong, and tailored immunosuppressive treatments, with adjustment or reduction of maintenance post-transplant therapy, may be necessary to reduce the risk of new cSCC [8,19]. Dermatologists, transplant physicians, and patients must collaborate to ensure adherence to dermatologic surveillance recommendations and must monitor suspicious lesions, thereby reducing the risk of cSCC in SOTRs [87,88].

It should be noted that although immunosuppression of cSCC patients is often described in the setting of SOTRs, other conditions should not be disregarded. As such, patients immunosuppressed due to human immunodeficiency virus (HIV) infection, allogenic bone marrow transplant, or autoimmune diseases should be carefully monitored.

8. Follow-Up and Monitoring

Patients with cSCC should be closely followed up for recurrences and development of new keratinocyte cancers and melanoma, particularly if they have a history of cSCC [8]. Evidence from cohort studies and registries shows that patients with at least one cSCC are at risk for developing additional cSCC and other skin cancers [20]. Therefore, it is recommended that follow-up for all patients should include regular clinical assessment, including inspection and palpation of the excision site, the in-transit route, and the regional lymph nodes [7,8]. Patients who have had cSCC should receive counselling regarding the risk for new primary skin cancers, such as BCC, cSCC, and melanoma, and the benefits of in-office and self-screening for early detection should be made clear [7,20].

The frequency of visits should be adjusted individually, depending on patient-specific underlying risk characteristics for cSCC [7,8]. The recently updated interdisciplinary European guidelines for the management of invasive cSCC recommend scheduling follow-up examination based on low- or high-risk common primary, advanced, or regional disease and the immunosuppressive setting [8]. While patients with low-risk cSCC may be assessed every 6–12 months for 5 years, high-risk patients should be examined in the clinic every 3–6 months for 2 years (including lymph node US), every 6–12 months from year 3–5, and annually thereafter. Closer follow-up is advised in patients with locally advanced/metastatic cSCC or after surgery for locoregional metastases, including clinical and US evaluation every 3 months for 5 years and then once or twice per year [8]. Similarly, the National Comprehensive Cancer Network (NCCN) guidelines for cSCC recommend that the frequency of patients' follow-up should depend on whether the tumour has spread to lymph nodes. Local cSCC should be monitored at least every 3–12 months during the first 2 years, every 6–12 months for another 3 years, and then annually for life. For regional cSCC, a more frequent history and clinical examination is advised: every 1–3 months for 1 year, every 2–4 months for 1 year, every 4–6 months for another 3 years, and then every 6–12 months for life [7].

Imaging tests (CT, MRI, or PET-CT) for non-palpable regional lymph nodes should also be performed in patients with regional cSCC [7] or locally advanced or metastatic disease [8]. The NCCN guidelines establish that the frequency of imaging scans should be tailored to individual clinical factors [7], whereas the European guidelines recommend examination every 3–6 months in the first 3 years and then based on individual symptoms and stage [8].

It should be noted that increased surveillance may be required for cSCC patients at very high risk of other primary tumours and recurrence, such as immunosuppressed patients or individuals with haematological comorbidities, genetic predisposition, or previous history of cSCC [7,8]. Their follow-up schedule, including clinical and imaging evaluation, should be modified depending on the characteristics of individual primary tumours (e.g., number and frequency of development) [8].

9. Summary and Conclusions

Tumour staging, management decisions, and monitoring of the cSCC are continuously improving thanks to the development of novel diagnostic tools and therapeutic agents. Multidisciplinary teams may ensure that the most appropriate strategy, tailored to the patient's individual profile, is undertaken from the initial diagnosis to follow-up. Patient-specific characteristics for risk including comorbidities, clinical parameters, histopathology, and molecular biomarkers should be considered to provide an accurate diagnosis. While surgery remains the cornerstone of treatment and RT is often used in the management of cSCC, ICIs and targeted therapies have changed the therapeutic landscape, showing improved outcomes in patients with locally advanced or metastatic disease. Considering the likelihood of recurrence and the increased risk of developing multiple cSCC, particularly in immunosuppressed patients, patients should be closely followed up in the first two years after starting treatment and continued long-term surveillance by clinical and imaging examination is warranted.

Supplementary Materials: The following are available online at https://www.mdpi.com/article/10.3390/cancers14030629/s1, Table S1: Methods and results of the Delphi survey.

Author Contributions: Conceptualization, J.G.-F., R.M., O.S., F.R., J.M., A.T.-V., S.M. and I.A.; writing—original draft preparation, J.G.-F., R.M., O.S., F.R., J.M., A.T.-V., S.M. and I.A.; writing—review and editing, J.G.-F., R.M., O.S., F.R., J.M., A.T.-V., S.M. and I.A.; project administration, J.G.-F. All authors have read and agreed to the published version of the manuscript.

Funding: This work was supported by Sanofi.

Institutional Review Board Statement: Not applicable.

Informed Consent Statement: Not applicable.

Data Availability Statement: The authors confirm that all data supporting the findings of this study are available within the article and its Supplementary Materials.

Acknowledgments: The authors thank Celia Miguel-Blanco from Medical Statistics Consulting S.L. (Valencia, Spain) for editorial support in the writing of this manuscript, and the 15 panellists who participated in the survey: A. Rueda Domínguez (H.U. Carlos Haya, Málaga), A. Soria (H.U. Ramón y Cajal, Madrid), J.L. Rodríguez Peralto (H. 12 de Octubre, Madrid), M. Iglesias (H. del Mar, Barcelona), E. Muñoz (H. Vall d'Hebron, Barcelona), A. García-Castaño (H. Marqués de Valdecilla, Santander), I. Márquez (H. Gregorio Marañón, Madrid), P. Cerezuela (H. Virgen de la Arrixaca, Murcia), J. Cañueto (H.U. Salamanca, Salamanca), R. Botella (H.U. La Fe, Valencia), S. Puig (H. Clinic, Barcelona), Y. Escobar (H. Gregorio Marañón, Madrid), F. Almeida (H. Ramón y Cajal, Madrid), B. Bellosillo (H. de Mar, Barcelona), and E. Espinosa (H. La Paz, Madrid).

Conflicts of Interest: J. Garcia-Foncillas discloses consulting/advisory/honoraria speaker roles with Abbott, Amgen, Astellas, Astra Zeneca, Biocartis, Boehringer Ingelheim, BMS, Bayer, Celgene, Eisai, Foundation Medicine, GSK, Hospira, Janssen, Lilly, Merck Serono, MSD, Novartis, Pharmamar, Pfizer, Roche, Sanofi, Servier, Sysmex, and Tesaro. R. Mesía discloses consulting/advisory/honoraria speaker role with Amgen, BMS, Merck Serono, MSD, Roche, and Sanofi. O. Sanmartín discloses consulting/advisory/honoraria speaker role with Roche, ISDIN, and Sanofi. F. Rojo has received honorary for consultancy from Roche, BMS, MSD, Merck, Novartis, Pfizer, Astra Zeneca, Abbvie, Pierre Fabre, InCyte, and Bayer. J. Mestre discloses consulting/advisory/honoraria speaker roles with Lavigor, Roche, and Sanofi. A. Tejera has received honoraria for consultancy from Sanofi. S. Martín discloses consulting/advisory/honoraria speaker roles with MSD, Sanofi/Regeneron, BMS, Novartis, Roche, and Pierre-Fabre. I. Azinovic discloses no conflicts of interest. The funder had no role in the design of the study; in the collection, analyses, or interpretation of data; in the writing of the manuscript; or in the decision to publish the results.

References

1. Karia, P.S.; Han, J.; Schmults, C.D. Cutaneous squamous cell carcinoma: Estimated incidence of disease, nodal metastasis, and deaths from disease in the United States, 2012. *J. Am. Acad. Dermatol.* **2013**, *68*, 957–966. [CrossRef] [PubMed]
2. Flohil, S.C.; van der Leest, R.J.; Arends, L.R.; de Vries, E.; Nijsten, T. Risk of subsequent cutaneous malignancy in patients with prior keratinocyte carcinoma: A systematic review and meta-analysis. *Eur. J. Cancer* **2013**, *49*, 2365–2375. [CrossRef] [PubMed]
3. Tejera-Vaquerizo, A.; Descalzo-Gallego, M.A.; Otero-Rivas, M.M.; Posada-García, C.; Rodríguez-Pazos, L.; Pastushenko, I.; Marcos-Gragera, R.; García-Doval, I. Skin Cancer Incidence and Mortality in Spain: A Systematic Review and Meta-Analysis. *Actas Dermosifiliogr.* **2016**, *107*, 318–328. [CrossRef] [PubMed]
4. Armstrong, B.K.; Kricker, A. The epidemiology of UV induced skin cancer. *J. Photochem. Photobiol. B* **2001**, *63*, 8–18. [CrossRef]
5. Stern, R.S. The risk of squamous cell and basal cell cancer associated with psoralen and ultraviolet A therapy: A 30-year prospective study. *J. Am. Acad. Dermatol.* **2012**, *66*, 553–562. [CrossRef]
6. Lindelöf, B.; Sigurgeirsson, B.; Gäbel, H.; Stern, R.S. Incidence of skin cancer in 5356 patients following organ transplantation. *Br. J. Dermatol.* **2000**, *143*, 513–519.
7. National Comprehensive Cancer Center. NCCN Clinical Practice Guidelines in Oncology; Squamous Cell Carcinoma (V2. 2018). Available online: https://oncolife.com.ua/doc/nccn/Squamous_Cell_Skin_Cancer.pdf (accessed on 13 September 2021).

8. Stratigos, A.J.; Garbe, C.; Dessinioti, C.; Lebbe, C.; Bataille, V.; Bastholt, L.; Dreno, B.; Concetta Fargnoli, M.; Forsea, A.M.; Frenard, C.; et al. European interdisciplinary guideline on invasive squamous cell carcinoma of the skin: Part 2. Treatment. *Eur. J. Cancer* **2020**, *128*, 83–102. [CrossRef]

9. Combalia, A.; Carrera, C. Squamous Cell Carcinoma: An Update on Diagnosis and Treatment. *Dermatol. Pract. Concept* **2020**, *10*, e2020066. [CrossRef]

10. Diamond, I.R.; Grant, R.C.; Feldman, B.M.; Pencharz, P.B.; Ling, S.C.; Moore, A.M.; Wales, P.W. Defining consensus: A systematic review recommends methodologic criteria for reporting of Delphi studies. *J. Clin. Epidemiol.* **2014**, *67*, 401–409. [CrossRef]

11. Norman, G. Likert scales, levels of measurement and the "laws" of statistics. *Adv. Health Sci. Educ. Theory Pract.* **2010**, *15*, 625–632. [CrossRef]

12. Fitch, K.; Bernstein, S.J.; Aguilar, M.D.; Burnanrd, B.; LaCalle, J.R.; Lazaro, P.; van het Loo, M.; McDonell, J.; Vader, J.; Kahan, J.P. *The Rand/UCLA Appropriateness Method User's Manual*; RAND Corporation: Santa Monica, CA, USA, 2001. Available online: https://www.rand.org/pubs/monograph_reports/MR1269.html (accessed on 3 March 2021).

13. Armin, M.B.; Edge, S.B.; Greene, F.; Byrd, D.R.; Brookland, R.K.; Washington, M.K.; Gershenwald, J.E.; Compton, C.C.; Hess, K.R.; Sullivan, D.C.; et al. (Eds.) *AJCC Cancer Staging Manual*, 8th ed.; Springer: New York, NY, USA, 2017.

14. Brantsch, K.D.; Meisner, C.; Schönfisch, B.; Trilling, B.; Wehner-Caroli, J.; Röcken, M.; Breuninger, H. Analysis of risk factors determining prognosis of cutaneous squamous-cell carcinoma: A prospective study. *Lancet Oncol.* **2008**, *9*, 713–720. [CrossRef]

15. Karia, P.S.; Jambusaria-Pahlajani, A.; Harrington, D.P.; Murphy, G.F.; Qureshi, A.A.; Schmults, C.D. Evaluation of American JoInt. Committee on Cancer, International Union against Cancer, and Brigham and Women's Hospital tumor staging for cutaneous squamous cell carcinoma. *J. Clin. Oncol.* **2014**, *32*, 327–334. [CrossRef]

16. Ruiz, E.S.; Karia, P.S.; Besaw, R.; Schmults, C.D. Performance of the American Joint. Committee on Cancer Staging Manual, 8th Edition vs the Brigham and Women's Hospital Tumor Classification System for Cutaneous Squamous Cell Carcinoma. *JAMA Dermatol.* **2019**, *155*, 819–825. [CrossRef] [PubMed]

17. Blechman, A.B.; Carucci, J.A.; Stevenson, M.L. Stratification of Poor Outcomes for Cutaneous Squamous Cell Carcinoma in Immunosuppressed Patients Using the American JoInt. Committee on Cancer Eighth Edition and Brigham and Women's Hospital Staging Systems. *Dermatol. Surg.* **2019**, *45*, 1117–1124. [CrossRef] [PubMed]

18. Tejera-Vaquerizo, A.; Canueto, J.; Llombart, B.; Martorell-Calatayud, A.; Sanmartin, O. Predictive Value of Sentinel Lymph Node Biopsy in Cutaneous Squamous Cell Carcinoma Based on the AJCC-8 and Brigham and Women's Hospital Staging Criteria. *Dermatol. Surg.* **2020**, *46*, 857–862. [CrossRef] [PubMed]

19. Stratigos, A.J.; Garbe, C.; Dessinioti, C.; Lebbe, C.; Bataille, V.; Bastholt, L.; Dreno, B.; Fargnoli, M.C.; Forsea, A.M.; Frenard, C.; et al. European interdisciplinary guideline on invasive squamous cell carcinoma of the skin: Part 1. epidemiology, diagnostics and prevention. *Eur. J. Cancer* **2020**, *128*, 60–82. [CrossRef]

20. Alam, M.; Armstrong, A.; Baum, C.; Bordeaux, J.S.; Brown, M.; Busam, K.J.; Eisen, D.B.; Iyengar, V.; Lober, C.; Margolis, D.J.; et al. Guidelines of care for the management of cutaneous squamous cell carcinoma. *J. Am. Acad. Dermatol.* **2018**, *78*, 560–578. [CrossRef]

21. Cañueto, J.; Martín-Vallejo, J.; Cardeñoso-Álvarez, E.; Fernández-López, E.; Pérez-Losada, J.; Román-Curto, C. Rapid growth rate is associated with poor prognosis in cutaneous squamous cell carcinoma. *Clin. Exp. Dermatol.* **2018**, *43*, 876–882. [CrossRef]

22. Gonzalez-Guerrero, M.; Martínez-Camblor, P.; Vivanco, B.; Fernández-Vega, I.; Munguía-Calzada, P.; Gonzalez-Gutierrez, M.P.; Rodrigo, J.P.; Galache, C.; Santos-Juanes, J. The adverse prognostic effect of tumor budding on the evolution of cutaneous head and neck squamous cell carcinoma. *J. Am. Acad. Dermatol.* **2017**, *76*, 1139–1145. [CrossRef]

23. Pickering, C.R.; Zhou, J.H.; Lee, J.J.; Drummond, J.A.; Peng, S.A.; Saade, R.E.; Tsai, K.Y.; Curry, J.L.; Tetzlaff, M.T.; Lai, S.Y.; et al. Mutational landscape of aggressive cutaneous squamous cell carcinoma. *Clin. Cancer Res.* **2014**, *20*, 6582–6592. [CrossRef]

24. Hussein, M.R. Ultraviolet radiation and skin cancer: Molecular mechanisms. *J. Cutan. Pathol.* **2005**, *32*, 191–205. [CrossRef]

25. Durinck, S.; Ho, C.; Wang, N.J.; Liao, W.; Jakkula, L.R.; Collisson, E.A.; Pons, J.; Chan, S.W.; Lam, E.T.; Chu, C.; et al. Temporal dissection of tumorigenesis in primary cancers. *Cancer Discov.* **2011**, *1*, 137–143. [CrossRef]

26. Botti, E.; Spallone, G.; Moretti, F.; Marinari, B.; Pinetti, V.; Galanti, S.; De Meo, P.D.; De Nicola, F.; Ganci, F.; Castrignanò, T.; et al. Developmental factor IRF6 exhibits tumor suppressor activity in squamous cell carcinomas. *Proc. Natl. Acad. Sci. USA* **2011**, *108*, 13710–13715. [CrossRef] [PubMed]

27. Moriyama, M.; Durham, A.D.; Moriyama, H.; Hasegawa, K.; Nishikawa, S.; Radtke, F.; Osawa, M. Multiple roles of Notch signaling in the regulation of epidermal development. *Dev. Cell* **2008**, *14*, 594–604. [CrossRef] [PubMed]

28. Su, F.; Viros, A.; Milagre, C.; Trunzer, K.; Bollag, G.; Spleiss, O.; Reis-Filho, J.S.; Kong, X.; Koya, R.C.; Flaherty, K.T.; et al. RAS Mutations in Cutaneous Squamous-Cell Carcinomas in Patients Treated with BRAF Inhibitors. *N. Engl. J. Med.* **2012**, *366*, 207–215. [CrossRef] [PubMed]

29. Küsters-Vandevelde, H.V.; Van Leeuwen, A.; Verdijk, M.A.; de Koning, M.N.; Quint, W.G.; Melchers, W.J.; Ligtenberg, M.J.; Blokx, W.A. CDKN2A but not TP53 mutations nor HPV presence predict poor outcome in metastatic squamous cell carcinoma of the skin. *Int. J. Cancer* **2010**, *126*, 2123–2132. [CrossRef]

30. Darr, O.A.; Colacino, J.A.; Tang, A.L.; McHugh, J.B.; Bellile, E.L.; Bradford, C.R.; Prince, M.P.; Chepeha, D.B.; Rozek, L.S.; Moyer, J.S. Epigenetic alterations in metastatic cutaneous carcinoma. *Head Neck* **2015**, *37*, 994–1001. [CrossRef]

31. Lui, V.W.; Peyser, N.D.; Ng, P.K.; Hritz, J.; Zeng, Y.; Lu, Y.; Li, H.; Wang, L.; Gilbert, B.R.; General, I.J.; et al. Frequent mutation of receptor protein tyrosine phosphatases provides a mechanism for STAT3 hyperactivation in head and neck cancer. *Proc. Natl. Acad. Sci. USA* **2014**, *111*, 1114–1119. [CrossRef]

32. Commandeur, S.; Ho, S.H.; de Gruijl, F.R.; Willemze, R.; Tensen, C.P.; El Ghalbzouri, A. Functional characterization of cancer-associated fibroblasts of human cutaneous squamous cell carcinoma. *Exp. Dermatol.* **2011**, *20*, 737–742. [CrossRef]

33. Moussai, D.; Mitsui, H.; Pettersen, J.S.; Pierson, K.C.; Shah, K.R.; Suárez-Fariñas, M.; Cardinale, I.R.; Bluth, M.J.; Krueger, J.G.; Carucci, J.A. The human cutaneous squamous cell carcinoma microenvironment is characterized by increased lymphatic density and enhanced expression of macrophage-derived VEGF-C. *J. Investig. Dermatol.* **2011**, *131*, 229–236. [CrossRef]

34. Ulrich, M.; Stockfleth, E.; Roewert-Huber, J.; Astner, S. Noninvasive diagnostic tools for nonmelanoma skin cancer. *Br. J. Dermatol.* **2007**, *157*, 56–58. [CrossRef]

35. Warszawik-Hendzel, O.; Olszewska, M.; Maj, M.; Rakowska, A.; Czuwara, J.; Rudnicka, L. Non-invasive diagnostic techniques in the diagnosis of squamous cell carcinoma. *J. Dermatol. Case Rep.* **2015**, *9*, 89–97. [CrossRef]

36. Zalaudek, I.; Giacomel, J.; Schmid, K.; Bondino, S.; Rosendahl, C.; Cavicchini, S.; Tourlaki, A.; Gasparini, S.; Bourne, P.; Keir, J.; et al. Dermatoscopy of facial actinic keratosis, intraepidermal carcinoma, and invasive squamous cell carcinoma: A progression model. *J. Am. Acad. Dermatol.* **2012**, *66*, 589–597. [CrossRef] [PubMed]

37. Yanofsky, V.R.; Mercer, S.E.; Phelps, R.G. Histopathological variants of cutaneous squamous cell carcinoma: A review. *J. Skin Cancer* **2011**, *2011*, 210813. [CrossRef] [PubMed]

38. Thompson, A.K.; Kelley, B.F.; Prokop, L.J.; Murad, M.H.; Baum, C.L. Risk Factors for Cutaneous Squamous Cell Carcinoma Recurrence, Metastasis, and Disease-Specific Death: A Systematic Review and Meta-analysis. *JAMA Dermatol.* **2016**, *152*, 419–428. [CrossRef] [PubMed]

39. Liao, L.J.; Lo, W.C.; Hsu, W.L.; Wang, C.T.; Lai, M.S. Detection of cervical lymph node metastasis in head and neck cancer patients with clinically N0 neck-a meta-analysis comparing different imaging modalities. *BMC Cancer* **2012**, *12*, 236. [CrossRef]

40. Jennings, L.; Schmults, C.D. Management of high-risk cutaneous squamous cell carcinoma. *J. Clin. Aesthetic Dermatol.* **2010**, *3*, 39–48.

41. Connolly, K.L.; Jeong, J.M.; Barker, C.A.; Hernandez, M.; Lee, E.H. A systematic review of comorbidity indices used in the nonmelanoma skin cancer population. *J. Am. Acad. Dermatol.* **2017**, *76*, 344–346.e2. [CrossRef]

42. Dhiwakar, M.; Khan, N.A.; McClymont, L.G. Surgery for head and neck skin tumors in the elderly. *Head Neck* **2007**, *29*, 851–856. [CrossRef]

43. Charles, A.J., Jr.; Otley, C.C.; Pond, G.R. Prognostic factors for life expectancy in nonagenarians with nonmelanoma skin cancer: Implications for selecting surgical candidates. *J. Am. Acad. Dermatol.* **2002**, *47*, 419–422. [CrossRef]

44. Delaney, A.; Shimizu, I.; Goldberg, L.H.; MacFarlane, D.F. Life expectancy after Mohs micrographic surgery in patients aged 90 years and older. *J. Am. Acad. Dermatol.* **2013**, *68*, 296–300. [CrossRef]

45. Reid, B.C.; Alberg, A.J.; Klassen, A.C.; Koch, W.M.; Samet, J.M. The American Society of Anesthesiologists' class as a comorbidity index in a cohort of head and neck cancer surgical patients. *Head Neck* **2001**, *23*, 985–994. [CrossRef]

46. Lee, E.H.; Nijhawan, R.I.; Nehal, K.S.; Dusza, S.W.; Levine, A.; Hill, A.; Barker, C.A. Comorbidity Assessment in Skin Cancer Patients: A Pilot Study Comparing Medical Interview with a Patient-Reported Questionnaire. *J. Skin Cancer* **2015**, *2015*, 953479. [CrossRef] [PubMed]

47. Porceddu, S.V.; Bressel, M.; Poulsen, M.G.; Stoneley, A.; Veness, M.J.; Kenny, L.M.; Wratten, C.; Corry, J.; Cooper, S.; Fogarty, G.B.; et al. Postoperative Concurrent Chemoradiotherapy Versus Postoperative Radiotherapy in High-Risk Cutaneous Squamous Cell Carcinoma of the Head and Neck: The Randomized Phase III TROG 05.01 Trial. *J. Clin. Oncol.* **2018**, *36*, 1275–1283. [CrossRef] [PubMed]

48. Carter, J.B.; Johnson, M.M.; Chua, T.L.; Karia, P.S.; Schmults, C.D. Outcomes of Primary Cutaneous Squamous Cell Carcinoma with Perineural Invasion: An 11-Year Cohort Study. *JAMA Dermatol.* **2013**, *149*, 35–42. [CrossRef] [PubMed]

49. Karia, P.S.; Morgan, F.C.; Ruiz, E.S.; Schmults, C.D. Clinical and Incidental Perineural Invasion of Cutaneous Squamous Cell Carcinoma: A Systematic Review and Pooled Analysis of Outcomes Data. *JAMA Dermatol.* **2017**, *153*, 781–788. [CrossRef] [PubMed]

50. Navarrete-Dechent, C.; Veness, M.J.; Droppelmann, N.; Uribe, P. High-risk cutaneous squamous cell carcinoma and the emerging role of sentinel lymph node biopsy: A literature review. *J. Am. Acad. Dermatol.* **2015**, *73*, 127–137. [CrossRef]

51. Tejera-Vaquerizo, A.; García-Doval, I.; Llombart, B.; Cañueto, J.; Martorell-Calatayud, A.; Descalzo-Gallego, M.A.; Sanmartín, O. Systematic review of the prevalence of nodal metastases and the prognostic utility of sentinel lymph node biopsy in cutaneous squamous cell carcinoma. *J. Dermatol.* **2018**, *45*, 781–790. [CrossRef]

52. Schmitt, A.R.; Brewer, J.D.; Bordeaux, J.S.; Baum, C.L. Staging for Cutaneous Squamous Cell Carcinoma as a Predictor of Sentinel Lymph Node Biopsy Results: Meta-analysis of American JoInt. Committee on Cancer Criteria and a Proposed Alternative System. *JAMA Dermatol.* **2014**, *150*, 19–24. [CrossRef]

53. Samsanavicius, D.; Kaikaris, V.; Cepas, A.; Ulrich, J.; Makstiene, J.; Rimdeika, R. Importance of sentinel lymphatic node biopsy in detection of early micrometastases in patients with cutaneous squamous cell carcinoma. *J. Plast. Reconstr. Aesthetic Surg.* **2018**, *71*, 597–603. [CrossRef]

54. Samsanavičius, D.; Kaikaris, V.; Norvydas, S.J.; Liubauskas, R.; Valiukevičienė, S.; Makštienė, J.; Maslauskas, K.; Rimdeika, R. Sentinel lymph node biopsy for high-risk cutaneous squamous cell carcinoma: Analysis of recurrence-free survival. *Medicina* **2016**, *52*, 276–282. [CrossRef]

55. Lee, C.T.; Lehrer, E.J.; Aphale, A.; Lango, M.; Galloway, T.J.; Zaorsky, N.G. Surgical excision, Mohs micrographic surgery, external-beam radiotherapy, or brachytherapy for indolent skin cancer: An international meta-analysis of 58 studies with 21,000 patients. *Cancer* **2019**, *125*, 3582–3594. [CrossRef]

56. Martin, T.; Moutrie, Z.; Tighe, D.; Haady, H.; Fogarty, G. Volumetric modulated arc therapy (VMAT) for skin field cancerisation of the nose—A technique and case report. *Int. J. Radiol. Radiat. Ther.* **2018**, *5*, 142–148. [CrossRef]

57. Likhacheva, A.; Awan, M.; Barker, C.A.; Bhatnagar, A.; Bradfield, L.; Brady, M.S.; Buzurovic, I.; Geiger, J.L.; Parvathaneni, U.; Zaky, S.; et al. Definitive and Postoperative Radiation Therapy for Basal and Squamous Cell Cancers of the Skin: Executive Summary of an American Society for Radiation Oncology Clinical Practice Guideline. *Pract. Radiat. Oncol.* **2020**, *10*, 8–20. [CrossRef] [PubMed]

58. Gunaratne, D.A.; Veness, M.J. Efficacy of hypofractionated radiotherapy in patients with non-melanoma skin cancer: Results of a systematic review. *J. Med. Imaging Radiat. Oncol.* **2018**, *62*, 401–411. [CrossRef] [PubMed]

59. Zaorsky, N.G.; Lee, C.T.; Zhang, E.; Keith, S.W.; Galloway, T.J. Hypofractionated radiation therapy for basal and squamous cell skin cancer: A meta-analysis. *Radiother. Oncol.* **2017**, *125*, 13–20. [CrossRef] [PubMed]

60. Nakamura, K.; Okuyama, R.; Saida, T.; Uhara, H. Platinum and anthracycline therapy for advanced cutaneous squamous cell carcinoma. *Int. J. Clin. Oncol.* **2013**, *18*, 506–509. [CrossRef] [PubMed]

61. Rogers, H.W.; Weinstock, M.A.; Feldman, S.R.; Coldiron, B.M. Incidence Estimate of Nonmelanoma Skin Cancer (Keratinocyte Carcinomas) in the U.S. Population, 2012. *JAMA Dermatol.* **2015**, *151*, 1081–1086. [CrossRef]

62. Shin, D.M.; Glisson, B.S.; Khuri, F.R.; Clifford, J.L.; Clayman, G.; Benner, S.E.; Forastiere, A.A.; Ginsberg, L.; Liu, D.; Lee, J.J.; et al. Phase II and biologic study of interferon alfa, retinoic acid, and cisplatin in advanced squamous skin cancer. *J. Clin. Oncol.* **2002**, *20*, 364–370. [CrossRef]

63. Cañueto, J.; Cardeñoso, E.; García, J.L.; Santos-Briz, Á.; Castellanos-Martín, A.; Fernández-López, E.; Blanco Gómez, A.; Pérez-Losada, J.; Román-Curto, C. Epidermal growth factor receptor expression is associated with poor outcome in cutaneous squamous cell carcinoma. *Br. J. Dermatol.* **2017**, *176*, 1279–1287. [CrossRef]

64. Gold, K.A.; Kies, M.S.; William, W.N., Jr.; Johnson, F.M.; Lee, J.J.; Glisson, B.S. Erlotinib in the treatment of recurrent or metastatic cutaneous squamous cell carcinoma: A single-arm phase 2 clinical trial. *Cancer* **2018**, *124*, 2169–2173. [CrossRef]

65. William, W.N., Jr.; Feng, L.; Ferrarotto, R.; Ginsberg, L.; Kies, M.; Lippman, S.; Glisson, B.; Kim, E.S. Gefitinib for patients with incurable cutaneous squamous cell carcinoma: A single-arm phase II clinical trial. *J. Am. Acad. Dermatol.* **2017**, *77*, 1110–1113.e2. [CrossRef]

66. Maubec, E.; Petrow, P.; Scheer-Senyarich, I.; Duvillard, P.; Lacroix, L.; Gelly, J.; Certain, A.; Duval, X.; Crickx, B.; Buffard, V.; et al. Phase II study of cetuximab as first-line single-drug therapy in patients with unresectable squamous cell carcinoma of the skin. *J. Clin. Oncol.* **2011**, *29*, 3419–3426. [CrossRef] [PubMed]

67. Foote, M.C.; McGrath, M.; Guminski, A.; Hughes, B.G.; Meakin, J.; Thomson, D.; Zarate, D.; Simpson, F.; Porceddu, S.V. Phase II study of single-agent panitumumab in patients with incurable cutaneous squamous cell carcinoma. *Ann. Oncol.* **2014**, *25*, 2047–2052. [CrossRef] [PubMed]

68. Montaudié, H.; Viotti, J.; Combemale, P.; Dutriaux, C.; Dupin, N.; Robert, C.; Mortier, L.; Kaphan, R.; Duval-Modeste, A.B.; Dalle, S.; et al. Cetuximab is efficient and safe in patients with advanced cutaneous squamous cell carcinoma: A retrospective, multicentre study. *Oncotarget* **2020**, *11*, 378–385. [CrossRef] [PubMed]

69. Vermorken, J.B.; Mesia, R.; Rivera, F.; Remenar, E.; Kawecki, A.; Rottey, S.; Erfan, J.; Zabolotnyy, D.; Kienzer, H.-R.; Cupissol, D.; et al. Platinum-Based Chemotherapy plus Cetuximab in Head and Neck Cancer. *N. Engl. J. Med.* **2008**, *359*, 1116–1127. [CrossRef] [PubMed]

70. Vermorken, J.B.; Trigo, J.; Hitt, R.; Koralewski, P.; Diaz-Rubio, E.; Rolland, F.; Knecht, R.; Amellal, N.; Schueler, A.; Baselga, J. Open-label, uncontrolled, multicenter phase II study to evaluate the efficacy and toxicity of cetuximab as a single agent in patients with recurrent and/or metastatic squamous cell carcinoma of the head and neck who failed to respond to platinum-based therapy. *J. Clin. Oncol.* **2007**, *25*, 2171–2177. [CrossRef]

71. Galbiati, D.; Cavalieri, S.; Alfieri, S.; Resteghini, C.; Bergamini, C.; Orlandi, E.; Platini, F.; Locati, L.; Giacomelli, L.; Licitra, L.; et al. Activity of platinum and cetuximab in cutaneous squamous cell cancer not amenable to curative treatment. *Drugs Context* **2019**, *8*, 212611. [CrossRef]

72. Sahni, S.; Valecha, G.; Sahni, A. Role of Anti-PD-1 Antibodies in Advanced Melanoma: The Era of Immunotherapy. *Cureus* **2018**, *10*, e3700. [CrossRef]

73. Goodman, A.M.; Kato, S.; Bazhenova, L.; Patel, S.P.; Frampton, G.M.; Miller, V.; Stephens, P.J.; Daniels, G.A.; Kurzrock, R. Tumor Mutational Burden as an Independent Predictor of Response to Immunotherapy in Diverse Cancers. *Mol. Cancer Ther.* **2017**, *16*, 2598–2608. [CrossRef]

74. European Medicines Agency. Libtayo: EPAR—Product Information. Available online: https://www.ema.europa.eu/en/documents/productinformation/libtayo-epar-product-information_en.pdf (accessed on 4 October 2021).

75. Regeneron Pharmaceuticals, Inc. LIBTAYO®[Cemiplimab-Rwlc] Injection Full US Prescribing Information. Available online: https://www.accessdata.fda.gov/drugsatfda_docs/label/2018/761097s000lbl.pdf (accessed on 14 September 2021).

76. Burova, E.; Hermann, A.; Waite, J.; Potocky, T.; Lai, V.; Hong, S.; Liu, M.; Allbritton, O.; Woodruff, A.; Wu, Q.; et al. Characterization of the Anti-PD-1 Antibody REGN2810 and Its Antitumor Activity in Human PD-1 Knock-In Mice. *Mol. Cancer Ther.* **2017**, *16*, 861–870. [CrossRef]
77. Migden, M.R.; Rischin, D.; Schmults, C.D.; Guminski, A.; Hauschild, A.; Lewis, K.D.; Chung, C.H.; Hernandez-Aya, L.; Lim, A.M.; Chang, A.L.S.; et al. PD-1 Blockade with Cemiplimab in Advanced Cutaneous Squamous-Cell Carcinoma. *N. Engl. J. Med.* **2018**, *379*, 341–351. [CrossRef]
78. Rischin, D.; Khushalani, N.I.; Schmults, C.D.; Guminski, A.D.; Chang, A.L.S.; Lewis, K.D.; Lim, A.M.L.; Hernandez-Aya, L.F.; Hughes, B.G.M.; Schadendorf, D.; et al. Integrated analysis of a phase 2 study of cemiplimab in advanced cutaneous squamous cell carcinoma: Extended follow-up of outcomes and quality of life analysis. *J. Immunother. Cancer* **2021**, *9*, e002757. [CrossRef] [PubMed]
79. Migden, M.R.; Khushalani, N.I.; Chang, A.L.S.; Lewis, K.D.; Schmults, C.D.; Hernandez-Aya, L.; Meier, F.; Schadendorf, D.; Guminski, A.; Hauschild, A.; et al. Cemiplimab in locally advanced cutaneous squamous cell carcinoma: Results from an open-label, phase 2, single-arm trial. *Lancet Oncol.* **2020**, *21*, 294–305. [CrossRef]
80. Migden, M.R.; Rischin, D.; Sasane, M.; Mastey, V.; Pavlick, A.; Schmults, C.D.; Chen, Z.; Guminski, A.D.; Hauschild, A.; Bury, D.; et al. Health-related quality of life (HRQL) in patients with advanced cutaneous squamous cell carcinoma (CSCC) treated with cemiplimab: Post hoc exploratory analyses of a phase II clinical trial. *J. Clin. Oncol.* **2020**, *38*, 10033. [CrossRef]
81. Hober, C.; Fredeau, L.; Ledard, A.P.; Boubaya, M.; Herms, F.; Aubin, F.; Benetton, N.; Dinulescu, M.; Jannic, A.; Cesaire, L.; et al. 1086P Cemiplimab for advanced cutaneous squamous cell carcinoma: Real life experience. *Ann. Oncol.* **2020**, *31*, S737. [CrossRef]
82. Grob, J.-J.; Gonzalez, R.; Basset-Seguin, N.; Vornicova, O.; Schachter, J.; Joshi, A.; Meyer, N.; Grange, F.; Piulats, J.M.; Bauman, J.R.; et al. Pembrolizumab Monotherapy for Recurrent or Metastatic Cutaneous Squamous Cell Carcinoma: A Single-Arm Phase II Trial (KEYNOTE-629). *J. Clin. Oncol.* **2020**, *38*, 2916–2925. [CrossRef] [PubMed]
83. Kim, C.; Cheng, J.; Colegio, O.R. Cutaneous squamous cell carcinomas in solid organ transplant recipients: Emerging strategies for surveillance, staging, and treatment. *Semin Oncol.* **2016**, *43*, 390–394. [CrossRef] [PubMed]
84. Lanz, J.; Bouwes Bavinck, J.N.; Westhuis, M.; Quint, K.D.; Harwood, C.A.; Nasir, S.; Van-de-Velde, V.; Proby, C.M.; Ferrándiz, C.; Genders, R.E.; et al. Aggressive Squamous Cell Carcinoma in Organ Transplant Recipients. *JAMA Dermatol.* **2019**, *155*, 66–71. [CrossRef]
85. Nuño-González, A.; Vicente-Martín, F.J.; Pinedo-Moraleda, F.; López-Estebaranz, J.L. High-risk cutaneous squamous cell carcinoma. *Actas Dermosifiliogr.* **2012**, *103*, 567–578. [CrossRef]
86. Manyam, B.V.; Garsa, A.A.; Chin, R.I.; Reddy, C.A.; Gastman, B.; Thorstad, W.; Yom, S.S.; Nussenbaum, B.; Wang, S.J.; Vidimos, A.T.; et al. A multi-institutional comparison of outcomes of immunosuppressed and immunocompetent patients treated with surgery and radiation therapy for cutaneous squamous cell carcinoma of the head and neck. *Cancer* **2017**, *123*, 2054–2060. [CrossRef]
87. Cheng, J.Y.; Li, F.Y.; Ko, C.J.; Colegio, O.R. Cutaneous Squamous Cell Carcinomas in Solid Organ Transplant Recipients Compared with Immunocompetent Patients. *JAMA Dermatol.* **2018**, *154*, 60–66. [CrossRef]
88. Rizvi, S.M.H.; Aagnes, B.; Holdaas, H.; Gude, E.; Boberg, K.M.; Bjørtuft, Ø.; Helsing, P.; Leivestad, T.; Møller, B.; Gjersvik, P. Long-term Change in the Risk of Skin Cancer After Organ Transplantation: A Population-Based Nationwide Cohort Study. *JAMA Dermatol.* **2017**, *153*, 1270–1277. [CrossRef] [PubMed]

Systematic Review

Treatment Options and Post-Treatment Malignant Transformation Rate of Actinic Cheilitis: A Systematic Review

Katerina Bakirtzi [1], **Ilias Papadimitriou** [1], **Dimitrios Andreadis** [2] **and Elena Sotiriou** [1,*]

[1] First Department of Dermatology and Venereology, School of Medicine, Aristotle University of Thessaloniki, P.O. Box 54643 Thessaloniki, Greece; mpakirtzi@auth.gr (K.B.); ipapadimi@auth.gr (I.P.)
[2] Department of Oral Medicine and Pathology, School of Dentistry, Aristotle University of Thessaloniki, P.O. Box 54624 Thessaloniki, Greece; dandrea@dent.auth.gr
* Correspondence: epsotiri@auth.gr; Tel.: +30-6977986818

Simple Summary: Actinic cheilitis is a precancerous condition that may evolve to a more aggressive type of skin cancer. Therefore, its therapy is crucial for the disease prognosis. In this systematic review, we tried to identify the best therapies of actinic cheilitis regarding safety, efficacy, recurrences, and the potential to progress to skin cancer. The therapeutic approach comprised invasive and topical treatments. The invasive therapies, such as partial surgery and laser treatments, had the best cosmetic and therapeutic results with few recurrences. Photodynamic therapy demonstrated satisfactory outcomes, while topical treatments were the least beneficial. Notably, the efficacy of photodynamic therapy was improved when combined with 5% imiquimod. However, except from photodynamic therapy, the other modalities were assessed in a limited number of patients. Finally, when actinic cheilitis is treated, no risk of cancer progression exists. Larger studies are necessary to confirm these results.

Citation: Bakirtzi, K.; Papadimitriou, I.; Andreadis, D.; Sotiriou, E. Treatment Options and Post-Treatment Malignant Transformation Rate of Actinic Cheilitis: A Systematic Review. *Cancers* **2021**, *13*, 3354. https://doi.org/10.3390/cancers13133354

Academic Editor: Francesca Ricci

Received: 30 May 2021
Accepted: 2 July 2021
Published: 4 July 2021

Publisher's Note: MDPI stays neutral with regard to jurisdictional claims in published maps and institutional affiliations.

Abstract: Actinic cheilitis is a premalignant condition that may evolve to squamous cell carcinoma. A consensus on its management has not been established, and large clinical trials are lacking. We aimed to review the existing data regarding the treatment of actinic cheilitis with various modalities regarding safety, efficacy, recursions, and post-treatment malignant transformation. A systematic review was conducted through Pubmed, Ovid and the Cochrane library for studies in English language and the references of included papers from inception to January 2021. Case series were considered if ≥6 patients were included. Of the 698 articles, 36 studies and, overall, 699 patients were eventually reviewed. Laser ablation and vermilionectomy provided the best clinical and aesthetic outcomes with few recurrences, while photodynamic therapy was linked to more relapses. Generally, the adverse events were minor and there was no risk of post-treatment malignant transformation. The limitations of our review include the heterogeneity and the small number of patients across studies. Conclusively, invasive treatments demonstrated superior therapeutic and safety profile. Nevertheless, high-quality head-to-head studies that assess different modalities for actinic cheilitis and report patient preferences are lacking.

Keywords: actinic cheilitis; treatment; imiquimod; photodynamic therapy; vermilionectomy; laser; diclofenac

1. Introduction

Actinic cheilitis (AC) is a premalignant lesion on the lips in patients who are overexposed to sunlight, and it has a significant chance of evolving into invasive squamous cell carcinoma (SCC). It primarily affects the lower lip of male individuals over the age of 50, and its clinical features include dryness, atrophy, scaling, erythema, ulceration, and a poorly demarcated border [1–4]. Most patients are of the Fitzpatrick I-II skin phototype [3]. It is also observed that people of lower education level and poor lifestyle conditions are more likely to develop AC [5]. The dermoscopic characteristics of AC comprise white structureless areas, scales, white halos of the vermilion of the lip, and erosions [6,7]. The slow progression of AC usually leads to a delay in the diagnosis, as it is often mistakenly regarded as a regular feature of aging [8].

The rate of malignant transformation of AC into SCC varies between 10 and 30%, while it is reported that 95% of SCCs on the lip occur on the ground of preexisting ACs [9,10]. Moreover, whereas up to 6% of cutaneous SCCs metastasize, the metastasis rate for a SCC located on the lips is four times higher than its peripheral counterpart [11]. Therefore, early detection and treatment of AC are of great value, since they could largely prevent SCC development.

Nevertheless, to date, no consensus on the proper management of AC exists. Surgical removal techniques exist, with vermilionectomy being the most commonly employed [12]. Yet, the surgical approach is destructive and is linked to various complications, such as scarring, persistent oedema and anaesthesia [13]. Conventional therapeutic approaches include topical chemo- or immunotherapy and radiation-based treatment, with the former being the less effective due to low patient adherence [14]. Topical application of fluorouracil FU, 5% imiquimod (5% IMI) and 3% diclofenac in 2.5% hyaluronic acid (DHA), 0.015% ingenol mebutate (IngMeb) and trichloroacetic acid (TCA), regarding topical regimens, photodynamic therapy (PDT) and CO_2 laser ablation (CO_2L), with or without aminolevulinic acid (ALA), or methyl-aminolevulinic acid (MAL), cryosurgery (CRYO) and electrodessication (ELD,)), regarding radiation-based and minimally invasive treatments, have been described [15]. Although there is some literature on the efficacy of AC treatments, there is still a lack of high-quality research to direct appropriate management decisions. The present systematic review aims to offer an overview of the efficacy of the current AC treatments with respect to clinical responses and, where available, histopathological or dermoscopic clearance, recurrence rates and the rate of post-treatment AC malignant transformation.

2. Search Strategy

This systematic review was conducted according to the Meta-analysis of Observational Studies in Epidemiology (MOOSE) proposal and the Preferred Reporting Items for Systematic Reviews and Meta-Analyses PRISMA guidelines [16,17]. The research was performed in MEDLINE through PubMed, Embase through Ovid and in the Cochrane Central Register of Controlled Trials (CENTRAL) library. The references of the suitable papers were screened for further relevant publications. A forward search was considered excessive since the searches already carried out yielded a plethora of results. The search was conducted in January 2021.

The search protocol was ((("actinic" OR "solar") AND ("cheilitis") OR ("cheilosis") AND ("treatment")) examining both medical subject headings (MeSH) and free text. Eligible studies were clinical trials, prospective and retrospective studies on human subjects and case series of at least six patients written in the English language. All studies needed to include cases with a pathological AC diagnosis, having received either surgical or noninvasive treatment, and report the clinical and/or pathological response as their outcome. Cross-sectional studies with no post-treatment follow-up visits were also excluded. Studies were also ruled out if they did not contain the results of AC therapies or did not fulfill the inclusion criteria.

3. Data Extraction

Two blinded reviewers (B.K. and P.I.) extracted data independently based on a standardized extraction form. Any discordance was resolved by consensus or by the involvement of a third investigator who was experienced in performing systematic reviews and meta-analyses (L.A.). The following data were collected from each eligible paper: year of the study, authors, publication date, type of manuscript, title, country, the treatment being studied (type, duration), study design, study population (sample size, age, sex, risk factors and comorbidities), means of diagnosis (histopathological or clinical), and sponsorship reported. Treatment evaluation was based on treatment response rate, healing time, recurrence rate, side effects, follow-up time, follow-up biopsy and aesthetic results. The risk of bias in every eligible study was classified as "low", "moderate", or "high" by the same reviewers per the PRISMA guidelines.

4. Results

The database search yielded 698 articles. After duplicates were removed, 281 articles were identified and, when titles and abstracts were reviewed, 91 papers were subsequently ruled out since they did not satisfy the inclusion and exclusion criteria. The 190 remaining articles were thoroughly reassessed for eligibility. After exclusion criteria were applied, 58 research papers remained. Comprehensive data extraction was performed for these 58 papers. Twenty-two studies were excluded based on the quality criteria, eventually leaving 36 papers for the data analysis, published between 1977 and 2019 [13,18–52]. High risk of bias levels was detected in all the included studies.

The shortlisted studies consisted of two RCTs, six retrospective and 22 prospective studies, and three case series. The number of patients in each study varied between six and fifty-two (699 in total) (Table 1). The gender was reported in 678 patients, most of whom were male (75.07%; 509/678), and the patients' ages ranged from 26 to 93 (mean: 63.18 ± 9.3). The localization of AC was recorded in 573 patients, with the lower lip being the most predominant site in 556 cases (97.03%), and the upper lip (1.05%; 6/573) or both lips (1.75%; 10/573) being significantly less affected. In nine studies, the percentage of the lip surface area affected by AC was also reported: >75% involvement in 24.77% of the patients (55/222), while in 70.72% (157/222) of patients, the lip involvement ranged between 50% and 75%. Data regarding the risk factors for AC development of the study population were collected from 30 articles. Seventy-two patients had fair skin, sixty-four had positive smoking and twenty-four positive alcohol drinking history, fifty-eight had suffered from skin tumors (non-melanoma skin cancer or melanoma: 49 patients; previous SCC of the lower lip: 9 patients), thirty-six were outdoor workers, four reported intense sunlight exposure and one was immunosuppressed.

Table 1. Studies and therapeutic process per treatment modality.

Treatment	Number of Studies	Number of Patients	Specific Treatment	Days of Treatment	Follow-Up Time (Range in Months)
Partial surgery [23,29,50]	3	28	CO_2 laser ablation, Electrodessication vs. CO_2 laser, Chemical peel	1	3–48
Laser [13,23–26,29,30,37,46,50–52]	12	278	CO_2 laser, Er: YAG laser	1–3	1.3–60
MAL PDT + 5%IMI [33]	1	34	-	1	12
Laser-mediated PDT [30,31]	2	33	ALA-dye, MAL-Er: YAG	1–3	1–12
FU [23,49,51]	3	28	1% and 5% fluorouracil	12–21	2–48
5% IMI [19,47]	2	25	-	12–30	1–18
PDT [18,20,31,32,34–36,39,41–45]	13	241	Daylight PDT, ALA PDT, MAL PDT	1–6	1–60
0.015% IngMeb [19,40]	2	17	-	1	1–10
50%TCA [23]	1	10	-	1	48
DHA [19,22,38,53]	4	62	-	1	1–6

MAL—methyl-aminolevulinic acid; ALA—aminolevulinic acid; Er: Yag—Erbium: YAG; IMI—imiquimod; PDT—photodynamic therapy; FU—fluorouracil; DHA—3% diclofenac in 2.5% hyaluronic acid; TCA—trichloroacetic acid; IngMeb—ingenol mebutate.

There is not a widely accepted clinical measurement tool for the severity and therapeutic outcomes for AC to date. These outcome measures used in the studies were comparable to the newly established core outcome set for actinic keratoses clinical trials [54]. Most articles presented results regarding the clearance—based on the clinical or histopathologic response, recurrence or progression rate, side effects, long-term follow-up, cosmetic outcome and patient satisfaction—and some reported on treatment discontinuation, patient adherence or healing time (Table 2).

Table 2. Study results per treatment modality.

Treatment	Complete Response Rate (%)	Recurrence Rate (%)	Adverse Events	Excellent Cosmetic Results (%)	Discontinuation Rate (%)
Partial surgery [23,29,50]	100	0.0	Scarring	N/A	N/A
Laser [13,23–26,29,30,37,46,50–52]	93.39	6.42	Scarring, pain, oedema, erosion, pruritus	100.0	N/A
MAL PDT + 5%IMI [33]	79.41	5.88	Pain, erythema, burning sensation, scarring, oedema, pruritus, rash	N/A	5.88
Laser-mediated PDT [30,31]	75.76	6.10	Erythema, burning sensation, oedema, erosions	N/A	N/A
FU [23,49,51]	75.0	31.80	N/A	N/A	10.0
5% IMI [19,47]	76.0	N/A	Erythema, oedema, induration, erosions, burning sensation	N/A	N/A
PDT [18,20,31,32,34–36,39,41–45]	66.67	14.07	Pain, erythema, oedema, scaling, rash, erosions, burning sensation, scarring	67.65	5.86
0.015% IngMeb [19,40]	41.18	0.0	Erythema, oedema, scaling, erosions, burning sensation	N/A	0.0
50%TCA [23]	30.0	70.0	N/A	N/A	N/A
DHA [19,22,38,53]	45.16	6.52	Erythema, oedema, burning sensation	100.0	15.22

N/A—not available; MAL—methyl-aminolevulinic acid; ALA—aminolevulinic acid; Er: Yag—Erbium: YAG; IMI—imiquimod; PDT—photodynamic therapy; FU—fluorouracil; DHA—3% diclofenac in 2.5% hyaluronic acid; TCA—trichloroacetic acid; IngMeb—ingenol mebutate.

4.1. Therapies for AC

Overall, 699 patients have been treated with the following therapies: laser-therapy (319; 45.64%) [13,23–26,29,30,37,46,50–52], PDT (241; 34.48%) [18,20,31,32,34–36,38,39,41–45], DHA (62; 8.87%) [19,22,38,53], MAL PDT + 5% IMI (34; 4.86%) [33], MAL or ALA plus laser (laser-PDT) (33; 4.72%) [31], 5% IMI (25; 3.58%) [19,47], FU (28; 4.0%) [23,48,49], partial surgery (28, 4.0%) [23,29,50], 0.015% IngMeb (17; 2.43%) [19,40], 50% TCA (10; 1.43%) [23] and ALA-PDT plus excimer dye laser (1; 0.14%) [30]. Of the shortlisted articles, 31 investigated one therapy alone, treating 533 patients in total [18,20,22,24–26,32,33,35–53]. In four articles [13,29–31], the efficacy of two modalities has been compared in 198 treated areas of 142 patients: one study compared laser-PDT (erbium-doped yttrium aluminium garnet (Er:YAG) and MAL-PDT) to MAL PDT alone [31], one study compared two different methods of CO_2 laser [13], one study compared dye laser to laser and ALA-PDT [30], and one study compared CO_2 laser to ELD with high energy [29]. Only one paper (30 patients) compared the efficacy of three different therapies (5% IMI vs. 0.015% IngMeb vs. DHA) [19] and only one study (40 patients) compared the efficacy of four therapeutic approaches (5% FU vs. 50% TCA vs. CO_2 laser vs. lip shave) [23].

4.2. Outcomes

It must be noted at the outset that all treatment options included a limited number of cases, except for studies regarding PDT and laser therapy. Furthermore, outcomes were often evaluated only on a certain number of individuals and not on the whole study population who underwent each treatment.

4.3. Therapeutic Response

Thirty-one out of 36 studies reported complete clinical clearance of the treated area: a rate of 85.93% of patients [13,18–20,22–26,30–36,38–43,45–51,53]. Partial clinical response was observed in 25.37% of cases. Poor treatment effect and clinical deterioration were only described in a limited number of patients. The overall clinical recurrence rate, as estimated in 21 articles [13,20,22–25,30–33,38,39,41,42,45,46,48,50,51], was 11.24% of the treated areas. In terms of the histopathologic outcomes, a post-treatment biopsy was performed in 23 studies [13,18–20,23–26,29–36,41,44,47,49–51,53]. In 88.43% of the treated areas that were biopsied, evidence of complete response was reported, with 64.07% obtaining complete clearance.

4.4. Invasive Treatments

All laser treatments demonstrated excellent efficacy, with 93.39% (226/242) of patients included in the studies achieving complete clinical response, varying from 93.04% (214/230) for CO_2 laser ablation to 100% for Er: YAG (12/12) [13,23–26,29,30,37,46,50,52]. Complete histopathological response was reported in 96% of patients (72/75), whereas the recurrence rate was 6.42% (14/218). The number of AEs per patient treated with laser therapy was less than one (0.42/case), and the aesthetic result was deemed exceptional in 100% of cases. All patients who underwent partial surgery obtained complete response (14/14, 100%) and did not report any recurrences [23,29,50]. More than 80% of patients with histopathological evaluation achieved the relevant complete response (10/12; 83.33%), and the number of adverse events per patient was minimal (0.3/case).

4.5. Non-Invasive Treatments

An almost 80% (27/34; 79.41%) complete clinical response was achieved among patients treated with MAL-PDT and 5% IMI, whereas complete histopathological response reached 64.71% (22/34). The recurrence rate was 5.88% (2/34), and the number of adverse events per patient was 5.4, with two patients (5.88%) discontinuing treatment due to side effects [33]. Regarding laser application in conjunction with PDT, the relevant studies reported complete clinical response in 75.76% (25/33) of patients, varying between 68.4% for ALA PDT-dye laser and 85.7% for MAL PDT-Er: YAG laser, while the clinical recurrence rate was calculated at 6.1% (2/33) [30,31]. When FU was applied, complete clinical response was obtained in a satisfactory percentage of patients (21/28; 75.0%), with 1% FU demonstrating excellent performance (100% complete response) compared to 5% FU (68.21%); the recurrence rate, however, was quite high (7/22; 31.8%). In the histopathological follow-up evaluation, 5/6 (83.33%) patients achieved partial clearance, and 1/6 (16.67%) reported poor response. Treatment discontinuation due to adverse events was noted in 10.0% of patients [23,48,49]. During 5% IMI therapy, complete clinical response was experienced by 76.0% (19/25) of subjects, while the number of AEs per patient was calculated at 3.1 [19,47]. Photodynamic therapy was a treatment option that was thoroughly investigated. A total of 148 cases out of 222 (66.67%) across the included studies achieved complete clinical response. The MAL daylight treatment outperformed all other approaches (82.63%), whereas methyl-aminoxopentanoate-PDT had the lowest scores (55.6%). Furthermore, PDT with aminolevulinic acid rather than with methyl-aminolevulinate produced better clinical results (73.49% vs. 63.81% complete response rate, respectively). Complete histopathological response was obtained in 49.48% (48/97) of patients. Reflecting the clinical outcomes, ALA-PDT performed better than MAL-PDT (53.24% and 23.41%, respectively) in the histopathological evaluation. The cumulative recurrence rate of the relevant studies was calculated to be 14.07% (19/135). The number of side effects per patient was estimated to be 2.4, whereas the treatment interruption due to AEs was reported in 5.86% (13/222) of patients. Excellent cosmetic results were recorded in 92 out of 136 (67.65%) subjects [18,20,31,32,34–36,39,41–45]. The complete response rate was 41.18% (7/17) through ingenol mebutate therapy. No recurrences were observed throughout the follow-up period. All patients experienced adverse events; however, none of them discontinued treatment, as the side effects reported were mild and resolved within a maximum of two weeks without any medical intervention [19,40]. When patients were treated with DHA, almost half of them (28/62; 45.16%) obtained complete clinical response, whereas complete histopathological response was observed in 66.67% (4/6) of the assessed cases. Regarding the aesthetic results, all respondents (6/6) rated the outcomes as excellent. The recurrence rate was estimated in 6.52% (3/46) of cases, whereas seven patients discontinued treatment due to AEs (15.22%) [19,22,38,53]. Finally, after 50% TCA application, only 30% of cases (3/10) achieved complete clinical clearance [23].

4.6. Cosmetic Outcome

Seventeen articles [24–28,30–32,34–36,39,43,45,51,53,55] assessed the cosmetic result, which was described as excellent in 74.63% of patients. The cosmetic outcome depicted the patients' perspective in two studies [28,36] and the physician's evaluation in 14 studies [24–27,30–32,34,35,39,43,45,51,53,55], while, in one study, both measurements were used [34]. All patients treated with laser ablation or DHA reported excellent aesthetic results with no scarring. In a study of two surgical techniques, W-plasty outperformed vermilionectomy in terms of scar retraction [55]. Excellent cosmetic outcomes ranged from 58% to 88% in patients who underwent vermilionectomy [27,28]. The relevant PDT results varied widely: one study reported excellent results in 60% of cases, while fair or poor outcomes were observed in 40% of cases [31], two papers reported excellent outcomes in nearly 80% of cases [32,34] and another study reported very good results in 33% of patients [36].

4.7. Healing Time in Different Studies

The mean healing time estimated in 13 articles [13,18,24–26,29,30,37,49–52] was 2.8 weeks (Range: 0.4–4 weeks). The healing time was primarily reported for CO_2 laser therapy, followed by MAL-PDT, and was reported less for 5% IMI, 1% FU, ELD, dermabrasion, a combination of dye-laser and ALA-PDT, and for Er: YAG laser.

4.8. Adverse Events

Twenty-four papers provided data on AEs [13,18–20,26,27,31–37,39,40,43–47,50,51,53] with a total of 1027 AEs experienced by 541 patients. The most common side effects included erythema, pain, edema and burning sensation. Moreover, based on information provided by three studies, a mean VAS pain value of 5.62 ± 1.75 in 69 patients was calculated (57 treated with PDT, 12 with Er: YAG) [26,35,44]. Overall, 16 patients discontinued treatment due to AEs, as reported in six studies [22,23,33,35,38,43]. Most AEs were mild to moderate in severity and subsided within two weeks post-treatment without therapeutic intervention [34,53]. Persistent AEs for up to one year were reported after surgical treatment and included labial tension and diminished sensitivity in nearly 36% of cases [27].

4.9. Post-Treatment Malignant Transformation

Malignant transformation after surgical treatment was examined in three longitudinal studies, and none of them reported any incidence [13,23,25]. Nevertheless, in a study where different approaches of CO_2 laser implementation were compared and in a case series of CO_2 vermilionectomy, low rates (1/43; 2.33% and 1/14; 7.14%) of malignant transformation in the treated areas were observed [29,37]. On the other hand, neither of the two reports [23,34] where patients had undergone non-invasive treatment for AC with long follow-up periods (1.5–4 years) could detect any case of post-treatment malignant transformation.

4.10. Assessment of Recurrence

Five articles reported the effect of laser ablation (four regarding CO_2 laser and one regarding Er: YAG laser ablation) [13,23–26]. Of those, in only one study of 40 cases, the recurrence rate reached 13% [13]. On the other hand, no recurrence was observed through vermilionectomy in any of the relevant studies [27,28,55]. Topical treatments presented a high recurrence rate with diclofenac 3% gel obtaining 33% [53] and FU application 55% [23] recurrence rates. Both studies regarding laser treatment combined with PDT presented recurrence rates of 8% [30,31]. Six papers presented the results of PDT monotherapy [18,31,32,34–36], with one those presenting the results of PDT combined with 5% IMI [33]; the latter achieved the lowest recurrence rate (12%). Although, in two papers on PDT monotherapy, no recurrences have been observed, in the remaining four articles, 25–60% of patients suffered from AC recurrence. Three randomized controlled trials compared the outcomes of various therapies. The first compared topical FU application, chemical peel, vermilionectomy and CO_2 laser ablation [23]. Among the four modalities, chemical peel obtained the highest recurrence rate, which reached 70%, while patients treated with CO_2 laser ablation or vermilionectomy did not experience any recurrence. The second RCT compared laser-assisted PDT to PDT alone, and it was proved that laser-assisted PDT outperformed PDT monotherapy in terms of recurrence rates (8% vs. 50%, respectively) [31]. Lastly, the third RCT suggested that W-plasty compared to classic vermilionectomy were equally satisfying when recurrence rates were concerned [55].

5. Risk of Bias and Quality of the Shortlisted Studies

The risk of bias in individual papers was determined as per the Cochrane Reviews recommendations using the updated RoB-2 tool [56]. As stated in the Cochrane handbook, "a bias is a systematic error, or deviation from the truth, in results or inferences, which means that multiple replications of the same study would reach the wrong answer on average". Six kinds of bias were assessed, namely: 1. Selection bias, when the study population does not represent the target population. 2. Performance bias, when the conduct of a study negligently introduces differences between randomized groups other than the intervention being investigated. 3. Attrition bias, when subjects are lost to follow-up, or they miss at least one measurement time points during the study period. 4. Reporting bias, when a trial reports only part of its estimated outcomes. 5. Other sources of bias. 6. Overall.

Bias analysis showed that all articles included in the review presented a high risk of bias. Only two studies' designs proceeded to patient randomization [13,31]. None of the studies followed the blinding process either for therapeutic intervention or evaluation. The follow-up time varied between 3 and 48 months across studies. All articles reporting a follow-up time of fewer than 8 months had histopathological verification of AC clearance. Most trials were non-randomized, observational cohort studies. Even though the number of randomized clinical trials was fairly limited, the data obtained from these papers suggest that laser ablation or vermilionectomy demonstrate the lowest recurrence rates. Finally, given the small number of trials testing each therapeutic approach and the heterogeneity of results and study design, statistical comparison or a meta-analysis was deemed inappropriate.

6. Discussion

In this systematic review, we demonstrated the results of 36 studies of AC treatment and assessed the relevant outcomes of the surgical (laser ablation and vermilionectomy) and topical therapeutic (5-FU, diclofenac gel and PDT) approach. We concluded that the best response was obtained through partial surgery and laser therapy, either alone or combined with PDT, with MAL PDT + 5% IMI, FU and PDT alone achieving lower clearance rates. Our findings derived from pool data analysis and individual studies whenever the number of studies for each aspect was not enough to accrue meaningful conclusions.

A recent consensus established an international core outcome set for clinical trials on actinic keratosis treatment based on physician and patient Delphi surveys [54]. The significance of the aesthetic result and adverse events were ranked, among other factors, lower by patients compared to physicians. This could be explained by the composition of the study population, which included patients with a history of skin cancer; therefore, recurrence was inevitably deemed more important than the cosmetic outcome. Furthermore, the localization of the actinic keratoses was not determined in the consensus. It may thus be possible for cosmetic outcomes to be of particular relevance for patients with AC due to the visibility of the lesions.

It should be noted that, so far, the studies on AC do not concentrate on patients' future treatment preferences; this is made apparent in the available literature. Only one study investigated patient treatment satisfaction, with 80% of cases considering the treatment as beneficial [21]. Patient satisfaction should be more closely examined, at least regarding surgical versus topical treatment, considering the invasive nature of vermilionectomy and laser-assisted therapies.

Even though the number of patients for each treatment was low in each study, the cumulative number of areas treated with each therapy was significant. Specifically, at least 200 cases were treated with laser therapies and PTD alone, and sixty-two areas with DHA. Laser therapy—mainly CO_2 laser treatment—outperformed the other therapeutic options in all aspects, including high rates of complete response and low recurrence rates. Carbon dioxide laser ablation seems to be linked to fewer side effects with a shorter time to resolution than vermilionectomy, although head-to-head trials are lacking [23,55]. Unlike non- or minimally invasive treatments, though, vermilionectomy has the added benefit of enabling the histopathological evaluation of the lesion. Novel surgical procedures such as W-plasty may provide comparable results to conventional vermilionectomy with better cosmetic outcomes. Our results regarding the beneficial effect of CO_2 laser therapy for actinic cheilitis are in line with the guidelines for the SCC from the National Comprehensive Cancer Network of the U.S.A., which suggest ablative laser vermilionectomy as a recommended treatment for extensive AC [57]. On the other hand, there are no relevant European and British guidelines for the management of AC.

Concerning other therapeutic options, PDT scored relatively lower in terms of complete response than other treatments such a laser therapy, FU and 5% IMI, while over 12% of patients suffered from a recurrence. Moreover, DHA showed an exceptionally low complete response rate, whereas the recurrence rate is as good as laser therapy. Both treatments seem to be less efficient, possibly due to crusts which may impede the therapeutic effect and patient compliance.

As regards the other therapies, the relevant results should be interpreted with caution, given the small study population of each paper. On the whole, the subsequent employment of two different modalities seems to have a synergistic effect on the therapeutic outcome of each treatment. By way of example, unlike the application of 5% IMI and MAL-PDT alone, which were characterized by modest results, the combination of MAL-PDT with 5% IMI yielded an almost 80% complete response. One possible explanation for this superior efficacy may lie in their different modes of action, since PDT selectively destroys cancer cells and IMI enhances the immune response. Sotiriou et al. suggested that the inflammation generated after PDT could promote the activation of innate immunity against the malignant cells through the action of 5% IMI [33]. Finally, the different evaluation criteria of the histopathologic response across studies and the post-treatment biopsies at different time points do not permit a valid clinicopathological correlation of the treatment outcome. Nevertheless, most therapies provided a satisfactory safety profile with few side effects and excellent aesthetic results.

Minimally invasive therapies have the additional advantage of higher patient compliance rates, unlike topical treatments, which are more prone to discontinuation. More sophisticated procedures that minimize side effects may make the surgical approach a more attractive and beneficial option given its high efficacy. Therefore, it would be recommended that the risk–benefit ratio should be assessed for each of the therapeutic modalities during the treatment decision process, offering tailored management. In high-risk patients, for example, such as immunosuppressed patients or patients with a history of skin cancer, where definite and timely cures are essential, the surgical approach should be preferrable over topical treatment. Nevertheless, in younger, low-risk individuals, non-invasive therapies may be more appropriate. Patients' preferences and well-being should also be considered before making management decisions.

It should be remembered that the original purpose of AC therapy is to minimize the risk of the pre-malignant lesions evolving into a squamous cell carcinoma in the future. Most SCC cases reported in a study were diagnosed clinically, even when the diagnostic confidence was low [46]. Therefore, in these cases, the malignant lesion could pre-exist prior to treatment initiation. The other case series describing the malignant transformation of AC included patients with mostly moderate and severe epithelial dysplasia [29,37]. A possible explanation could be that the features of epithelial dysplasia create the conditions for potential malignancy. Whatever the case, well-designed studies exploring the malignization incidence are essential to provide accurate data for the various treatments, especially for topical therapies.

The results of our review should be viewed in light of some limitations. First, the sample size of most studies was small, while the substantial diversity in the quality and characteristics of presented evidence did not allow any direct comparison. Moreover, it should be noted that the data provided by the relevant studies were restricted only to positive outcomes, and the negative or ambiguous results of AC therapies could be ruled out. Third, there was heterogeneity in the histopathologic post-treatment evaluation and follow-up time points across the published studies. Lastly, none of the studies assessed cryotherapy as a therapeutic option, a treatment that has become a mainstay for the management of AC in everyday clinical practice.

7. Conclusions

Our review highlights the need for higher quality and more comprehensive studies in the field of AC management. However, given the available data, our review suggests that laser treatment alone or combined with PDT seems to offer the best clinical outcomes, while FU, 5% IMI and PDT alone or combined have a satisfactory therapeutic profile. Large, randomized controlled studies are necessary to validate the kinds of conclusions drawn from this review in terms of the efficacy and safety of the traditional therapies for AC so that dermatologists can select the optimum therapeutic approach for these patients.

Author Contributions: Conceptualization, E.S.; methodology, E.S. and K.B.; software, K.B.; validation, E.S. and D.A.; formal analysis, K.B.; investigation, K.B. and I.P.; resources, E.S. and D.A.; data curation, K.B. and I.P.; writing—original draft preparation, K.B.; writing—review and editing, E.S.; visualization, K.B.; supervision, E.S.; project administration, E.S.; funding acquisition, None. All authors have read and agreed to the published version of the manuscript.

Funding: This research received no external funding.

Institutional Review Board Statement: Not applicable.

Informed Consent Statement: Not applicable.

Conflicts of Interest: The authors declare no conflict of interest.

References

1. Jadotte, Y.T.; Schwartz, R.A. Solar cheilosis: An ominous precursor: Part I. Diagnostic insights. *J. Am. Acad. Dermatol.* **2012**, *66*, 173–184. [CrossRef]
2. Vieira, R.A.; Minicucci, E.M.; Marques, M.E.; Marques, S.A. Actinic cheilitis and squamous cell carcinoma of the lip: Clinical, histopathological and immunogenetic aspects. *An. Bras. Dermatol.* **2012**, *87*, 105–114. [CrossRef]
3. Rodríguez-Blanco, I.; Flórez, Á.; Paredes-Suárez, C.; Rodríguez-Lojo, R.; González-Vilas, D.; Ramírez-Santos, A.; Paradela, S.; Suárez Conde, I.; Pereiro-Ferreirós, M. Actinic Cheilitis Prevalence and Risk Factors: A Cross-sectional, Multicentre Study in a Population Aged 45 Years and Over in North-west Spain. *Acta Derm. Venereol.* **2018**, *98*, 970–974. [CrossRef]
4. Silva, L.V.O.; de Arruda, J.A.A.; Abreu, L.G.; Ferreira, R.C.; da Silva, L.P.; Pelissari, C.; Silva, R.N.F.; Nóbrega, K.H.S.; de Andrade, B.A.B.; Romañach, M.J.; et al. Demographic and Clinicopathologic Features of Actinic Cheilitis and Lip Squamous Cell Carcinoma: A Brazilian Multicentre Study. *Head Neck Pathol.* **2020**, *14*, 899–908. [CrossRef] [PubMed]
5. de Souza Lucena, E.; Costa, D.; Da Silveira, E.; Lima, K. Prevalence and factors associated to actinic cheilitis in beach workers. *Oral Dis.* **2012**, *18*, 575–579. [CrossRef] [PubMed]
6. Ito, T.; Natsuga, K.; Tanimura, S.; Aoyagi, S.; Shimizu, H. Dermoscopic features of plasma cell cheilitis and actinic cheilitis. *Acta Derm. Venereol.* **2014**, *94*, 593–594. [CrossRef]
7. Benati, E.; Pampena, R.; Bombonato, C.; Borsari, S.; Lombardi, M.; Longo, C. Dermoscopy and reflectance confocal microscopy for monitoring the treatment of actinic cheilitis with ingenol mebutate gel: Report of three cases. *Dermatol. Ther.* **2018**, *31*, e12613. [CrossRef]
8. Sober, A.J.; Burstein, J.M. Precursors to skin cancer. *Cancer* **1995**, *75*, 645–650. [CrossRef]
9. Lopes, M.L.; Silva, F.L., Jr.; Lima, K.C.; Oliveira, P.T.; Silveira, É.J. Clinicopathological profile and management of 161 cases of actinic cheilitis. *An. Bras. Dermatol.* **2015**, *90*, 505–512. [CrossRef]
10. Dancyger, A.; Heard, V.; Huang, B.; Suley, C.; Tang, D.; Ariyawardana, A. Malignant transformation of actinic cheilitis: A systematic review of observational studies. *J. Investig. Clin. Dent.* **2018**, *9*, e12343. [CrossRef] [PubMed]
11. Karia, P.S.; Morgan, F.C.; Ruiz, E.S.; Schmults, C.D. Clinical and incidental perineural invasion of cutaneous squamous cell carcinoma: A systematic review and pooled analysis of outcomes data. *JAMA Dermatol.* **2017**, *153*, 781–788. [CrossRef] [PubMed]
12. Holmkvist, K.A.; Roenigk, R.K. Squamous cell carcinoma of the lip treated with Mohs micrographic surgery: Outcome at 5 years. *J. Am. Acad. Dermatol.* **1998**, *38*, 960–966. [CrossRef]
13. de Peres, F.F.G.; Brandão, A.A.H.; Carvalho, Y.R.; Filho, U.D.; Plapler, H. A study of actinic cheilitis treatment by two low-morbidity CO_2 laser vaporization one-pass protocols. *Lasers Med. Sci.* **2009**, *24*, 375–385. [CrossRef]
14. Grada, A.; Feldman, S.R.; Bragazzi, N.L.; Damiani, G. Patient-reported outcomes of topical therapies in actinic keratosis: A systematic review. *Dermatol. Ther.* **2021**, *34*, e14833. [CrossRef]
15. de Carvalho, M.V.; de Moraes, S.L.D.; Lemos, C.A.A.; Santiago, J.F., Jr.; do Vasconcelos, B.C.E.; Pellizzer, E.P. Surgical versus non-surgical treatment of actinic cheilitis: A systematic review and meta-analysis. *Oral Dis.* **2019**, *25*, 972–981. [CrossRef] [PubMed]
16. Stroup, D.F.; Berlin, J.A.; Morton, S.C.; Olkin, I.; Williamson, G.D.; Rennie, D.; Moher, D.; Becker, B.J.; Sipe, T.A.; Thacker, S.B. Meta-analysis of observational studies in epidemiology: A proposal for reporting. *JAMA* **2000**, *283*, 2008–2012. [CrossRef]
17. Moher, D.; Liberati, A.; Tetzlaff, J.; Altman, D.G.; Group, P. Preferred reporting items for systematic reviews and meta-analyses: The PRISMA statement. *PLoS Med.* **2009**, *6*, e1000097. [CrossRef] [PubMed]
18. Chaves, Y.N.; Torezan, L.A.; Lourenço, S.V.; Neto, C.F. Evaluation of the efficacy of photodynamic therapy for the treatment of actinic cheilitis. *Photodermatol. Photoimmunol. Photomed.* **2017**, *33*, 14–21. [CrossRef]
19. Husein-ElAhmed, H.; Almazan-Fernandez, F.M.; Husein-ElAhmed, S. Ingenol mebutate versus imiquimod versus diclofenac for actinic cheilitis: A 6-month follow-up clinical study. *Clin. Exp. Dermatol.* **2019**, *44*, 231–234. [CrossRef]
20. Levi, A.; Hodak, E.; Enk, C.D.; Snast, I.; Slodownik, D.; Lapidoth, M. Daylight photodynamic therapy for the treatment of actinic cheilitis. *Photodermatol. Photoimmunol. Photomed.* **2019**, *35*, 11–16. [CrossRef]
21. de Bezerra, H.I.O.; Gonzaga, A.K.G.; da Silveira, É.J.D.; de Oliveira, P.T.; de Medeiros, A.M.C. Fludroxycortide cream as an alternative therapy for actinic cheilitis. *Clin. Oral Investig.* **2019**, *23*, 3925–3931. [CrossRef]
22. Gda, S.L.; Silva, G.F.; Gomes, A.P.; de Araújo, L.M.; Salum, F.G. Diclofenac in hyaluronic acid gel: An alternative treatment for actinic cheilitis. *J. Appl. Oral Sci.* **2010**, *18*, 533–537. [CrossRef]
23. Robinson, J.K. Actinic cheilitis: A prospective study comparing four treatment methods. *Arch. Otolaryngol. Head Neck Surg.* **1989**, *115*, 848–852. [CrossRef]
24. David, L.M. Laser vermilion ablation for actinic cheilitis. *J. Dermatol. Surg. Oncol.* **1985**, *11*, 605–608. [CrossRef]
25. Whitaker, D.C. Microscopically proven cure of actinic cheilitis by CO_2 laser. *Lasers Surg. Med.* **1987**, *7*, 520–523. [CrossRef]
26. Orenstein, A.; Goldan, O.; Weissman, O.; Winkler, E.; Haik, J. A new modality in the treatment of actinic cheilitis using the Er: YAG laser. *J. Cosmet. Laser Ther.* **2007**, *9*, 23–25. [CrossRef]
27. Sanchez-Conejo-Mir, J.; Bernal, A.P.; Moreno-Gimenez, J.; Camacho-Martinez, F. Follow-up of vermilionectomies: Evaluation of the technique. *J. Dermatol. Surg. Oncol.* **1986**, *12*, 180–184. [CrossRef] [PubMed]
28. Lustig, J.; Librus, H.; Neder, A. Bipedicled myomucosal flap for reconstruction of the lip after vermillionectomy. *Oral Surg. Oral Med. Oral Pathol.* **1994**, *77*, 594–597. [CrossRef]
29. Laws, R.A.; Wilde, J.L.; Grabski, W.J. Comparison of electrodessication with CO_2 laser for the treatment of actinic cheilitis. *Dermatol. Surg.* **2000**, *26*, 349–353. [CrossRef] [PubMed]

30. Alexiades-Armenakas, M.R.; Geronemus, R.G. Laser-mediated photodynamic therapy of actinic cheilitis. *J. Drugs Dermatol. JDD* **2004**, *3*, 548–551.

31. Choi, S.; Kim, K.; Song, K.H. Efficacy of ablative fractional laser-assisted photodynamic therapy for the treatment of actinic cheilitis: 12-month follow-up results of a prospective, randomized, comparative trial. *Br. J. Dermatol.* **2015**, *173*, 184–191. [CrossRef] [PubMed]

32. Suárez-Pérez, J.A.; López-Navarro, N.; Herrera-Acosta, E.; Aguilera, J.; Gallego, E.; Bosch, R.; Herrera, E. Treatment of actinic cheilitis with methyl aminolevulinate photodynamic therapy and light fractionation: A prospective study of 10 patients. *Eur. J. Dermatol.* **2015**, *25*, 623–624. [CrossRef] [PubMed]

33. Sotiriou, E.; Lallas, A.; Goussi, C.; Apalla, Z.; Trigoni, A.; Chovarda, E.; Ioannides, D. Sequential use of photodynamic therapy and imiquimod 5% cream for the treatment of actinic cheilitis: A 12-month follow-up study. *Br. J. Dermatol.* **2011**, *165*, 888–892. [CrossRef]

34. Sotiriou, E.; Apalla, Z.; Koussidou-Erremonti, T.; Ioannides, D. Actinic cheilitis treated with one cycle of 5-aminolaevulinic acid-based photodynamic therapy: Report of 10 cases. *Br. J. Dermatol.* **2008**, *159*, 261–262. [CrossRef] [PubMed]

35. Sotiriou, E.; Apalla, Z.; Chovarda, E.; Panagiotidou, D.; Ioannides, D. Photodynamic therapy with 5-aminolevulinic acid in actinic cheilitis: An 18-month clinical and histological follow-up. *J. Eur. Acad. Dermatol. Venereol.* **2010**, *24*, 916–920. [CrossRef]

36. Berking, C.; Herzinger, T.; Flaig, M.J.; Brenner, M.; Borelli, C.; Degitz, K. The efficacy of photodynamic therapy in actinic cheilitis of the lower lip: A prospective study of 15 patients. *Dermatol. Surg.* **2007**, *33*, 825–830. [CrossRef] [PubMed]

37. Zelickson, B.D.; Roenigk, R.K. Actinic cheilitis. Treatment with the carbon dioxide laser. *Cancer* **1990**, *65*, 1307–1311. [CrossRef]

38. Gonzaga, A.K.G.; de Oliveira, P.T.; da Silveira, É.J.D.; Queiroz, L.M.G.; de Medeiros, A.M.C. Diclofenac sodium gel therapy as an alternative to actinic cheilitis. *Clin. Oral Investig.* **2018**, *22*, 1319–1325. [CrossRef] [PubMed]

39. Radakovic, S.; Tanew, A. 5-aminolaevulinic acid patch-photodynamic therapy in the treatment of actinic cheilitis. *Photodermatol. Photoimmunol. Photomed.* **2017**, *33*, 306–310. [CrossRef]

40. Flórez, Á.; Batalla, A.; de la Torre, C. Management of actinic cheilitis using ingenol mebutate gel: A report of seven cases. *J. Dermatol. Treat.* **2017**, *28*, 149–151. [CrossRef]

41. Fai, D.; Romanello, E.; Brumana, M.B.; Fai, C.; Vena, G.A.; Cassano, N.; Piaserico, S. Daylight photodynamic therapy with methyl-aminolevulinate for the treatment of actinic cheilitis. *Dermatol. Ther.* **2015**, *28*, 355–368. [CrossRef] [PubMed]

42. Kim, S.K.; Song, H.S.; Kim, Y.C. Topical photodynamic therapy may not be effective for actinic cheilitis despite repeated treatments. *Eur. J. Dermatol.* **2013**, *23*, 917–918. [CrossRef]

43. Calzavara-Pinton, P.G.; Rossi, M.T.; Sala, R. A retrospective analysis of real-life practice of off-label photodynamic therapy using methyl aminolevulinate (MAL-PDT) in 20 Italian dermatology departments. Part 2: Oncologic and infectious indications. *Photochem. Photobiol. Sci.* **2013**, *12*, 158–165. [CrossRef] [PubMed]

44. Ribeiro, C.F.; Souza, F.H.; Jordão, J.M.; Haendchen, L.C.; Mesquita, L.; Schmitt, J.V.; Faucz, L.L. Photodynamic therapy in actinic cheilitis: Clinical and anatomopathological evaluation of 19 patients. *An. Bras. Dermatol.* **2012**, *87*, 418–423. [CrossRef] [PubMed]

45. Fai, D.; Romano, I.; Cassano, N.; Vena, G.A. Methyl-aminolevulinate photodynamic therapy for the treatment of actinic cheilitis: A retrospective evaluation of 29 patients. *G Ital. Dermatol. Venereol.* **2012**, *147*, 99–101.

46. Castiñeiras, I.; Del Pozo, J.; Mazaira, M.; Rodríguez-Lojo, R.; Fonseca, E. Actinic cheilitis: Evolution to squamous cell carcinoma after carbon dioxide laser vaporization. A study of 43 cases. *J. Dermatol. Treat.* **2010**, *21*, 49–53. [CrossRef]

47. Smith, K.J.; Germain, M.; Yeager, J.; Skelton, H. Topical 5% imiquimod for the therapy of actinic cheilitis. *J. Am. Acad. Dermatol.* **2002**, *47*, 497–501. [CrossRef]

48. Epstein, E. Treatment of lip keratoses (actinic cheilitis) with topical fluorouracil. *Arch. Dermatol.* **1977**, *113*, 906–908. [CrossRef]

49. Warnock, G.R.; Fuller, R.P., Jr.; Pelleu, G.B., Jr. Evaluation of 5-fluorouracil in the treatment of actinic keratosis of the lip. *Oral Surg. Oral Med. Oral Pathol.* **1981**, *52*, 501–505. [CrossRef]

50. Dufresne, R.G., Jr.; Garrett, A.B.; Bailin, P.L.; Ratz, J.L. Carbon dioxide laser treatment of chronic actinic cheilitis. *J. Am. Acad. Dermatol.* **1988**, *19*, 876–878. [CrossRef]

51. Conejo-Mir, J.; Garciandia, C.; Casals, M.; Artola, J.; Navarrete, M. CO_2 laser treatment of actinic cheilitis. Comparison with surgical vermib'onectomy and evaluation of the technique. *J. Dermatol. Treat.* **1995**, *6*, 85–88. [CrossRef]

52. Neder, A.; Nahlieli, O.; Kaplan, I. CO_2 Laser Used in Surgical Treatment of Actinic Cheilitis. *J. Clin. Laser Med. Surg.* **1992**, *10*, 373–375. [CrossRef] [PubMed]

53. Ulrich, C.; Forschner, T.; Ulrich, M.; Stockfleth, E.; Sterry, W.; Termeer, C. Management of actinic cheilitis using diclofenac 3% gel: A report of six cases. *Br. J. Dermatol.* **2007**, *156*, 43–46. [CrossRef] [PubMed]

54. Reynolds, K.A.; Schlessinger, D.I.; Vasic, J.; Iyengar, S.; Qaseem, Y.; Behshad, R.; DeHoratius, D.M.; Denes, P.; Drucker, A.M.; Dzubow, L.M.; et al. Core Outcome Set for Actinic Keratosis Clinical Trials. *JAMA Dermatol.* **2020**, *156*, 326–333. [CrossRef]

55. Rossoe, E.W.T.; Tebcherani, A.J.; Sittart, J.A.; Pires, M.C. Actinic cheilitis: Aesthetic and functional comparative evaluation of vermilionectomy using the classic and W-plasty techniques. *An. Bras. Dermatol.* **2011**, *86*, 65–73. [CrossRef] [PubMed]

56. Sterne, J.A.C.; Savović, J.; Page, M.J.; Elbers, R.G.; Blencowe, N.S.; Boutron, I.; Cates, C.J.; Cheng, H.-Y.; Corbett, M.S.; Eldridge, S.M.; et al. RoB 2: A revised tool for assessing risk of bias in randomised trials. *BMJ* **2019**, *366*, l4898. [CrossRef] [PubMed]

57. Fisher, K.; Gastman, B.; Ghosh, K.; Grekin, R.C.; Grossman, K.; Ho, A.L.; Lewis, K.D.; Loss, M.; Lydiatt, D.D.; Messina, J.; et al. NCCN Guidelines Index Table of Contents Discussion NCCN Guidelines Version 2.2018 Panel Members Squamous Cell Skin Cancer. Available online: http://oncolife.com.ua/doc/nccn/Squamous_Cell_Skin_Cancer.pdf (accessed on 17 April 2021).

22, 14, 5614

MDPI

Article

Brachytherapy in the Treatment of Non-Melanoma Skin Peri-Auricular Cancers—A Retrospective Analysis of a Single Institution Experience

Mateusz Bilski [1,2,3], Paweł Cisek [1,2], Izabela Baranowska [4], Izabela Kordzińska-Cisek [5], Nina Komaniecka [6], Anna Hymos [7], Ewelina Grywalska [7] and Paulina Niedźwiedzka-Rystwej [6,*]

1 Department of Radiotherapy, Medical University of Lublin, 20-093 Lublin, Poland
2 Department of Brachytherapy, St. John's Cancer Centre, 20-090 Lublin, Poland
3 Department of Radiotherapy, St. John's Cancer Centre, 20-090 Lublin, Poland
4 Department of Medical Physics, St. John's Cancer Centre, 20-090 Lublin, Poland
5 Department of Clinical Oncology, St. John's Cancer Centre, 20-090 Lublin, Poland
6 Institute of Biology, University of Szczecin, 71-412 Szczecin, Poland
7 Department of Experimental Immunology, Medical University of Lublin, 20-093 Lublin, Poland
* Correspondence: paulina.niedzwiedzka-rystwej@usz.edu.pl

Simple Summary: Skin cancer is one of the most common cancers worldwide. Non-melanoma skin neoplasms in the head and neck area, the location of lesions around the ear, accounts for approximately 13–15% of all cases. The problematic location of neoplasms within the auricle and around the ear often causes many problems in surgical treatment, which can lead to unsatisfactory cosmetic effects or the presence of positive surgical margins. The presence of positive surgical margins leads to recurrence in about 10–67% of such patients. The aim of the study was to analyse the effectiveness, toxicity profile, and cosmetic effect of two different brachytherapy techniques (contact and interstitial brachytherapy). In our study, we analysed the results of 33 patients treated with HDR contact or interstitial brachytherapy. We showed that this is a highly effective, short, and relatively low burden on patients with cancer of the outer ear, involving the auricle and the skin of the adjacent area. The toxicity of the treatment was low.

Citation: Bilski, M.; Cisek, P.; Baranowska, I.; Kordzińska-Cisek, I.; Komaniecka, N.; Hymos, A.; Grywalska, E.; Niedźwiedzka-Rystwej, P. Brachytherapy in the Treatment of Non-Melanoma Skin Peri-Auricular Cancers—A Retrospective Analysis of a Single Institution Experience. *Cancers* **2022**, *14*, 5614. https://doi.org/10.3390/cancers14225614

Academic Editor: Primož Strojan

Received: 8 September 2022
Accepted: 7 November 2022
Published: 15 November 2022

Publisher's Note: MDPI stays neutral with regard to jurisdictional claims in published maps and institutional affiliations.

Abstract: The location of skin neoplasms in the area of the ears qualifies patients to the so-called high-risk group. The location of neoplasms within the auricle and around the ear often causes many problems in surgical treatment. This is due to the presence of cartilage, the difficulty of performing procedures with obtaining a visually satisfactory cosmetic effect, especially in the presence of extensive lesions and can lead to positive surgical margins which leads to a high risk of recurrence. In such cases, the use of brachytherapy, both as an independent method and as a complementary method after surgery, may be an effective method of local control with an acceptable risk of radiation complications. However, there are no large retrospective studies on the use of brachytherapy in this anatomical region. The aim of the study was to analyse the effectiveness, toxicity profile, and cosmetic effect of two different brachytherapy techniques (contact and interstitial brachytherapy). Methods: This paper presents the results of a retrospective analysis of 33 patients treated with contact or interstitial high-dose-rate (HDR) brachytherapy for skin cancers of the outer ear, involving the auricle and the skin of the adjacent area. Brachytherapy was used both as a definitive treatment (15 patients—43%) and adjuvant treatment after surgery (18 patients—57%). The basic criterion for adjuvant treatment was a positive or narrow (<1 mm) resection margin. Fraction doses from 3 to 7 Gy per fraction were used at intervals from six hours (interstitial brachytherapy) to a maximum of seven days (contact brachytherapy). The treatment time ranged from 1 to 42 days, and the total dose range was 7 to 49 Gy. The follow-up was 29.75 months (range 2–64). Results: In the group of patients treated with adjuvant therapy, in the patients with post-radiation reaction, the mean time from surgery to the start of brachytherapy was 7.72 ± 3.05 weeks, the median was 8 (6–12) weeks, and in the group without post-radiation reaction, the mean time was 11.13 ± 4.41 weeks, the median time was 11 weeks (8–14). The risk of a post-radiation reaction increased significantly more often in patients with more advanced disease. In the case of contact brachytherapy, the post-radiation reaction occurred significantly more often (14/21 patients—43%) than in the case of interstitial brachytherapy (3/11 patients—9.4%). In patients with post-radiation reactions, a significantly larger volume of the skin receiving a dose of 200% was found, and the volume receiving a dose of 150% was close to statistical significance. The mean volume of the skin receiving a 200% dose in the group with post-radiation reactions was 28.05 ± 16.56 cm^3, the median was 24.86 (0.5–52.3) cm^3, and the mean volume in the group without post-radiation reaction was 17.98 ± 10.96 cm^3, median 14.95 (3.9–44.96) cm^3. The result was statistically significant ($Z = 2.035$, $p = 0.041$). Conclusion: Interstitial HDR (high-dose-rate) brachytherapy for non-melanoma skin cancers around the ear is highly effective, short, and has a relatively low burden on the patient. The toxicity of the treatment was low. In the case of contact brachytherapy, the toxicity profile is slightly higher but acceptable for patients. This method is preferred in patients in whom interstitial brachytherapy is impossible to perform due to anatomical and logistical reasons. The unquestionable advantage of contact brachytherapy is its ability to be performed on an outpatient basis without the need to stay in the hospital. No severe and late CTCAE $\geq$III and late RTOG $\geq$III toxicity was observed. In patients after surgery, in order to minimise the risk of radiation reaction, it is optimal to start treatment at least eight weeks after surgery. In the presence of extensive lesions, the use of interstitial brachytherapy seems to be more advantageous, especially when the expected volume of healthy skin in the dose range of 200% and 150% is above 15 cm^3 and 50 cm^3, respectively.

Keywords: HDR brachytherapy; non-melanoma skin cancer; cancer around the ear; radiation therapy

---

1. Introduction

Skin cancer is one of the most common cancers worldwide [1]. The term non-melanoma skin cancers includes the diagnosis of basal cell carcinoma (BCC) and squamous cell carcinoma (SCC) in nearly 99% of cases. Data on the prevalence of its non-melanoma forms are underestimated, but the World Cancer Research Fund reports that in 2020, nearly 1.2 million new cases were diagnosed, of which, over 722,000 were diagnosed in men and over 475,000 were diagnosed in women [2]. These neoplasms are most frequently diagnosed in Australia and the United States of America, and the most diagnoses per 100,000 inhabitants are in Australia and New Zealand, respectively, 140 and 127.5 [1]. Skin cancer in the US is the most common cancer with nearly 3.5 million newly diagnosed patients each year [3]. The latest survey conducted by the European Academy of Dermatology and Venereology (EADV) shows that approximately 7,304,000 Europeans are diagnosed with skin cancer, which is 1.71% of the adult European population [4]. A special group is patients with the location of neoplasms in the area of the ears as well as the central part of the face, which qualifies them for the so-called high-risk groups [5]. In non-melanoma skin neoplasms in the head and neck area, the location of lesions around the ear accounts for approximately 13–15% of all cases [6–8]. The problematic location of neoplasms within the auricle and around the ear often causes many problems in surgical treatment. This is due to the presence of cartilage, the difficulty of performing procedures with obtaining a visually satisfactory cosmetic effect, especially in the presence of extensive lesions, and an increased risk of the presence of positive surgical margins. This can lead to relapses and potential wound-healing problems. In patients with localisation of non-melanoma skin cancers within the head and neck, the percentage of incomplete resections ranges from 3–10.8%, depending on the surgical technique [6,7]. According to various sources, the presence of positive surgical margins leads to recurrence in about 10–67% of such patients, compared to 5–14% when the resection is complete [9–11]. The use of brachytherapy in this localisation as a method of conservative treatment seems to be a particularly attractive method. Data on the use of brachytherapy in this area are limited and refer to individual cases. Few reports from recent years also include the use of teleradiotherapy in this area [12–17]. The potential use of interstitial brachytherapy and/or contact brachytherapy in this group of patients provides issues for comparative analysis and facilitates the selection of the optimal technique. The aim of the study was to analyse the effectiveness, toxicity profile, and cosmetic effect of two different brachytherapy techniques (contact and interstitial brachytherapy).

2. Materials and Methods

2.1. Patient Characteristics

A group of 33 patients (14 men and 19 women) with skin cancer involving the ear and peri-auricular localisation including the skin of the adjacent area was included in the study (Figure 1). These patients were treated with high-dose-rate (HDR) brachytherapy at the Brachytherapy Department of the Centre of Oncology of the Lublin Region from 2010–2020 (Lublin, Poland). Histopathological confirmation of skin cancer was obtained in all patients. Squamous cell carcinoma dominated (20 patients—57%), followed by basal cell carcinoma (nine patients—26%), and further undifferentiated (four patients—11%). Brachytherapy was used both as a definitive treatment (15 patients—43%) and adjuvant treatment after surgery (18 patients—57%). The basic criterion for adjuvant treatment was a positive or narrow (<1 mm) resection margin. Eligibility for definitive brachytherapy resulted from the patient's lack of consent to resection (eight patients), the inoperability of the neoplasm (three patients), or a health condition that prevented surgery (four patients). At the time of this analysis, no systemic treatment with checkpoint inhibitors and the Hedgehog pathway was available in Poland.

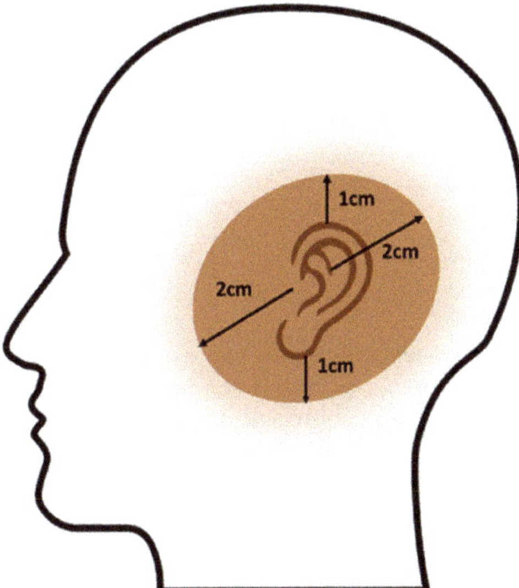

Figure 1. Lesion localisation area in all patients included in the study.

The tumour stage was determined on the TNM scale of the seventh edition. Lymph node and distant metastases were excluded in all patients. The majority of patients were T2 (15 patients—43%), then T3 (11 patients—31%), and then T1 (seven patients—20%). T4 patients were not eligible for brachytherapy. In the analysed group of patients, contact brachytherapy was dominant (21 patients—66%), and interstitial brachytherapy was used less frequently (11 patients—33%). Patient characteristics are presented in Table 1.

Table 1. Clinical and histopathological characteristics of patients undergoing brachytherapy.

Clinical and Histopathological Factor	Number of Patients (Percentage)	Mean Value	Median Value (Range)
Age (years)	33	80.79 +/− 9.76	83 (63–98)
Gender			
Men	14 (42%)	-	-
Women	19 (58%)	-	-
Histopathological type			
Squamous cell carcinoma	20 (57%)	-	-
Basal cell carcinoma	9 (26%)	-	-
Undifferentiated carcinoma	4 (11%)	-	-
Stage			
T1N0M0	7 (20%)	-	-
T2N0M0	15 (43%)	-	-
T3N0M0	11 (31%)	-	-
Largest lesion size (cm)	-	3.46 +/− 1.79	3 (0–7.5)
Definitive brachytherapy	15 (43%)	-	-
Brachytherapy after surgery	18 (57%)	-	-
Application of brachytherapy			
Contact brachytherapy	21 (66%)	-	-
Interstitial brachytherapy	11 (33%)	-	-
Time from surgery to the start of brachytherapy treatment (weeks)	19	9.16 +/− 2.54	8 (6–14)
Number of applicators	-	6.42 +/− 2.61	6 (3–15)

2.2. Procedure, Planning, and Treatment

Contact brachytherapy was performed by making an individual applicator from a thermoplastic mask and elastic applicators (by Varian Medical System) sewn or glued on it. The applicators were placed parallel to each other at 0.7–1.5 cm intervals covering the area of the neoplastic infiltration or postoperative scar with a 2 cm margin. The interstitial brachytherapy procedure was performed under local infiltration anaesthesia with the use of 1% Lignocaine. Flexible applicators of 35 cm in length from the Varian Medical System were used. They were inserted 2–5 mm under the skin, parallel to each other according to the Paris system assumptions so that they encompassed the tumour or tumour bed with a healthy skin margin of 1–2 cm. The applicators were placed parallel to each other at a distance of 0.7–1.5 cm. After the contact applicator or the interstitial procedure was performed, a treatment planning tomography was performed with the use of a 32-row Siemens computer tomograph. The layer thickness was from 1–3 mm. Based on the position of the applicators, the CTV (clinical target volume) area was drawn on the tomography scans for treatment planning. In the case of adjuvant brachytherapy, the dose was specified in accordance with the recommendations of GEC ESTRO/ACROP—5 mm from the applicator [18]. In case the thickness of the auricle was less than 5 mm, the dose was applied to its entire thickness. If possible, the applicators were placed on both sides of the auricle. This has allowed to avoid overdose and ensured a homogeneity of the dose. The critical organs were skin, bone, and brain.

2.3. Treatment Planning and Dosimetry Analysis

Patients were treated according to various fractionation schedules (Table 2). Fraction doses from 3 to 7 Gy per fraction were used at intervals from six hours (interstitial brachytherapy) to a maximum of seven days (contact brachytherapy). The treatment time ranged from 1 to 42 days, and the total dose range was 7 to 49 Gy. All doses were converted according to the linear-quadratic model into the biologically effective dose (BED) where the alpha/beta ratio was 10 Gy for the tumour and early radiation reaction and 3 Gy for the late radiation reaction. The dosimetric analysis is presented in Table 3. The choice of the scheme was dictated by a number of factors. The shorter schedule was used in elderly patients, in worse general conditions, and in less advanced disease. The shorter interval between fractions concerned interstitial procedure; the longer concerned contact procedure. Contact brachytherapy was selected in lesions with a maximum depth of 3 mm. Interstitial brachytherapy was chosen in lesions with a depth of more than 3 mm. Contact brachytherapy was used preferentially in the case of the auricle. Treatment planning was carried out using the Brachyvision treatment planning system. An Ir 192 source of 0.6 mm diameter and an average activity of 10 Ci was used. Treatment was performed with the 24-channel Gammamed or Gammamed Plus (Varian) apparatus. Figure 2 shows examples of CTV volume, critical organ contouring, and dose distribution with the histogram of one of the treated patients.

Table 2. Dose fractionation schedules applied in the treatment of all patients.

Dose Fractionation Schedules			
Total Number of Fractions	**Number of Fractions per Day**	**Dose per Fraction**	**Total Dose**
15	1	3 Gy	45 Gy
9	2 (interval between fractions of minimum 6 h)	5 Gy	45 Gy
14	2 (interval between fractions of minimum 6 h)	3 Gy	42 Gy
5	1	4 Gy	20 Gy
3	1 every 7 days	7 Gy	21 Gy

Table 3. Dosimetry characteristics of doses in particular target volumes and in the brain and associated bone. CTV—clinical target volume, V—volume, D—dose, BED—biologically effective dose.

	Variable	Mean Value	Median Value (Range)
General prescription	Total dose (Gy)	37.73 ± 12.31	45 (7–49)
	Fraction dose (Gy)	3.65 ± 0.88	3.5 (3–7)
	Number of fractions	11.18 ± 4.63	14 (1–15)
	Number of irradiation days	16.15 ± 10.4	19 (1–42)
Target dose	CTV D0.10cc [Gy]	18.14 ± 18.44	12.39 (5.78–88.01)
	CTV D0.10cc [Gy] × number of fractions	164.02 ± 113.61	147 (28.9–645.21)
	CTV D90 [Gy]	4.09 ± 0.87	3.8 (3.1–6.2)
	CTV D90 [Gy] × number of fractions	43.67 ± 17.34	49.5 (6.2–92.55)
	CTV D90 [Gy] × number of fractions BED 10	61.42 ± 25.96	68.78 (10.04–149.65)
	CTV D100[Gy]	2.75 ± 0.56	2.58 (2.04–4.44)
	CTV D100 [Gy] × number of fractions	29.72 ± 12.5	34.35 (3.74–66.6)

Table 3. *Cont.*

	Variable	Mean Value	Median Value (Range)
	CTV D100 [Gy] × number of fractions	37.93 ± 16.92	42.22 (5.14–96.17)
	CTV V100 [% of volume]	96.76 ± 2.72	97.71 (90.05–99.8)
	CTV V150 [% of volume]	53.6 ± 13.33	53.9 (25.3 −77.65)
	CTV V200 [%of volume]	22.86 ± 14.66	19.49 (0.5–52.31)
OAR's dose	Brain D1.00cc [Gy]	1.53 ± 0.69	1.49 (0.3–2.94)
	Brain D1.00cc [Gy] × number of fractions	16.46 ± 9.65	14.40 (2.7–33.6)
	Brain D1.00cc [Gy] × number of fractions BED 3	25.95 ± 17.21	21.77 (2.97–58.66)
	Brain D1.00cc [Gy] × number of fractions BED 10	19.31 ± 11.88	17.76 (2.78–41.12)
	Brain D0.10cc [Gy]	1.79 ± 0.83	1.72 (0.34–4.07)
	Brain D0.10cc [Gy] × number of fractions	19.04 ± 10.86	17.25 (3.06–36.9)
	Brain D0.10cc [Gy] × number of fractions BED 3	31.62 ± 20.42	27.77 (3.41–67.13)
	Brain D0.10cc [Gy] × number of fractions BED 10	22.81 ± 13.68	20.54 (3.16–45.98
	Bone D1.00cc [Gy]	2.92 ± 0.93	2.95 (0.82–5.44)
	Bone D1.00cc [Gy] × number of fractions	31.42 ± 14.49	37.20 (4.73–51.6)
	Bone D1.00cc [Gy] × number of fractions BED 3	63.25 ± 32.89	67.92 (9.4–137.65)
	Bone D1.00cc [Gy] × number of fractions BED 10	40.97 ± 19.83	46.42 (6.97–75.59)
	Bone D0.10cc [Gy]	3.63 ± 1.37	3.51 (1.05–8.14)
	Bone D0.10cc [Gy] × number of fractions	38.39 ± 17.91	42.6 (8.14–71.4)
	Bone D0.10cc [Gy] × number of fractions BED 3	86.50 ± 47.15	95.51 (12.75–184.57)
	Bone D0.10cc [Gy] × number of fractions BED 10	52.83 ± 26.38	56.01 (10.44–105.38)

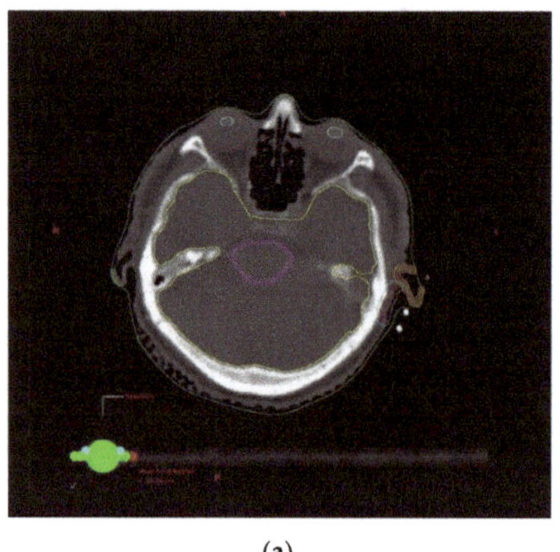

(a)

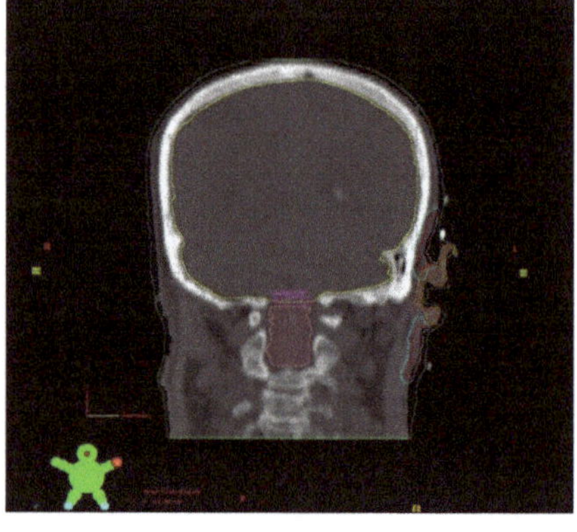

(b)

Figure 2. *Cont.*

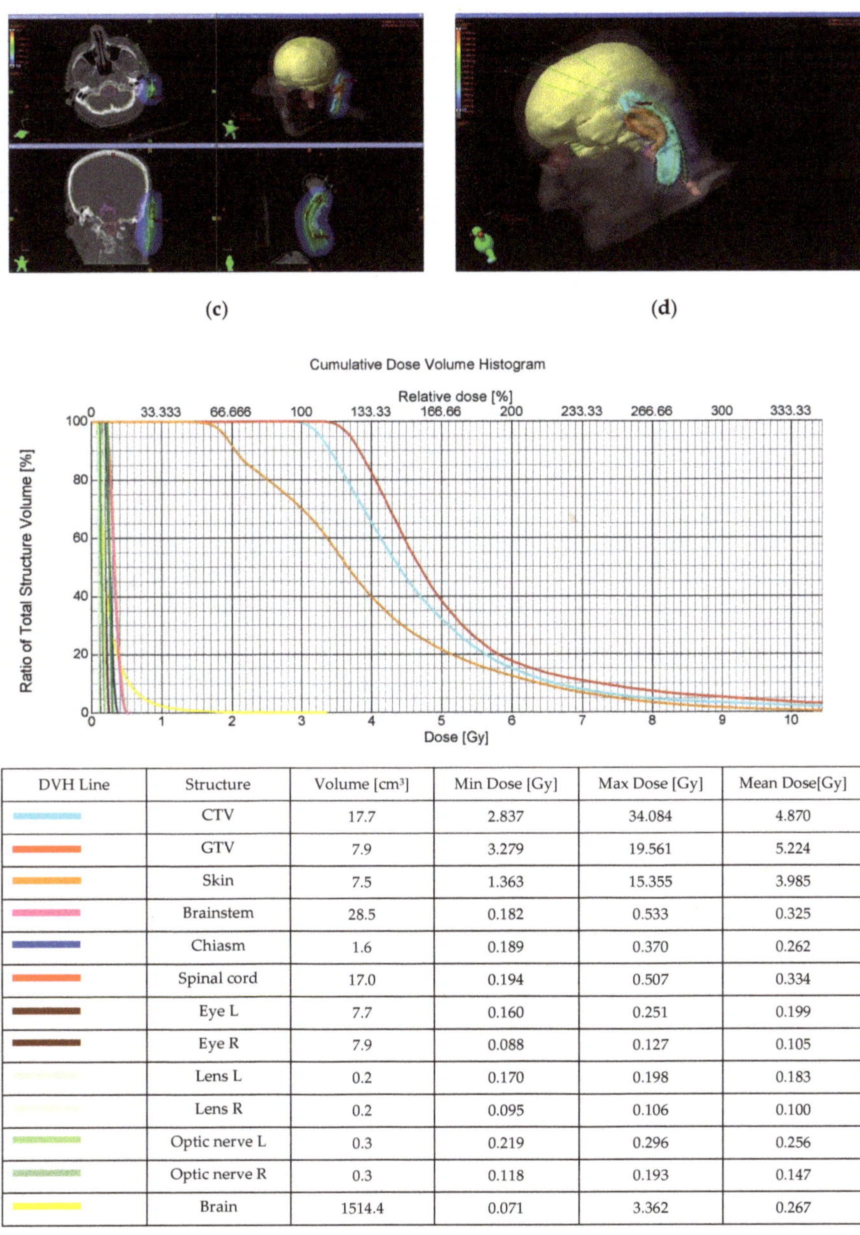

(c) (d)

Cumulative Dose Volume Histogram

DVH Line	Structure	Volume [cm³]	Min Dose [Gy]	Max Dose [Gy]	Mean Dose[Gy]
	CTV	17.7	2.837	34.084	4.870
	GTV	7.9	3.279	19.561	5.224
	Skin	7.5	1.363	15.355	3.985
	Brainstem	28.5	0.182	0.533	0.325
	Chiasm	1.6	0.189	0.370	0.262
	Spinal cord	17.0	0.194	0.507	0.334
	Eye L	7.7	0.160	0.251	0.199
	Eye R	7.9	0.088	0.127	0.105
	Lens L	0.2	0.170	0.198	0.183
	Lens R	0.2	0.095	0.106	0.100
	Optic nerve L	0.3	0.219	0.296	0.256
	Optic nerve R	0.3	0.118	0.193	0.147
	Brain	1514.4	0.071	3.362	0.267

(e)

Figure 2. (**a**) Delineation of the brachytherapy target volumes and organs at risk for a patient with non-melanoma skin cancer around the ear, transverse plane; (**b**) Delineation of the brachytherapy target volumes and organs at risk for a patient with non-melanoma skin cancers around the ear, coronal plane; (**c**) The 2D and 3D dose distribution of brachytherapy application; (**d**) A 3D reconstruction of the clinical target volume and CTV surface applicators; (**e**) Dose-volume histogram for target volumes and OARs.

2.4. Follow-Up after Treatment

Patients in the post-treatment period underwent cyclical clinical evaluation, initially monthly for up to three months; some evaluations also included ENT checking. In selected patients, periodic ultrasounds of the neck and abdominal cavity and chest X-rays were also performed. The patients were assessed for local recurrence, lymph node metastasis, and distant metastasis. The frequency and degree of the severity of early and late radiation reactions were also assessed. Due to the number of patients and the retrospective nature of the data, only the acute reaction was reported. The stage I reaction concerned all patients. In 16 patients (48%), stage II and III reactions were found. No stage IV radiation complications were found in the analysed group of patients. The RTOG scale was used to assess toxicity. Figure 3 illustrates the therapeutic process and the achieved effect during the first six months of follow-up after the completion of brachytherapy treatment in one of the patients.

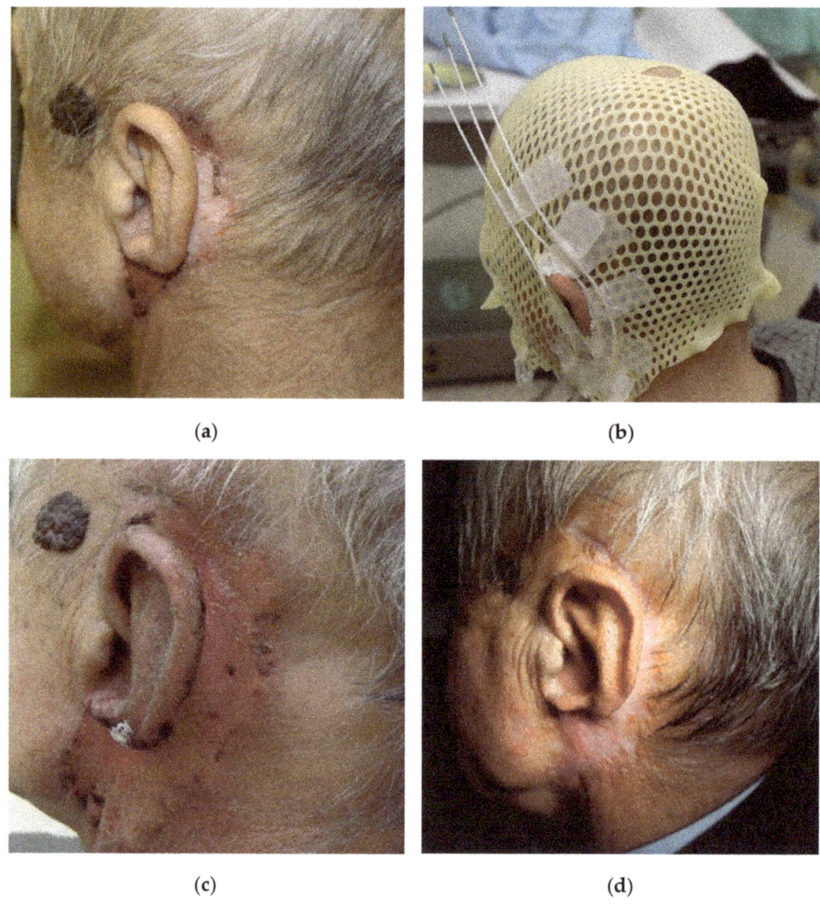

(a)   (b)

(c)   (d)

Figure 3. Sample photos of one of the treated patients (the same patient whose plan is presented in Figure 2). (**a**) the image of the primary lesion before the start of treatment; (**b**) the patient with a thermoplastic mask and sewn/glued applicators in the area of the neoplastic lesion; (**c**) the image of an acute radiation reaction, four weeks after the end of brachytherapy; (**d**) the effect obtained six months after the end of brachytherapy.

2.5. Statistical Analysis

Depending on the occurrence of a grade $\geq$II reaction, a statistical analysis of selected parameters for the risk of reaction was performed. When examining the relationship between the variables, the non-parametric Mann–Whitney U test (for independent variables, to compare the differences between the two groups of patients) was used. In order to compare the frequency of the analysed categories depending on the parameters tested, a non-parametric chi-square test was used. The significance level in all tests was $p = 0.05$. Statistical analysis was performed in the Statistica ver. 13.1 and 13.3 (StatSoft Poland).

3. Results

In the mean follow-up period of 29.75 months (range 2–64), in one patient (3%), there was a relapse of the disease. It was a local recurrence in the irradiated area. The treatment tolerance was good; only 16 (48%) patients had a clinically significant reaction (grade II and above).

The analysis of the influence of the tested parameters on the occurrence of the post-radiation reaction showed a statistically significant effect of the time from the surgical treatment to the start of irradiation. In the group of patients treated with adjuvant therapy, for the patients with post-radiation reactions, the mean time from surgery to the start of brachytherapy was 7.72 ± 3.05 weeks, and the median was 8 (6–12) weeks, and for the group without post-radiation reactions, the mean time was 11.13 ± 4.41 weeks, and the median time was 11 weeks (8–14). The result was statistically significant ($Z = 3.011$, $p = 0.002$) (Figure 4). The severity of the post-radiation reaction was also influenced by the tumour stage on the TNM scale (T parameter). The risk of a post-radiation reaction increased significantly more often in patients with more advanced disease ($Z = -2.447$, $p = 0.014$). The type of brachytherapy used also had an influence on the occurrence of the reaction. In the case of contact brachytherapy, the post-radiation reaction occurred significantly more often (14/21 patients—43%) than in the case of interstitial brachytherapy (3/11 patients—9.4%) $Z = 4.50$, $p = 0.339$. In the case of contact brachytherapy, the maximum depth of the treated lesion was 5 mm, and the average depth was 3.5 mm. The maximum isodose to the skin was 467.75% (in one pixel), the average isodose was 256%, and the median dose did not exceed 200% isodose.

In patients with post-radiation reaction, a significantly larger volume of the skin receiving the dose of 200% was found, and the volume receiving the dose of 150% was close to statistical significance. The mean volume of the skin receiving the 150% dose in the group with post-radiation reactions was 57.8 ± 13.17 cm^3, and the median was 59.53 (25.3–77.65) cm^3, and in the no-post-radiation reaction group, the mean volume was 49.64 ± 12.59 cm^3, and the median volume was 52.35 (30.1–71.69) cm^3 ($Z = 1.297$, $p = 0.053$). The mean volume of the skin receiving the 200% dose in a group with post-radiation reactions was 28.05 ± 16.56 cm^3, and the median was 24.86 (0.5–52.3) cm^3, and the mean volume in the group without post-radiation reactions was 17.98 ± 10.96 cm^3, and the median was 14.95 (3.9–44.96) cm^3. The result was statistically significant ($Z = 2.035$, $p = 0.041$) (Figures 5 and 6). The individual results and their statistical significance are presented in Tables 4 and 5.

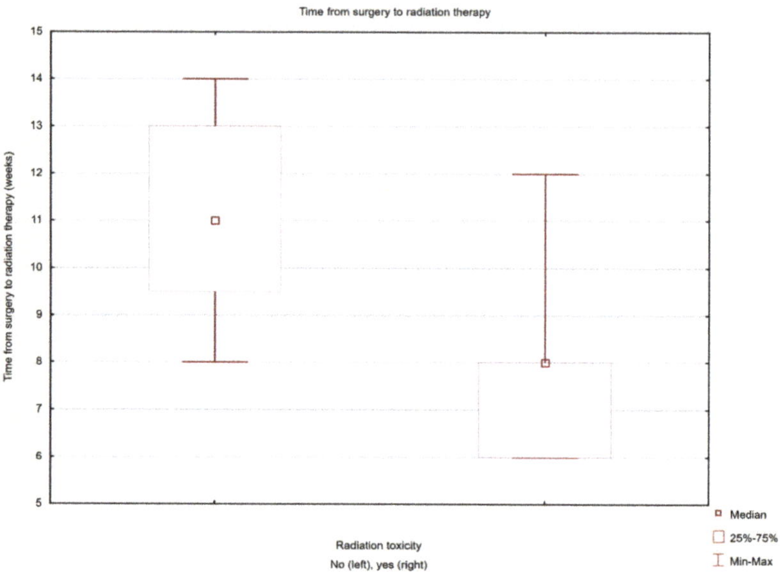

Figure 4. Analysis of the influence of time from surgery on the severity of post-radiation reaction.

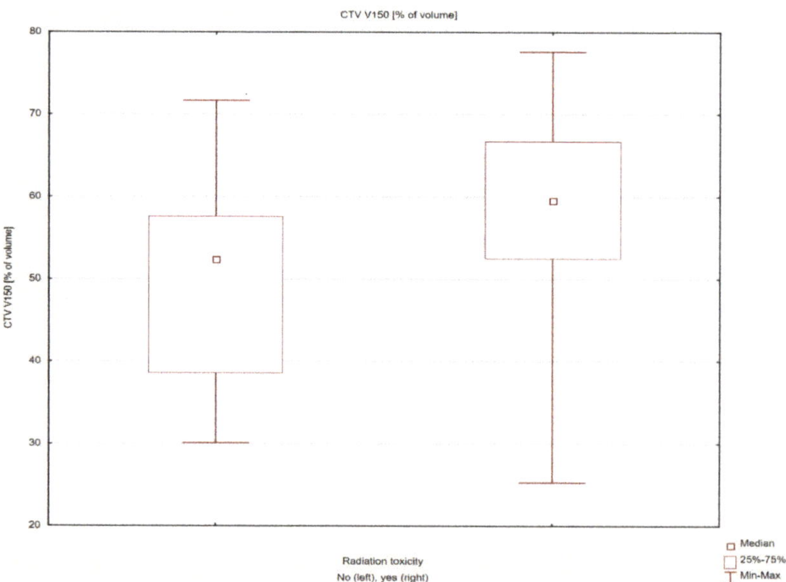

Figure 5. Analysis of the influence of the mean skin volume receiving a dose of 150% on the severity of post-radiation reaction.

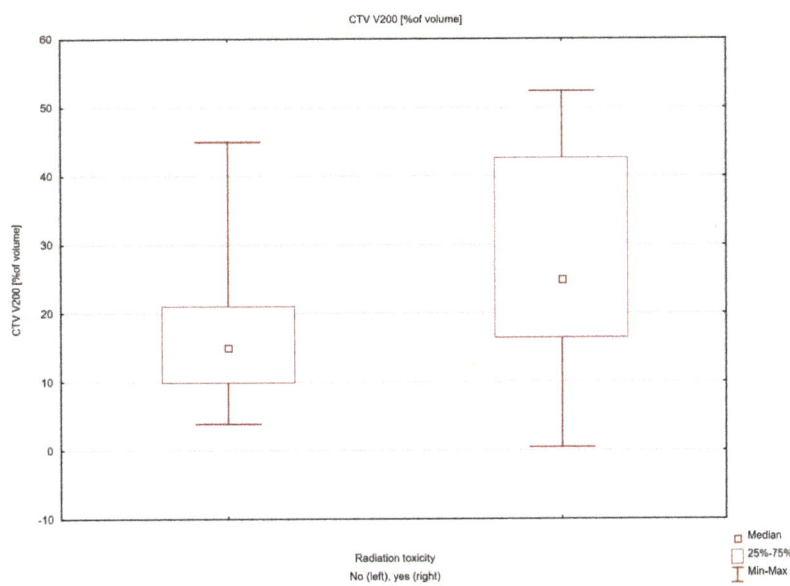

Figure 6. Analysis of the influence of the mean skin volume receiving the 200% dose on the severity of post-radiation reaction.

Table 4. Influence of selected factors on the severity of radiation reaction on the basis of the Mann–Whitney non-parametric U test. CTV—clinical target volume, D—dose, BED—biologically effective dose.

	Factor	**Z**	*p*
General information	Age	0.66713	0.504689
	Lesion size	−1.43029	0.152634
	Time from surgery to irradiation	3.01115	0.002603
	Number of applicators	−0.62662	0.530910
Prescription dose	Total dose	−0.34345	0.731262
	Fraction dose	−0.64187	0.520960
	Number of fractions	0.36689	0.713702
	Number of irradiation days	0.54918	0.582880
Target dose	CTV D0.10cc [Gy]	−0.88261	0.377447
	CTV D0.10cc [Gy] × number of fractions	−1.45901	0.144564
	CTV D90 [Gy]	−0.14411	0.885412
	CTV D90 [Gy] × number of fractions	0.21617	0.828857
	CTV D90 [Gy] × number of fractions BED 10	−0.14411	0.885412
	CTV D100 [Gy]	0.55848	0.576517
	CTV D100 [Gy] × number of fractions	0.30621	0.759444
	CTV D100 [Gy] × number of fractions BED 10	0.12609	0.899663

Table 4. *Cont.*

	Factor	*Z*	*p*
	CTV V100 [% of volume]	1.67530	0.093877
	CTV V150 [% of volume]	−1.92733	0.053939
	CTV V200 [%of volume]	−2.03541	0.041811
OAR's dose	Brain D1.00cc [Gy]	1.49528	0.134841
	Brain D1.00cc [Gy] × number of fractions	1.53106	0.125756
	Brain D1.00cc [Gy] × number of fractions BED 3	1.74721	0.080602
	Brain D1.00cc [Gy] × number of fractions BED 10	1.53106	0.125756
	Brain D0.10cc [Gy]	1.44112	0.149553
	Brain D0.10cc [Gy] × number of fractions	1.49503	0.134907
	Brain D0.10cc [Gy] × number of fractions BED 3	1.60311	0.108912
	Brain D0.10cc [Gy] × number of fractions BED 10	1.49503	0.134907
	Bone D1.00cc [Gy]	0.68459	0.493605
	Bone D1.00cc [Gy] × number of fractions	1.04481	0.296112
	Bone D1.00cc [Gy] × number of fractions BED 3	1.04481	0.296112
	Bone D1.00cc [Gy] × number of fractions BED 10	1.11687	0.264053
	Bone D0.10cc [Gy]	0.1802	0.856996
	Bone D0.10cc [Gy] × number of fractions	0.9368	0.34886
	Bone D0.10cc [Gy] × number of fractions BED 3	0.86474	0.387181
	Bone D0.10cc [Gy] × number of fractions BED 10	0.86474	0.387181

Table 5. Influence of selected factors on the severity of radiation reaction on the basis of the chi-square test. TNM—tumour, node, metastasis.

Feature	*Z*	*p*
Gender	0.75	0.387
Histopathological type	12.833	0.615
TNM	13	0.602
Indications for brachytherapy—individual vs. adjuvant after surgery	2.53	0.120
Application—contact vs. interstitial	4.5	0.034
Location—auricle vs. adjacent area of the auricle	0.76	0.383

Toxicity

The early and late toxicity of radiotherapy and the application procedure were analysed. Toxicity data were collected for all patients. CTCAE and RTOG scales were used. CTCAE (Common Terminology Criteria for Adverse Events) is an international standard for defining and categorizing adverse events, while RTOG (Toxicity criteria of the Radiation Therapy Oncology Group) is a scale that determines the degree of skin damage after radiotherapy.

Among the 33 analysed patients, 17 (52%) people according to CTCAE and 15 (45.5%) according to RTOG did not present any radiation reaction, and their skin remained unchanged. A total of 14 subjects (42%) had CTCAE grade I, and grade I RTOG was observed in 15 (45.5%) patients. Moderate adverse events occurred in 2 patients (6%) according to CTCAE. RTOG grade II reactions occurred in 3 patients (9%). None of the patients presented a reaction of grade III or higher in both the CTCAE and RTOG scores. A summary of the data on radiation reactions after the applied brachytherapy is presented in Tables 6 and 7.

Table 6. Characteristics of the early toxicity of radiotherapy and the brachytherapy procedure. CTCAE—Common Terminology Criteria for Adverse Events.

CTCAE (Degree of Severity)	Number of Patients (Percentage)
Grade 0 (none)	17 (52%)
Grade 1 (mild)	14 (42%)
Grade 2 (moderate)	2 (6%)
Grade 3 (severe)	-
Grade 4 (life-threatening)	-
Grade 5 (death)	-

Table 7. Characteristics of the toxicity of radiotherapy and the brachytherapy procedure. RTOG—toxicity criteria of the Radiation Therapy Oncology Group.

RTOG	Number of Patients (Percentage)
Grade 0	15 (45.5%)
Grade 1	15 (45.5%)
Grade 2	3 (9%)
Grade 3	-
Grade 4	-

4. Discussion

Patients with neoplasms of the head and neck region are at risk of late radiation complications, such as tissue fibrosis, oedema, cartilage and/or bone necrosis, and dry mucosa. The risk of radiation complications depends on the irradiated region, the total dose, and the duration of treatment [19]. Radiation complications that arise after radiotherapy of the outer ear occur after the administration of a conventionally fractionated total dose above 60 Gy [20]. These complications include ear pain, tinnitus, hearing impairment, and otitis exudate [21]. There are no new data on radiation complications from the external ear; the old data indicate a risk of cartilage necrosis, especially with lesions greater than 4 cm [22].

In the case of brachytherapy, due to the method of dose distribution, the real range of radiation is small, and it is not possible to exceed the tolerance dose for the middle and inner ear. The heterogeneity of the dose in the area of the auricle still remains a problem as it increases the risk of hot spots in the vicinity of the applicators. For this reason, the use of most of the available applicators (Valencia, Leipzig) or skin electronic brachytherapy does not allow to achieve a good dose distribution [23–26]. The solution to this problem is the use of individual applicators. In the work of Kuncman et al., individual mould applicators were used, composed of a mixture of paraffin and wax, obtaining a homogeneous dose distribution similar to IMRT [27].

Another solution is to use individual applicators attached to a thermoplastic mask [28]. The application of this method provides, with small and irregular changes, a more conformal dose distribution than electron brachytherapy with high treatment efficacy [29]. In the case of deeper lesions with a very irregular surface and different depths of infiltration, interstitial brachytherapy is used, using plastic tubes in one or more areas [30]. In the analysed group, patients with different severity of acute radiation reactions in a particularly difficult location—the outer ear and adjacent tissues—were compared. A group with a mild reaction and a group with an intense acute radiation reaction were separated and analysed. Factors that influence the presence of a reaction are the time from surgery to irradiation and the volume of high-dose areas, i.e., 150 and 200% of the dose. The technique of brachytherapy also turned out to be important in patients undergoing interstitial brachytherapy; the radiation reaction is lower.

The study also showed that the treatment was well tolerated; it was also characterised by low toxicity. There are no data in the literature describing in such a detailed manner brachytherapy in the area of the outer ear. In the study by Kuncman et al. [27], 10 patients with tumours of the auricle and external nose were irradiated. Dosimetry analysis showed good dose homogeneity in this area, comparable to EBRTF, which may indicate the low toxicity of this treatment. Similarly, in the study by Brovchuk et al., using various brachytherapy techniques, 751 patients were irradiated in the facial area, of which, 36 patients were affected by skin cancer in the area of the outer ear [26]. The treatment tolerance in the whole group was good, acute toxicity of grade II and higher concerned only 48% of patients, and cosmetic effects assessed as good and very good concerned 98% of patients. Contrary to the group of patients analysed in our study, the greater severity of the radiation reaction was observed in patients with squamous cell carcinoma compared to patients with basal cell carcinoma. Additionally, in contrast to the authors' analysis, the study of Brovchuk et al. did not show any differences in clinical outcomes depending on the applied brachytherapy technique (interstitial, contact) [26].

Similar studies with the use of brachytherapy in skin cancers also indicate the low toxicity of treatment and a good cosmetic effect, although they cover different locations within the head and neck [24,26,30]. The only study on external ear neoplasms deals with LDR (low-dose-rate) brachytherapy and does not include dosimetry analysis. Nevertheless, the results of this study indicate excellent local control and a good cosmetic effect, especially in the case of tumours smaller than 4 cm [22].

5. Conclusions

Interstitial HDR (high-dose-rate) brachytherapy for non-melanoma skin cancers around the ear is highly effective and short and has a relatively low burden on the patient. The toxicity of the treatment was low. In the case of contact brachytherapy, the toxicity profile is slightly less favourable but acceptable for patients. This method is preferred in patients in whom interstitial brachytherapy is impossible to perform due to anatomical and logistical reasons. The unquestionable advantage of contact brachytherapy is its ability to be performed on an outpatient basis without the need to stay in the hospital. No severe and late CTCAE $\geq$ III and late RTOG $\geq$ III toxicity was observed. In patients after surgery, in order to minimise the risk of radiation reaction, it is optimal to start treatment at least eight weeks after surgery. In the presence of extensive lesions, the use of interstitial brachytherapy seems to be more advantageous, especially when the expected volume of healthy skin in the dose range of 200% and 150% is above 15 cm^3 and 50 cm^3, respectively.

Author Contributions: M.B., P.C., I.B., I.K.-C., N.K., A.H., E.G. and P.N.-R., conceptualisation; M.B., P.C., I.K.-C. and I.B., data acquisition and interpretation; M.B., P.C., I.K.-C., I.B. and N.K., visualisation; M.B., P.C., I.K.-C., I.B. and N.K., writing—original draft preparation; E.G. and P.N.-R.; writing—review and editing; E.G., A.H. and P.N.-R., funding acquisition. All authors have read and agreed to the published version of the manuscript.

Funding: This research was funded by the Medical University of Lublin, grants no. PBmb164 and no. DS640.

Institutional Review Board Statement: The study was conducted according to the guidelines of the Declaration of Helsinki and approved by the Ethics Committee of the Medical University of Lublin, Poland (approval no. KE-0254/236/2020).

Informed Consent Statement: Informed consent was obtained from all subjects involved in the study. Written informed consent has been obtained from the patients to publish this paper.

Data Availability Statement: Due to privacy and ethical concerns, the data that support the findings of this study are available on request from the corresponding author, M.B.

Conflicts of Interest: The authors declare no conflict of interest.

References

1. Zhang, W.; Zeng, W.; Jiang, A.; He, Z.; Shen, X.; Dong, X.; Feng, J.; Lu, H. Global, regional and national incidence, mortality and disability-adjusted life-years of skin cancers and trend analysis from 1990 to 2019: An analysis of the Global Burden of Disease Study 2019. *Cancer Med.* **2021**, *10*, 4905–4922. [CrossRef] [PubMed]
2. World Cancer Research Fund International. Available online: https://www.wcrf.org/cancer-trends/skin-cancer-statistics/ (accessed on 12 May 2022).
3. Schmalbach, C.E.; Malloy, K.M. Head and Neck Cutaneous Cancer. *Otolaryngol. Clin. N. Am.* **2021**, *54*, 15–16. [CrossRef] [PubMed]
4. Richard, M.-A.; Paul, C.; Nijsten, T.; Gisondi, P.; Salavastru, C.; Taieb, C.; Trakatelli, M.; Puig, L.; Stratigos, A. The Burden of Skin Diseases [BOSD] in Europe: Preliminary results about skin cancers diagnosis and care pathway. In Proceedings of the Abstract no 353 Presented at EADV Spring Symposium 2022, Ljubljana, Slovenia, 12–14 May 2022.
5. Newlands, C.; Currie, R.; Memon, A.; Whitaker, S.; Woolford, T. Non-melanoma skin cancer: United kingdom national multidisciplinary guidelines. *J. Laryngol. Otol.* **2016**, *130*, S125–S132. [CrossRef] [PubMed]
6. Van Lee, C.B.; Roorda, B.M.; Wakkee, M.; Voorham, Q.; Mooyaart, A.L.; de Vijlder, H.C.; Nijsten, T.; van den Bos, R.R. Recurrence rates of cutaneous squamous cell carcinoma of the head and neck after Mohs micrographic surgery vs. standard excision: A retrospective cohort study. *Br. J. Dermatol.* **2019**, *181*, 338–343.
7. Dalal, A.J.; Ingham, J.; Collard, B.; Merrick, G. Review of outcomes of 500 consecutive cases of non-melanoma skin cancer of the head and neck managed in an oral and maxillofacial surgical unit in a District General Hospital. *Br. J. Oral. Maxillofac. Surg.* **2018**, *56*, 805–809. [CrossRef]
8. De Freitas, C.; Santos, A.N.; Bittner, G.C.; Sanabria, B.D.; Levenhagen, M.; Hans-Filho, G. Nonmelanoma Skin Cancer at Critical Facial Sites: Results and Strategies of the Surgical Treatment of 102 Patients. *J. Skin Cancer* **2019**, *2019*, 4798510. [CrossRef]
9. Rieger, K.E.; Linos, E.; Egbert, B.M.; Swetter, S.M. Recurrence rates associated with incompletely excised low-risk nonmelanoma skin cancer. *J Cutan Pathol.* **2010**, *37*, 59–67. [CrossRef]
10. Kumar, P.; Orton, C.I.; McWilliam, L.J.; Watson, S. Incidence of incomplete excision in surgically treated basal cell carcinoma: A retrospective clinical audit. *Br. J. Plast Surg.* **2000**, *53*, 563–566. [CrossRef]
11. Lara, F.; Santamaría, J.R.; Garbers, L.E. Recurrence rate of basal cell carcinoma with positive histopathological margins and related risk factors. *An. Bras. Dermatol.* **2017**, *92*, 58–62. [CrossRef]
12. Spigariolo, C.B.; Berti, E.; Brambilla, R.; Piccinno, R. Radiation therapy of non-melanoma skin cancer of the pinna: An Italian 35-year experience. *Ital. J. Dermatol. Venerol.* **2022**, *157*, 92–100. [CrossRef]
13. Foley, H.; Hopley, S.; Brown, E.; Bernard, A.; Foote, M. Conformal orbit sparing radiation therapy: A treatment option for advanced skin cancer of the parotid and ear region. *J. Med. Radiat. Sci.* **2016**, *63*, 186–194. [CrossRef] [PubMed]
14. Cisek, P.; Kieszko, D.; Bilski, M.; Dębicki, R.; Grywalska, E.; Hrynkiewicz, R.; Bębnowska, D.; Kordzińska-Cisek, I.; Rolińska, A.; Niedźwiedzka-Rystwej, P.; et al. Interstitial HDR Brachytherapy in the Treatment of Non-Melanocytic Skin Cancers around the Eye. *Cancers* **2021**, *13*, 1425. [CrossRef] [PubMed]
15. Tanous, A.; Tighe, D.; Bartley, J.; Gottschalk, G.; Gilmour, T.; Lotz, N.; Fogarty, G.B. Lesion-based radiotherapy of the ears, lips and eyelids for skin cancer. *Int. J. Radiol. Radiat. Ther.* **2021**, *8*, 32–42. [CrossRef]
16. Kim, J.W.; Yun, B.M.; Shin, M.S.; Kang, J.K.; Kim, J.; Kim, Y.S. Effectiveness of radiotherapy for head and neck skin cancers: A single-institution study. *Radiat. Oncol. J.* **2019**, *37*, 293–300. [CrossRef] [PubMed]
17. Tagliaferri, L.; Giarrizzo, I.; Fionda, B.; Rigante, M.; Pagliara, M.M.; Casà, C.; Parrilla, C.; Lancellotta, V.; Placidi, E.; Salvati, A.; et al. ORIFICE (Interventional Radiotherapy for Face Aesthetic Preservation) Study: Results of Interdisciplinary Assessment of Interstitial Interventional Radiotherapy (Brachytherapy) for Periorificial Face Cancer. *J. Pers. Med.* **2022**, *12*, 1038. [CrossRef] [PubMed]
18. Guinot, J.L.; Rembielak, A.; Perez-Calatayud, J.; Rodríguez-Villalba, S.; Skowronek, J.; Tagliaferri, L.; Guix, B.; Gonzalez-Perez, V.; Valentini, V.; Kovacs, G.; et al. GEC-ESTRO ACROP recommendations in skin brachytherapy. *Radiother. Oncol.* **2018**, *126*, 377–385. [CrossRef] [PubMed]
19. Brook, I. Late side effects of radiation treatment for head and neck cancer. *Radiat. Oncol. J.* **2020**, *38*, 84–92. [CrossRef] [PubMed]
20. Landier, W. Ototoxicity and cancer therapy. *Cancer* **2016**, *122*, 1647–1658. [CrossRef]
21. Nader, M.E.; Gidley, P.W. Challenges of hearing rehabilitation after radiation and chemotherapy. *J. Neurol. Surg. B Skull. Base* **2019**, *80*, 214–224. [CrossRef]
22. Mazeron, J.-J.; Ghalie, R.; Zeller, J.; Marinello, G.; Marin, L.; Raynal, M.; Le Bourgeois, J.-P.; Pierquin, B. Radiation therapy for carcinoma of the pinna using iridium 192 wires: A series of 70 patients. *Int. J. Radiat. Oncol. Biol. Phys.* **1986**, *12*, 1757–1763. [CrossRef]
23. Guix, B.; Finestres, F.; Tello, J.-I.; Palma, C.; Martinez, A.; Guix, J.-R.; Guix, R. Treatment of skin carcinomas of the face by high-dose-rate brachytherapy and custom-made surface molds. *Int. J. Radiat. Oncol. Biol. Phys.* **2000**, *47*, 95–102. [CrossRef]
24. Gauden, R.; Pracy, M.; Avery, A.-M.; Hodgetts, I.; Gauden, S. HDR brachytherapy for superficial non-melanoma skin cancers. *J. Med. Imaging Radiat. Oncol.* **2013**, *57*, 212–217. [CrossRef] [PubMed]

25. Pons-Llanas, O.; Ballester-Sánchez, R.; Celada-Álvarez, F.J.; Candela-Juan, C.; García-Martínez, T.; Llavador-Ros, M.; Botella-Estrada, R.; Barker, C.; Ballesta, A.; Tormo-Micó, A.; et al. Clinical implementation of a new electronic brachytherapy system for skin brachytherapy. *J. Contemp. Brachytherapy* **2015**, *6*, 417–423. [CrossRef] [PubMed]

26. Tormo, A.; Celada, F.; Rodriguez, S.; Botella, R.; Ballesta, A.; Kasper, M.; Ouhib, Z.; Santos, M.; Perez-Calatayud, J. Non-melanoma skin cancer treated with HDR Valencia applicator: Clinical outcomes. *J.Contemp. Brachytherapy* **2014**, *6*, 167–172. [CrossRef] [PubMed]

27. Kuncman, Ł.; Kozłowski, S.; Pietraszek, A.; Pietrzykowska-Kuncman, M.; Danielska, J.; Sobotkowski, J.; Łuniewska-Bury, J.; Fijuth, J. Highly conformal CT based surface mould brachytherapy for non-melanoma skin cancers of earlobe and nose. *J. Contemp. Brachytherapy* **2016**, *8*, 195–200. [CrossRef] [PubMed]

28. Sen, S.; Bandyopadhyay, A.; Pal, J.; Ghosh, A.; Deb, A. A dosimetric study of electron beam therapy vs. high-dose-rate mould brachytherapy in adjuvant treatment of non-melanoma skin carcinomas of the head and neck region. *J. Contemp. Brachytherapy* **2019**, *11*, 547–555.

29. Maroñas, M.; Guinot, J.L.; Arribas, L.; Carrascosa, M.; Tortajada, M.I.; Carmona, R.; Estornell, M.; Muelas, R. Treatment of facial cutaneous carcinoma with high-dose rate contact brachytherapy with customized molds. *Brachytherapy* **2011**, *10*, 221–227. [CrossRef]

30. Arguis, M.; Murcia-Mejía, M.; Henríquez, I.; Gómez, D.; Lafuerza, A.; Esteban, C.; León, R.; Polo, Y.; Grau, I.; Arenas, M. The role of brachytherapy in non-melanoma skin cancer treatment. *Radiother. Oncol.* **2012**, *103*, S477. [CrossRef]

Article

Basal Cell Carcinoma Treated with High Dose Rate (HDR) Brachytherapy—Early Evaluation of Clinical and Dermoscopic Patterns during Irradiation

Tomasz Krzysztofiak [1], Grażyna Kamińska-Winciorek [2,*], Andrzej Tukiendorf [3], Magdalena Suchorzepka [4] and Piotr Wojcieszek [1]

[1] Brachytherapy Department, Maria Sklodowska-Curie National Research Institute of Oncology (MSCNRIO), Gliwice Branch, ul. Wybrzeże Armii Krajowej 15, 44-100 Gliwice, Poland; tomasz.krzysztofiak@io.gliwice.pl (T.K.); piotr.wojcieszek@io.gliwice.pl (P.W.)

[2] Skin Cancer and Melanoma Team, Department of Bone Marrow Transplantation and Haematology-Oncology, Maria Sklodowska-Curie National Research Institute of Oncology, Gliwice Branch, ul. Wybrzeże Armii Krajowej 15, 44-100 Gliwice, Poland

[3] Department of Public Health, Wrocław Medical University, ul. Bartla 5, 51-618 Wrocław, Poland; andrzej.tukiendorf@gmail.com

[4] Pathology Department, Maria Sklodowska-Curie National Research Institute of Oncology, Gliwice Branch, ul. Wybrzeże Armii Krajowej 15, 44-100 Gliwice, Poland; m.suchorzepka@gmail.com

* Correspondence: dermatolog.pl@gmail.com

Citation: Krzysztofiak, T.; Kamińska-Winciorek, G.; Tukiendorf, A.; Suchorzepka, M.; Wojcieszek, P. Basal Cell Carcinoma Treated with High Dose Rate (HDR) Brachytherapy—Early Evaluation of Clinical and Dermoscopic Patterns during Irradiation. *Cancers* **2021**, *13*, 5188. https://doi.org/10.3390/cancers13205188

Academic Editor: Aimilios Lallas

Received: 28 September 2021
Accepted: 10 October 2021
Published: 16 October 2021

Publisher's Note: MDPI stays neutral with regard to jurisdictional claims in published maps and institutional affiliations.

Simple Summary: Basal cell carcinoma (BCC) is the most frequent malignancy of the Caucasian population. High dose rate (HDR) brachytherapy is a re-emerging treatment method for various skin cancers. Dermoscopy is an acknowledged and widely used diagnostic tool providing the bridge between histopathology and clinical examination. Current literature lacks data reporting on the dermoscopic observation of basal cell carcinomas undergoing brachytherapy. In this article, the authors describe clinical and dermoscopic patterns of basal cell carcinomas from 23 patients treated with HDR brachytherapy, and analyse the evolution of BCC structures.

Abstract: Basal cell carcinoma (BCC) is the most frequent malignancy of the Caucasian population. Dermoscopy is an established diagnostic method providing the bridge between clinical and pathological examination. Surface skin high dose rate (HDR) brachytherapy is an organ sparing treatment method used for non-surgical candidates. This prospective study aimed to observe clinical and dermoscopic features and their evolution in 23 patients with pathologically confirmed BCC that have been treated with HDR brachytherapy. In all cases, custom-made surface moulds were used. HDR brachytherapy was performed with 192Ir, dose 45Gy was delivered to the tumour in nine fractions of 5Gy, three times a week. The evolution of clinical and dermoscopic features was followed up at the beginning of treatment, and on the day of every fraction (t1–t9). Dermoscopic evaluation of neoplastic and non-neoplastic structures was based on current diagnostic criteria according to current literature. Univariate logistic regression showed a decreasing number of clinical and pathological features of basal cell carcinoma with every treatment fraction. The effect was more strongly pronounced for cancer-related dermoscopic structures compared with non-neoplastic features. We used multivariate ordinal logistic regression with random effects to prove that the patients' age corresponds with the tumour's response to radiation—which may implicate a better response to treatment among older patients. High dose rate brachytherapy decreases the number of clinical and dermoscopic features typical for basal cell carcinoma. The effect is more pronounced among older patients.

Keywords: HDR brachytherapy; basal cell carcinoma; dermoscopy

1. Introduction

Basal cell carcinoma (BCC) is the most frequent skin cancer and most frequent malignancy among Caucasians worldwide [1–3]. It is characterized by slow progression and low mortality. Most lesions (80%) are localized on the head region, among which 90% are located on the face [2,4,5]. Diagnostics include anamnesis, clinical and pathological examination. Dermoscopy (epiluminescence microscopy and skin surface microscopy) is an in vivo examination of epidermis and dermis structures. It is an easy to perform, painless, non-invasive, repetitive diagnostic technique allowing doctors to observe neoplastic and non-neoplastic skin lesions in at least $10\times$ magnification in polarized and non-polarized light [6,7]. It gives additional information in the preliminary assessment of tumour morphologic type [8–10]. Moreover, this examination helps to determine surgical margins preoperatively [11,12]. It may also be beneficial in ex-vivo examinations for resected lesions [13]. The current literature lacks data concerning dermoscopic feature evaluation of basal cell carcinomas treated with brachytherapy.

Skin surface brachytherapy uses radioactive sources placed close to the tumour, allowing the delivery of high radiation doses to the treatment area with a steep gradient dose in the normal tissues (i.e., conformity). Nowadays, the most commonly used radioactive source in surface brachytherapy is Iridium 192. It delivers a dose over 12Gy per hour, which is defined as a high-dose-rate brachytherapy (HDR BT). One fraction of the treatment usually takes several minutes [14].

In recent years brachytherapy has gained more attention because of its unique features: high conformity—rapid dose drop-off outside of the target area and the possibility of optimizing the radiation dose distribution. Due to the wide range of commercial applicators available or individual applicators (custom made or 3D printed), sub-millimeter precision in dose optimization can be achieved [14].

Surgery remains the primary treatment method of non-melanoma skin cancer (NMSC's) [15,16]. Mohs surgery is superior to standard excision with the lowest 5 and 10-year recurrence rate [17,18]. Radiotherapy remains an option only for non-surgical candidates, but most guidelines refer only to external beam radiotherapy [15,16]. Other treatment options such as topical therapies (5% imiquimod or 5% fluorouracil) and destructive approaches (curettage, electrocautery, cryotherapy, laser ablation) should be considered only in patients with low-risk superficial BCC. Photodynamic therapy was found to be effective only for superficial BCC and in cases of thin nodular BCC [15,16].

Skin surface HDR BT is recommended for elderly patients, patients with comorbidities and non-surgical candidates. Furthermore, for patients with tumours involving facial aesthetic units, for whom radical surgery would cause massive deformations, may be advised to choose brachytherapy. There is a need for a standardised, non-invasive, fast and easy use method of follow-up, presumably a dermoscopy.

2. Materials and Methods

This study aimed to assess dermoscopic features of basal cell carcinoma among patients qualified for brachytherapy, with a subsequent analysis of clinical and dermoscopic patterns of the treated area. Researchers evaluated lesions clinically, obtained macroscopic and dermoscopic photographs of tumours for further comparison. Dermoscopic features occurring, disappearing and changing during exposition to ionizing irradiation were analysed. Acute radiation dermatitis was also monitored and stratified, which allows objective post-treatment observation.

2.1. Patients

Twenty-three T1 and T2 (according to AJCC ed. 8th edition) patients with a median age of 72 years (SD:9.7 years) who received HDR brachytherapy between September 2020 and March 2021 were included in the study. Patients were disqualified from surgery because of age (ranged from 86 to 95 years, 3/23), the poor expected cosmetic outcome of surgery—tumour localization in the central face region (16/23), or due to patient's choice

(4/23). Tumours were localized on the central facial region in 16 cases, lateral face in four cases, scalp in one case and neck in two cases. Details of patients' clinical, histopathological and therapeutic characteristics are shown in Table 1.

Table 1. Clinical, histopathological and therapeutic characteristics of observed patients' group.

Sex (M/F)	Median Age	Location of Tumor	Tumour Size According to AJCC [19]	Clinical Subtype [8]	Primary or Recurrent Type	High-Risk and Low-Risk BCC According to NCCN [15]	Histopathological Type (WHO Classification) [20]	Acute Morbidity According to RTOG [21]
10/13	72 (58–95)	Forehead (2) Ear (1) Temporal region (1) Infraorbital region (2) Buccal region (3) Neck (2) Nose (9) Orbital region (2)	T1 (20) T2 (3)	Superficial (5) Nodular (12) Sclerodermiform (3) Infiltrating (3)	Primary (20) Recurrent (3)	High (23)	Nodular (12) Nodular (adenoid) (3) Superficial (4) Micronodular (1) Infiltrating (3)	G0–1 (11) G2 (2) G3 (3) G4 (7)

2.2. Treatment

The treatment protocol of the Brachytherapy Department of Maria Sklodowska-Curie National Research Institute of Oncology (MSCNRIO), Gliwice Branch, was developed over the last three decades of experience in the field. It has commonalities with The Groupe Européen de Curiethérapie of the European Society for Radiotherapy & Oncology (GEC-ESTRO) recommendations [22], but there are some differences, i.e., maximal surface dose, mould thickness and fractionation. The schedule is nine fractions of 5Gy delivered to the tumour with a 5 mm margin. The treatment was prescribed three times a week to a total dose of 45 Gy. Custom-made surface mould polyacrylamide applicators precisely covering the treatment area were prepared. On the surface of the applicator, plastic tubes were glued for Iridium 192 HDR radioactive source loading. The position of the source and time of treatment was planned by a medical physicist with OncentraBrachy (Elekta AB, Stockholm, Sweden) software (version 4.6.0), using computed tomography images of the treatment area with an attached applicator. The study protocol was approved by the Local Ethics Committee (KB/430-41/20). All patients gave written informed consent.

2.3. Clinical Evaluation

Clinical evaluation of basal cell carcinomas was performed according to 12 clinical features at the beginning of the treatment (t1), and before every treatment fraction(t2–t9). Digital photographic images were obtained using DermLite Cam (3Gen,. San Juan Capistrano, CA, USA) digital dermoscopy camera. Presence (1) or absence (0) of 12 clinical features of BCC according to Tognetti et al. classification [23] were described. They are presented in Table 2.

Moreover, acute radiodermatitis (according to the RTOG/EORTC scale) [21] was observed. Twelve patients experienced grade 0–1 toxicity, two experienced grade 2, two experienced grade 3, and six experienced grade 4 skin radiation toxicity, called acute radiodermatitis.

Table 2. Clinical structures of BCC according to Tognetti classification [23].

Clinical Structures	Description	Significance
Pink macules	Well-demarcated pink macules	Superficial BCC
Erythematous papules	Red papules	Nodular BCC
Erythematous nodules	Red nodules	Nodular BCC
Pearly-shiny papules	Translucent papules	Superficial BCC
Pearly-shiny nodules	Translucent nodules	Nodular BCC
Scar-like plaque	Ivory-white plaque resembling scar or morpheaform plaque	Scleroderma-like BCC
Erosions	Multiple small de-epithelized areas	Superficial/nodular BCC
Ulcerations	One or more large red to blackish red areas representing hematogenous crusts	Superficial/nodular BCC
Telangiectasia	Large dilated vessels	Superficial/nodular/scleroderma-like BCC
Short vessels	Short fine vessels	Superficial/nodular/scleroderma-like BCC
Pigmented structures	Hyperpigmented macule, papule, or nodule	Superficial/nodular/scleroderma-like BCC
Crust/scale	Small brown-red to brown-yellow crusts	Superficial/nodular/scleroderma-like BCC

2.4. Histopathologic Assessment

All lesions before brachytherapy were confirmed in histopathological examination. Standard stains with hematoxylin and eosin were used in all cases, and the specimens were assessed by a qualified pathologist (M.S). Due to the presence of heterogeneous tumours, a dominant pattern in the microscopic image was reported according to the WHO classification of tumours (2018) [20].

2.5. Dermoscopic Procedure and Image Data Collection

Dermoscopy was performed by a certified medical doctor who is an expert in dermoscopy and an integral member of the skin cancer brachytherapy team. Dermoscopic assessment of skin lesions was performed using the polarized DermLiteFoto dermoscope (3Gen, LLC, San Juan Capistrano, CA, USA) at tenfold magnification. Dermoscopic images (n = 603) of 23 BCCs were acquired nine times (t1–t9)—at the beginning of the treatment and on the day of each fraction. Dermoscopic images were captured and saved using the DermLiteCam digital dermoscopy camera with polarized light and then independently analysed by two certified dermoscopists (T.K and G.K-W), who were blinded to any patient/protocol data following the methods of the study. Dermoscopic evaluation of the tumour was performed based on the third consensus of the International Society of Dermoscopy (61 features) [7], and non—neoplastic features according to an expert consensus of the International Dermoscopy Society (overall 31 features) [6] were described as present (1) and absent (0). Moreover, the evaluators were required to score the presence (1) or absence (0) of dermoscopic features including BCC-associated criteria (16) (Table 3) that were selected based current knowledge [7] in modification of Lallas et al. [8] and Tognetti's et al. classification [23].

Non-neoplastic patterns of the surrounding area of the tumour were described according to the expert consensus of the International Dermoscopy Society [6]. Since the score is not described in Table 3, while analysing the non-neoplastic surroundings of tumour, scale covering tumour was analysed as a non-neoplastic feature. Correlation between clinically observed radiodermatitis and non-neoplastic dermoscopic patterns of area surrounding tumor was analysed.

Table 3. Dermoscopic structures of BCC-associated criteria (16 features) ranked according to Lallas et al. [8] and Tognetti's et al. [23] in own modification based on dermoscopic terminology from IDS [7].

Descriptive Terminology	Metaphoric Terminology	Description	Significance
Lines, white, perpendicular	Shiny white streaks (former: chrysalis, chrysalids, crystalline)	Short thick shiny orthogonal crossing lines	Melanoma, BCC, Spitz nevus, dermatofibroma
Lines, radial, connected to a common base	Leaf-like areas	Greyish/bluish brown peripheral globular extensions arising from pigmented network or adjacent confluent pigmented areas	BCC
Lines, radial, converging to a central dot or clod	Spoke wheel area	Brown or greyish well-circumscribed radial projections, usually around a dark brown/black, bluish central axis	BCC
Clods, brown or blue, concentric (clod within a clod)	Concentric globules	Gray/brown/black/blue globular structures with darker central areas	BCC
Clods, blue, large, clustered	Blue-grey ovoid nests	Pigmented ovoid or elongated structures well circumscribed and separate from pigmented tumor body	BCC
Clods, blue, small	Blue globules	Numerous loosely arranged round to oval well circumscribed structures, smaller than nests	BCC
Clods, white, shiny	Shiny white blotches and strands	White structures in the form of circles, oval structures, or large structureless areas, bright white.	BCC
Gray dots	In focus dots	Small well defined loosely arranged grey dots in focus at dermoepidermal junction	BCC
Structureless zone, polychromatic	Rainbow pattern	Many different colours of the rainbow ranging from red to violet	Various diagnoses
Structureless zone, blue	Blue-white veil-like structures	Irregularly margined confluent blue pigmentation with overlying white ground glass haze	Melanoma
Structureless zone, white	Scar-like depigmentation	White structureless areas	BCC presence, tumour fibrotic stroma
Vessel Morphology			
Serpentine	Linear vessels with multiple bends	Linear vessels with multiple bends	Flat BCC, melanoma
Linear	Superficial fine vessels	Telangiectatic vessels in papillary dermis	BCC, inflammation
Vessel Arrangement			
Branched	Arborizing vessels	Bright-red, sharply in focus, large of thick diameter vessels dividing into smaller vessels	BCC
Others	Erosions	Thin crusts overlaying superficial loss of epidermis	BCC
	Ulcerations	Loss of epidermis with/without haematogenous crusts	Various diagnoses

2.6. Statistical Analysis

The entire photographic databaseoff 21,526 observations including all clinical and dermoscopic structures was analysed in the final statistical assessment. In the statistical analysis we used standard methods. Univariate logistic regression was applied to evaluate the impact of the BT fractions on binary skin diagnostic outcomes. In turn, to estimate the influence of the collected risk factors on the 16 available completed (ranked) number of Lallas/Tognetti's features and non-neoplastic lesions, a multivariate ordinal logistic model was used. Due to repeated measures with consecutive BT fractions for each patient, the regressions were extended for random effects. The statistical outcomes were expressed by a classical odds ratio (OR) together with a confidence 95% interval (CI 95%) and a p-value. The computation was performed in the R platform [24].

3. Results

All patients suffered from high-risk basal cell carcinomas (23/23). Twenty tumours were stratified as T1, three as T2 according to AJCC 8th edition [19]. All tumours were observed nine times (t1–t9) before treatment and before every treatment fraction among all patients (the completeness of the Lallas/Tognetti's features in the consecutive t1–t9 times were: 40%, 41%, 42%, 43%, 40%, 40%, 35%, 32%, and 29%, respectively), while in case of non-neoplastic lesions, analogously: 9%, 9%, 8%, 9%, 9%, 9%, 8%, 8% and 8%.

The list of statistically significant ORs ($p < 0.05$) of the influence of BT fractions on the skin diagnostic outcomes (univariate logistic regression with random effects) is reported in Table 4.

Table 4. Statistically significant ORs ($p < 0.05$) of the influence of BT fractions on the skin diagnostic outcomes (univariate logistic regression with random effects).

Clinical Outcome	OR (CI 95%), *p*-Value
Erythematous nodules	0.51 (0.43,0.59), <0.0001
Pearly, shiny nodules	0.67 (0.58,0.78), <0.0001
Scar-like plaque	0.61 (0.52,0.70), <0.0001
Erosion	0.84 (0.75,0.94), 0.0029
Ulceration	1.49 (1.27,1.74), <0.0001
Telangiectasia	0.83 (0.73,0.95), 0.0055
Pigmented structures	1.15 (1.00,1.32), 0.0465
Dermoscopic Outcome	**OR (CI 95%), *p*-Value**
Clods, white, shiny	0.69 (0.60,0.80), <0.0001
Gray dots	0.66 (0.56,0.77), <0.0001
Lines, white, perpendicular	0.84 (0.74,0.95), 0.0046
Vessels arrangement—branched	0.67 (0.57,0.79), <0.0001
Vessels—linear	0.81 (0.70,0.94), 0.0049
Vessels—serpentine	0.61 (0.51,0.73), <0.0001
Vessels—polymorphous	0.73 (0.63,0.84), <0.0001
Structureless zone, white	0.65 (0.53,0.79), <0.0001
Erosions	0.83 (0.74,0.94), 0.0028
Ulcerations	1.75 (1.50,2.05), <0.0001

The statistical interpretation of the ORs listed in Table 4 is as follows: each subsequent BT fraction statistically generates a lower chance of occurrence of the erythematous nodules on the skin; the use of one fraction reduces the risk of their occurrence by nearly half, and the use of two doses by $(1 - 0.51^2) \times 100\% = 74\%$, i.e., almost three-quarters. In turn, the administered radiotherapy increases the risk of ulceration; the difference of one dose statistically generates an increased risk by almost a half, and for two doses: $1.49^2 = 2.22$, i.e., more than twice. Statistical interpretations of the remaining results reported in Table 4 is analogous. The graphical presentation of ORs of the influence of BT fractions on the skin diagnostic outcomes is shown in Figure 1.

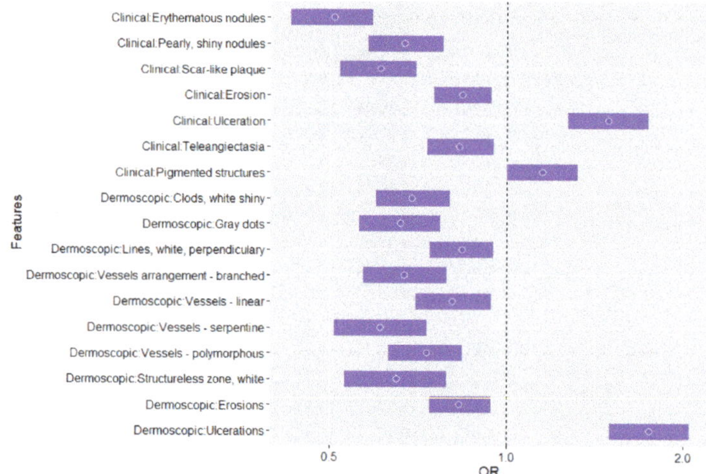

Figure 1. Influence of brachytherapy fractions on skin diagnostic outcomes.

The statistically significant ORs ($p < 0.05$) of the influence of BT fractions and available risk factors on ranked Tognetti's diagnostic clinical (23) and Lallas/Tognetti's dermoscopic (8) features (multivariate ordinal logistic regression with random effects) is reported in Table 5 and graphically in Figure 2.

Table 5. The influence of BT fractions and available risk factors on ranked Tognetti's clinical (23) and Lallas/Tognetti's dermoscopic (8) features of BCCs.

Neoplastic:	Risk Factor	OR (CI 95%), *p*-Value
Yes	BT fraction	0.77 (0.76,0.78), <0.0001
	Age	0.98 (0.97,0.99), 0.0005
No	BT fraction	0.94 (0.85,1.04), 0.2190
	Age	1.01 (0.87,1.16), 0.9440

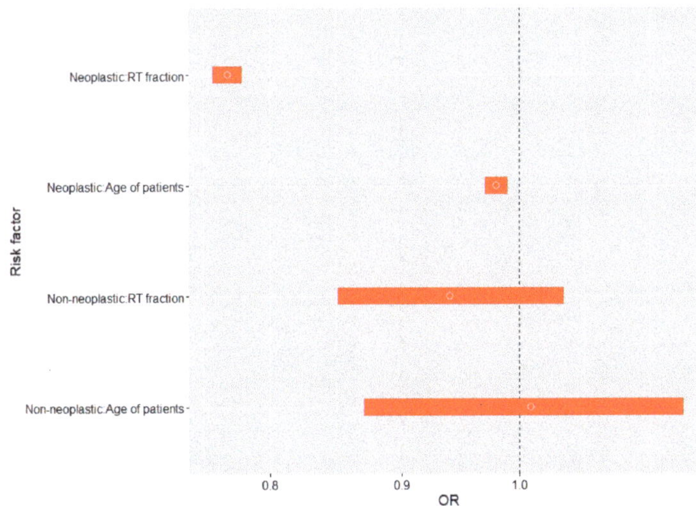

Figure 2. The influence of BT fractions and available risk factors on ranked Tognetti's clinical (23) and Lallas/Tognetti dermoscopic (8) features of BCCs in correlation with non-neoplastic features (10).

The statistical interpretation of the ORs reported in Table 5 is as follows: each subsequent BT fraction statistically decreases the Lallas/Tognetti's diagnostic rank, i.e., the first dose of BT by 23%, and two fractions by nearly $(1 - 0.77^2) \times 100\% \approx 40$. In addition, a 10-year difference in age of patients generates a $(1 - 0.98^{10}) \times 100\% = 18\%$ reduction in the cumulative number of clinical and dermoscopic features characterizing tumor, while in case of non-neoplastic features, ORs for the same risk factors were statistically non-significant ($p > 0.05$). The results are shown graphically in Figure 2.

Non-neoplastic dermoscopic patterns of the area surrounding the tumour are presented in Table 6. There is a confirmed negative, statistically significant correlation between intensity (grade) of clinically described acute radiodermatitis (according to RTOG/EORTC criteria) [21] and the number of its non-neoplastic dermoscopic features (OR = 0.65 (0.45,0.93), $p = 0.017$). The number of dermoscopic features decreases with increasing clinical grade of radiodermatitis ($p = 0.017$).

Table 6. Percentage occurrence of dermoscopic non-neoplastic features [6] depending on grade of radiodermatitis according to RTOG/EORTC [21].

TERMINOLOGY		G0	G1	G2	G3	G4
VESSELS	Dotted	0.0%	0.0%	0.0%	0.0%	0.0%
MORPHOLOGY	Linear	24.3%	35.7%	0.0%	50.0%	8.3%
	Branched	24.3%	31.4%	0.0%	50.0%	8.3%
	Curved	4.9%	4.3%	0.0%	0.0%	8.3%
VESSELS	Uniform	2.9%	0.0%	0.0%	0.0%	4.2%
DISTRIBUTION	Clustered	1.9%	0.0%	0.0%	0.0%	0.0%
	Peripheral	1.0%	2.9%	0.0%	0.0%	0.0%
	Reticular	1.9%	7.1%	0.0%	0.0%	4.2%
	Unspecific	19.6%	30.0%	0.0%	50.0%	8.3%
SCALE	White	46.6%	45.7%	25.0%	0.0%	12.5%
COLOUR	Yellow	38.2%	34.3%	100.0%	83.3%	8.3%
	Brown	12.6%	22.9%	50.0%	50.0%	0.0%
SCALES	Diffuse	30.1%	27.1%	100.0%	16.7%	4.2%
DISTRIBUTION	Central	1.9%	11.4%	0.0%	0.0%	0.0%
	Peripheral	20.4%	22.9%	0.0%	0.0%	8.3%
	Patchy	13.6%	11.4%	0.0%	66.7%	0.0%
FOLLICULAR	Plugs	13.6%	15.7%	0.0%	0.0%	8.3%
FINDINGS	Red dots	0.0%	0.0%	0.0%	0.0%	0.0%
	Peripheral White colour	3.9%	2.9%	0.0%	16.7%	0.0%
	Peripheral Pigmentation	5.8%	4.3%	0.0%	0.0%	0.0%
OTHER	White	0.0%	0.0%	0.0%	0.0%	0.0%
STRUCTURES	Brown	0.0%	0.0%	0.0%	0.0%	0.0%
COLOR	Grey	0.0%	0.0%	0.0%	0.0%	0.0%
	Blue	0.0%	0.0%	0.0%	0.0%	0.0%
	Orange	0.0%	0.0%	0.0%	0.0%	0.0%
	Yellow	0.0%	0.0%	0.0%	0.0%	0.0%
	Purple	0.0%	0.0%	0.0%	0.0%	0.0%
OTHER	Structureless	0.0%	0.0%	0.0%	0.0%	0.0%
STRUCTURES	Dots	0.0%	0.0%	0.0%	0.0%	0.0%
MORPHOLOGY	Lines	0.0%	0.0%	0.0%	0.0%	0.0%
	Circles	0.0%	0.0%	0.0%	0.0%	0.0%

4. Discussion

In general, the population nodular subtype of BCC is dominating especially on the head, with 60–80% occurrence. It is followed by a superficial BCC subtype with 10–30% occurrence [25]. Among our group, multiple clinical subtypes of BCC were observed—superficial (5/23), nodular (12/23), scleroderma-like (3/23) and infiltrating (3/23).

The broad spectrum of clinical patterns of basal cell carcinomas translates into multiple histopathological variants. All share common morphological features: basaloid cell aggregates with hyper-coloured cell nuclei and sparse cytoplasm, characteristic stroma and retraction artifact. Recent WHO classification divides basal cell carcinomas into two groups [20]. The first group includes tumours with low recurrence risk: nodular, superficial, pigmented, infundibulocystic with adnexal differentiation and fibroepithelial (Pinkus tumour). The second group contains tumours with high recurrence rate: basosquamous, sclerosing/morpheic, infiltrating, BCC with sarcomatoid differentiation and micronodular. The nodular variant is the most common basal cell carcinoma. It is characterized by large islands of neoplastic cells with peripheral palisading. It is not a homogenous group of tumours. It divides into keratotic nodular BCC—with features of mature keratosis in central parts of neoplastic cell islands, nodulocystic BCC—with retrograde cystic changes along islands, adenoid nodular BCC—forming reticular islands with pseudo adenoidal structures. In the studied material nodular variant was dominant (15 cases), among which in three cases pseudo adenoidal structures characterizing adenoid cystic subtype were found. In four cases low recurrence risk superficial type was present. In this variant in histopathological examination connection between neoplastic cell nests and the epidermal layer is present. Microscopically cancer foci spread into several locations was also observed, which explains this tumour's other description—superficial multifocal BCC. High recurrence-risk tumours in the studied group were represented by one micronodular and three infiltrating BCCs. Morphologically micronodular BCC is built from islands of neoplastic cells, however islands are considerably smaller (even <0.15 mm) and are more dispersed. They tend to invade skin nerves, subdermal tissues and muscles. Invasive BCC represents narrow strings and nests of neoplastic cells in rich collagen stroma, especially in front of the tumour. Invasive growth patterns often include neuroinvasion. It shares many similarities with scleroderma-like BCC and is often wrongly recognized as such [20].

In searched literature, only one data concerning dermoscopy in monitoring BCC's treatment effects using high dose ionizing radiation therapy was reported [26]; Moreover, dermoscopy was also used in the monitoring of changes in the course of lentigo maligna radiotherapy [27], or dermoscopic margin delineation in radiotherapy planning for superficial or nodular basal cell carcinoma [28]. Until the present data concerning dermoscopic features of BCC's undergoing brachytherapy has not been reported. Dermoscopic follow-up helped to monitor the therapeutic response to selected topical therapies including ingenolmebutate in BCC [29], and systemic therapy with vismodegib in BCC [23].

Obtained results prove that with every fraction of ionizing radiation the number of clinically visible tumour features diminish. The characteristic clinical appearance of the tumour (erythematous/pearl nodule with e.g., scar-like plaque—depending on the clinical and pathological variant) is replaced by ulceration (Figures 3A–I and 4A–I).

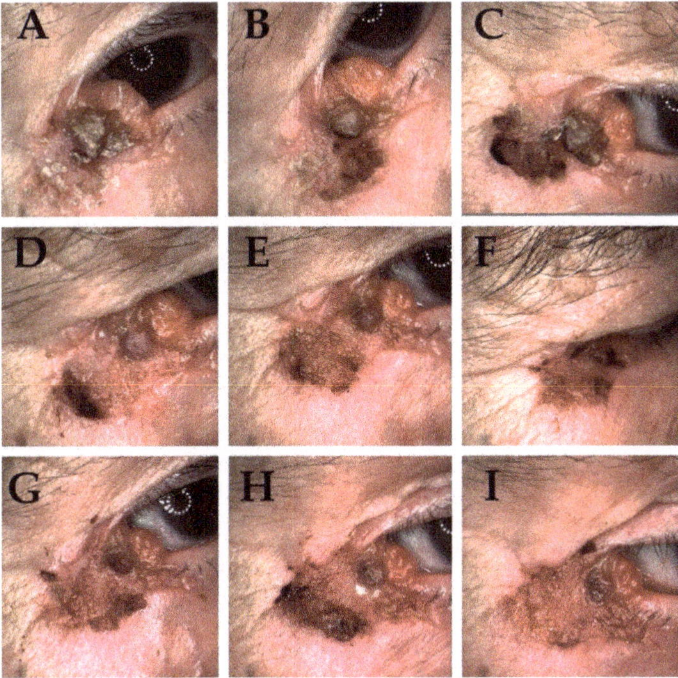

Figure 3. Clinical presentation of basal cell carcinoma of the orbital area of 95-year-old patient, infiltrating type. (**A**)—Tumour prior to brachytherapy, visible—Erythematous nodule, scar-like plaque, pigmented structures, telangiectasia, short vessels, erosion, ulceration. (**B–I**) (t2–t9) During subsequent treatment fractions visible gradual vanishing of nodular structures and vessels with formation of ulceration have been noticed.

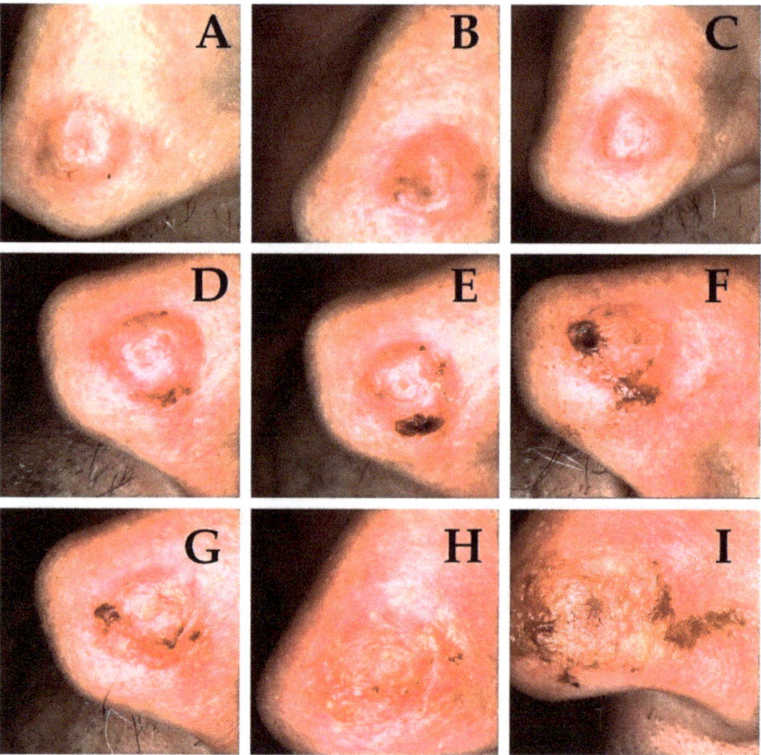

Figure 4. Clinical presentation of basal cell carcinoma of the nose of 83-year-old patient, nodular type. (**A**)—Tumour before treatment, with visible—erythematous nodule, short vessels and telangiectasia. (**B–I**)—evolution of tumour destruction with erosion arising in 4th fraction (t4) (**D**), evolving into ulceration with disappearing vessels in 6th fraction (t6) (**F**), and crust appearing in 7th fraction (t7) (**G**), with finally forming ulceration covered by scale and haemorrhagic crusts (t8–9) (**H,I**).

Destruction of the tumour corresponds with exposure of the underlying stroma, which may explain the increased number of pigmented structures in the irradiated areas. Increased number of pigmented structures indicates exposition of deeper layers of skin after destruction of tumour above. Only further dermoscopic follow up allows to assess the efficacy of the brachytherapy taking into account the persistence of the BCC (total clinical and dermoscopic clearance of the dermoscopic structures of the tumour). Persistence of ovoid nests and bluish dots was observed during monitoring of not efficient treatments reported in the previous publications (i.e., imiquimod applied topically in nodular BBC). Occurrence of these structures during irradiation has an important clinical significance—it raises the need for dermoscopic follow up, but also is an expected symptom of some BCC variants including nodular type, pigmented type previously covered by necrotic tissue or ulceration. Our study exemplifies that dermoscopy is a valuable diagnostic aid that helps to assess potential tumour recurrence/residual disease and to monitor post-brachytherapy response in the future. The presence of pigmented structures such as blue grey ovoid nests as a BCC related features in the previous studies concerning BCC treated with 5% imiquimod suggested poor therapeutic response of BCC [30]. Moreover, the blue-white veil areas and rainbow pattern were only observed in non-responding BCC lesions [31]. The opposite of the above, dermoscopy of responding lesions showed a higher frequency of lesions within focus grey dots [30]. In the final conclusion, the time of the disappearance of blue structures should be taken into account. In the previous study of Husein-ElAhmed [31] blue-grey globules were the fastest to exhibit clearance (50% at week 4), followed by leaf-like areas and large blue-grey ovoid nests [31]. Our short term dermoscopic follow up is a pioneer observation, therefore further investigations are necessary in this matter in the nearest future.

Dermoscopically higher correlation with the destruction of structures typical for neoplastic tumors (e.g., serpentine vessels) (Figure 5A–I) was observed, than structures also observed in healthy tissue (e.g., linear vessels). According to Reiter's BCC dermoscopic review, the occurrence of linear vessels in BCC corresponds to all histopathological subtypes of BCC [10]. We found no significant correlation between ongoing brachytherapy and changes in non-neoplastic features, which comparing to strong correlation with changes characteristic for carcinoma brings the conclusion that brachytherapy induces significant effect on the tumour. Clinical and dermoscopic erosion present in tumours often in the beginning of treatment is replaced by ulceration often present in the entire treated area which corresponds with clinical destruction of the treated tumour (Figures 4A–I and 6A–I).

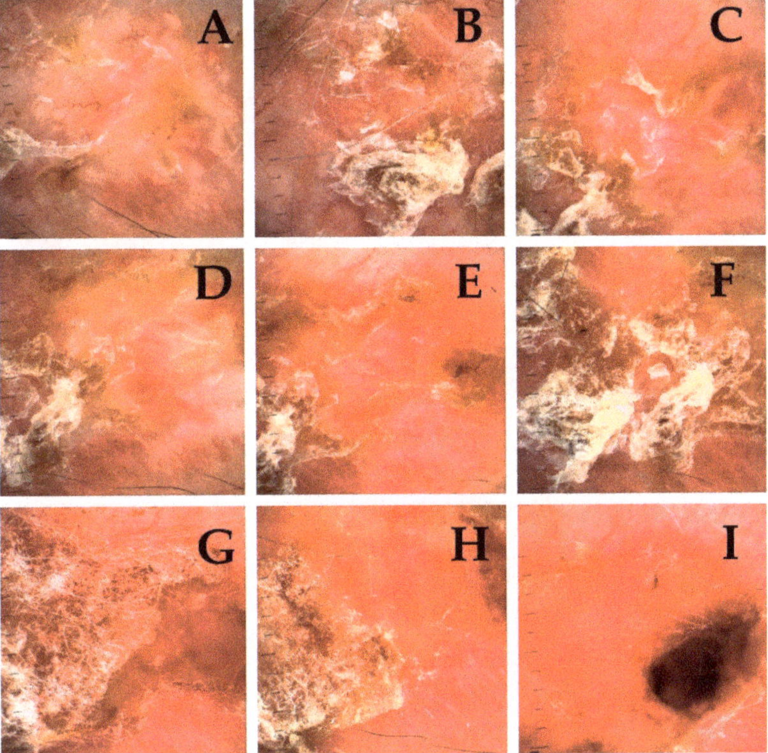

Figure 5. Dermoscopic presentation of basal cell carcinoma, nodular (adenoid type), localized on the forehead of 76-year-old patient. (**A**)—Dermoscopic image before brachytherapy showed the presence of—polymorphous, branched, linear and serpentine vessels, as well as white, shiny clods, perpendicular white lines and white and polychromatic structureless zones, white scale. (**B–H**)—Dermoscopy of the BCC during subsequent treatment fractions (t2–t8) revealed diminishing number of serpentine, branched vessels as well as formation of the ulceration arising (t9) (**I**).

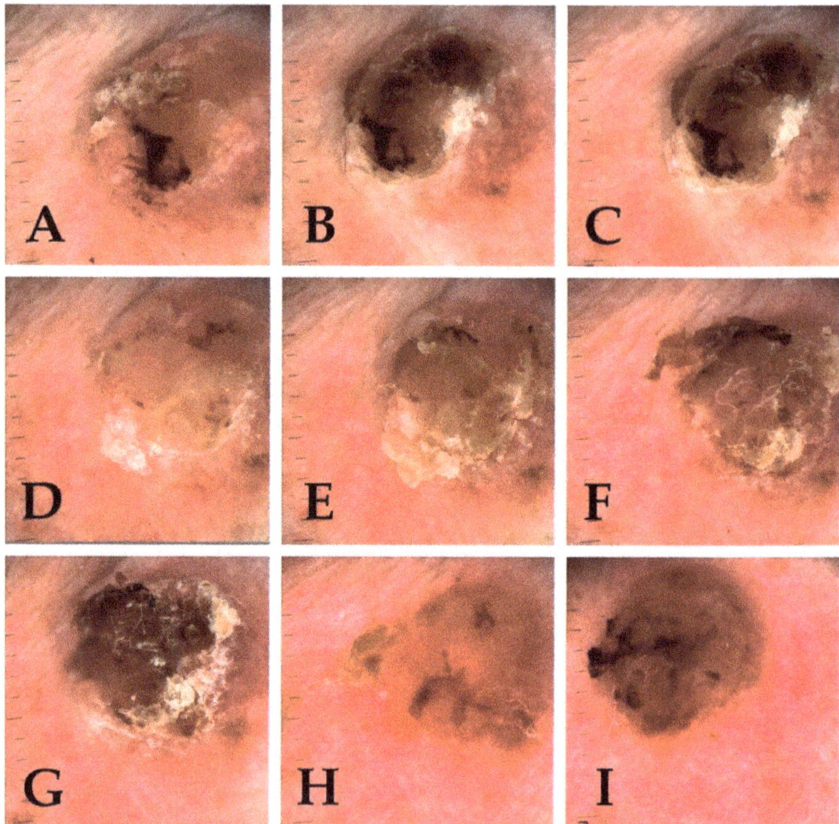

Figure 6. Dermoscopic picture of basal cell carcinoma of the infraorbital region, nodular type, of 67-year-old patient. (**A**)—Tumour prior to brachytherapy (t1) exhibit the presence of white clods, white structureless zones and erosion. (**B–I**)—White clods and structureless zones gradually vanishing from 5th fraction (t5), with developing predominance of ulceration.

Radiation also reduces the number of structures corresponding with fibrosis including perpendicular white lines, white, shiny clods and white structureless zones (Figure 7A–I). The diminished number of these structures may be the result of image overlapping from ulceration and inflammation.

Observed correlations correspond with the typical radiobiological response of tissue to ionizing radiation [32]. Skin is a hierarchic tissue, and regenerates from a population of stem cells in its deeper layers. Ionizing radiation from a brachytherapy source damages the DNA of stem cells of skin and tumour, without causing much damage to the population of mature cells of upper layers of skin. With time damaged stem cells cannot reproduce a diminishing number of mature cells, leading to ulceration in the irradiated area. Ulceration after tumor destruction heals within three months, from edges into the centre from non-damaged stem cells of healthy skin outside irradiated area [32]. A diminishing number of vessels in the treated area may be caused by radiation damage, but also because ulceration and inflammation may blur the dermoscopic image. Moreover, it is important to distinguish destruction of the tumour from ulceration of healthy skin surrounding BCC. The first effect is desirable, occurs in the second week of treatment. Ulceration of the surrounding healthy skin is an adverse effect defined as an acute radiodermatitis according to RTOG definitions [21]. Exact borders between ulceration created by tumour destruction and radiodermatitis of surrounding skin are often difficult to establish. Dermoscopic non-neoplastic features also correspond with clinically assessed radiodermatitis—i.e., in G2 acute radiodermatitis (according to RTOG/EORTC) diffuse scale is visible among 100% patients. Occurrence of ulceration (G4) corresponds with diminished number of all dermoscopic patterns (Figure 8).

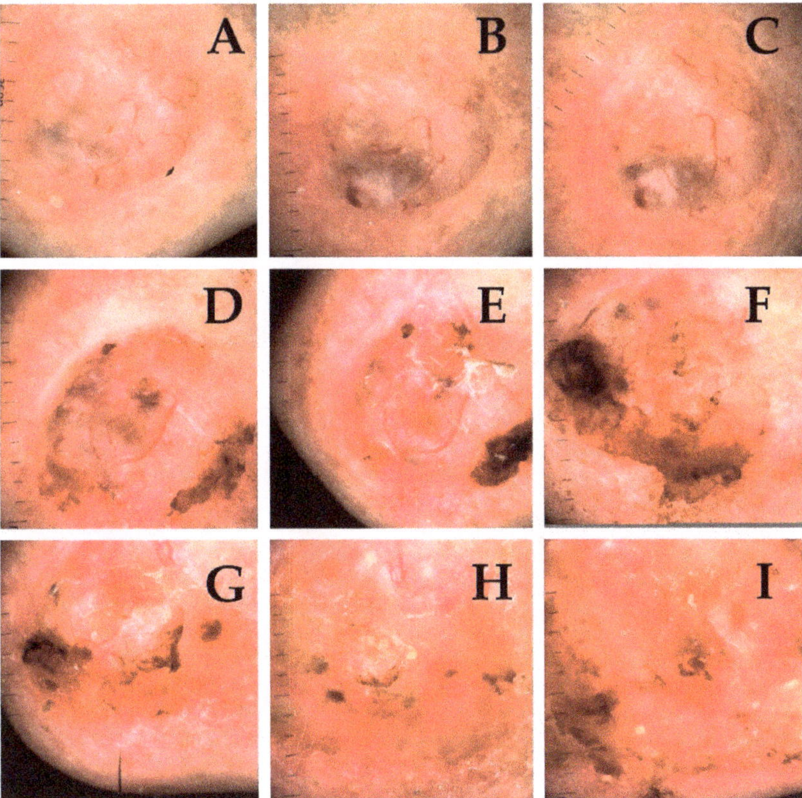

Figure 7. Dermoscopic image of basal cell carcinoma of the nose, nodular type, of 83-year-old patient (clinically presented in Figure 4). (**A**)—tumour prior to brachytherapy (t1), exhibit the presence of multiple white structures including: perpendicular white lines, structureless white zones, white shiny clods, accompanied by multiple polymorphous vessels (serpentine, in branched arrangement), with addition of blue, small clods. (**B,C**)—During the consecutive treatment fractions (t2–t3) mentioned structures remain visible. (**D–I**) Development of necrosis of the tumour (t4–t9) leads to gradual disappearance of dermoscopic structures with formation of ulceration.

No correlation between non–neoplastic features and brachytherapy in comparison to strong correlation between treatment and changes in BCC specific features indicate that response to radiation is tumour-specific, which also reflects radiobiology—tumouris more susceptible to radiation than healthy tissue [32].

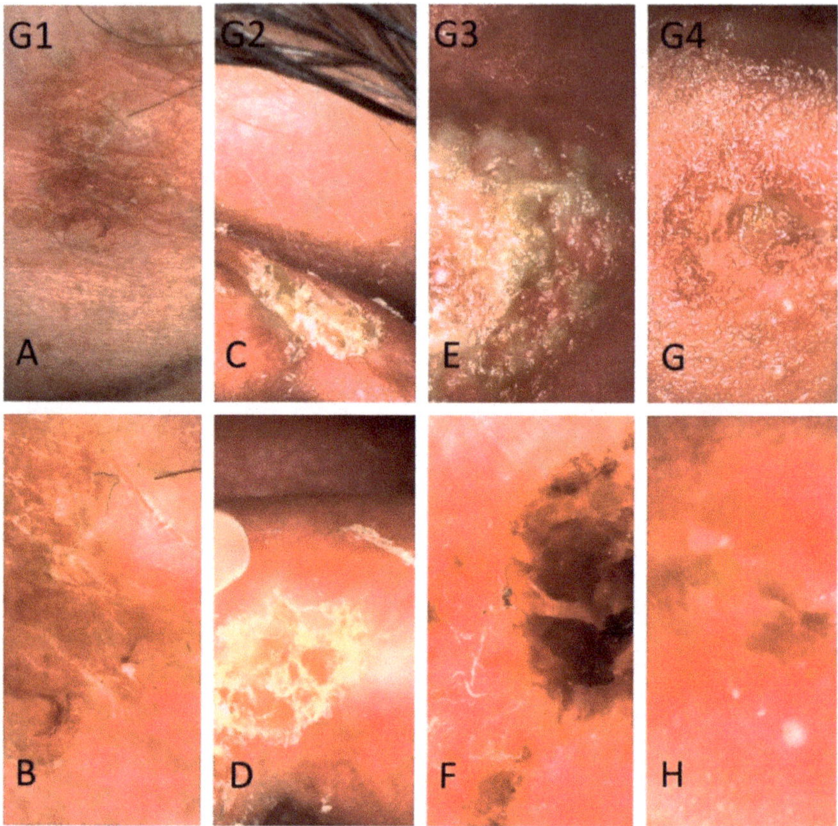

Figure 8. Dermoscopic images of various basal cell carcinomas with coexisting acute radiodermatitis in stage from G1 to G4 were clinically assessed according to RTOG criteria [21] (**A**)—minor erythema of the surrounding area with moderate dry desquamation (G1), (**B**)—dermoscopic image (G1) acute radiodermatitis reveals yellow scale with linear vessels in non-specific distribution in the surrounding area, (**C**)—bright erythema of the tumour's surroundings with dry desquamation (G2), (**D**)—dermoscopic image of acute radiodermatitis (G2) shows yellow scale (**E**)—confluent, moist desquamation (G3), (**F**)—dermoscopic image (G3) with linear vessels in unspecific distribution, yellow and brown crust. (**G**)—ulceration in acute radiodermatitis (G4), (**H**)—dermoscopic image (G4) reveals the absence of any dermoscopic features.

It is worth underlying that all patients in this study suffered from high-risk BCC's located in the facial ,H'-zone. According to European expert consensus for management of such lesions [16] primary treatment method should be surgery or radiotherapy for non-surgical candidates. Other destructive therapies including cryosurgery, curettage, electrodessication and laser ablation are dedicated only for small, low-risk non-facial BCC and for multiple small BCCs, whereas photodynamic therapy (PTD) can be considered in non-aggressive, low-risk BCC, i.e., small superficial and nodular types, not exceeding 2 mm tumour thickness, recurrent small and large BCC [16]. Less common histologic variants of BCC, morphoeic, pigmented and micronodular types, as well as areas with higher risk of tumour survival and deep penetration (facial 'H'-zone), should not be treated with PDT [16].

Correlation between age and higher tissue response to radiation up to now was unexpected. Older patients may be better candidates for brachytherapy than younger since their response to radiation is more pronounced. The correlation between dermoscopic response to radiation and the relapse rate of treated tumours requires further study.

5. Limitations of the Study

Dermoscopic evaluation presented in the study did not include post treatment long-term follow up, therefore no data on efficacy of brachytherapy was presented. The observations published above indicate which dermoscopic features diminished during brachytherapy, and which persisted. Further observation is needed to assess clinical significance of those findings and to establish further correlation between the persistence of selected dermoscopic features and the relapse rate of brachytherapy.

6. Conclusions

High dose rate brachytherapy decreases the number of clinical and dermoscopic features typical for basal cell carcinoma in the early observation. The effect is more pronounced among older patients.

Author Contributions: Conceptualization, T.K. and G.K.-W.; Data curation, T.K. and M.S.; Formal analysis, T.K., G.K.-W., A.T. and M.S.; Funding acquisition, G.K.-W.; Investigation, T.K., G.K.-W. and M.S.; Methodology, T.K., G.K.-W. and A.T.; Project administration, T.K., G.K.-W. and P.W.; Resources, T.K., G.K.-W., A.T., M.S. and P.W.; Software, A.T.; Supervision, G.K.-W. and P.W.; Validation, T.K., G.K.-W. and A.T.; Visualization, T.K., G.K.-W. and A.T.; Writing—original draft, T.K., G.K.-W., A.T., M.S. and P.W.; Writing—review & editing, T.K., G.K.-W., A.T., M.S. and P.W. All authors have read and agreed to the published version of the manuscript.

Funding: This research received no external funding.

Institutional Review Board Statement: The study protocol was approved by the Local Ethics Committee (KB/430-41/20). Informed consent was obtained from all subjects involved to the study.

Informed Consent Statement: Informed consent was obtained from all subjects involved in the study.

Data Availability Statement: All data used in the study are available from the corresponding author upon request.

Conflicts of Interest: The authors declare no conflict of interest.

References

1. Rogers, H.W.; Weinstock, M.A.; Harris, A.R.; Hinckley, M.R.; Feldman, S.R.; Fleischer, A.B.; Coldiron, B.M. Incidence estimate of nonmelanoma skin cancer in the United States, 2006. *Arch. Dermatol.* **2010**, *146*, 283–287. [CrossRef] [PubMed]
2. Wojciechowska, U.; Didkowska, J.; Michałek, J.; Olasek, P.; Ciuba, A. *Cancer in Poland 2018, Krajowy Rejestr Nowotworów*; Ministerstwo Zdrowia: Warszawa, Poland, 2020; pp. 16–21.
3. Sng, J.; Koh, D.; Siong, W.C.; Choo, T.B. Skin cancer trends among Asians living in Singapore from 1968 to 2006. *J. Am. Acad. Dermatol.* **2009**, *61*, 426–432. [CrossRef] [PubMed]
4. Zanetti, R.; Rosso, S.; Martinez, C.; Nieto, A.; Miranda, A.; Mercier, M.; Loria, D.I.; Østerlind, A.; Greinert, R.; Navarro, C.; et al. Comparison of risk patterns in carcinoma and melanoma of the skin in men: A multi-centre case-case-control study. *Br. J. Cancer* **2006**, *94*, 743–751. [CrossRef] [PubMed]
5. Schmults, C.D.; Karia, P.S.; Carter, J.B.; Han, J.; Qureshi, A.A. Factors predictive of recurrence and death from cutaneous squamous cell carcinoma: A 10-year, single-institution cohort study. *JAMA Dermatol.* **2013**, *149*, 541–547. [CrossRef] [PubMed]
6. Errichetti, E.; Zalaudek, I.; Kittler, H.; Apalla, Z.; Argenziano, G.; Bakos, R.; Blum, A.; Braun, R.P.; Ioannides, D.; Lacarrubba, F.; et al. Standardization of dermoscopic terminology and basic dermoscopic parameters to evaluate in general dermatology (non-neoplastic dermatoses): An expert consensus on behalf of the International Dermoscopy Society. *Br. J. Dermatol.* **2020**, *182*, 454–467. [CrossRef] [PubMed]
7. Kittler, H.; Marghoob, A.A.; Argenziano, G.; Carrera, C.; Curiel-Lewandrowski, C.; Hofmann-Wellenhof, R.; Malvehy, J.; Menzies, S.; Puig, S.; Rabinovitz, H.; et al. Standardization of terminology in dermoscopy/dermatoscopy: Results of the third consensus conference of the International Society of Dermoscopy. *J. Am. Acad. Dermatol.* **2016**, *74*, 1093–1106. [CrossRef] [PubMed]
8. Lallas, A.; Apalla, Z.; Argenziano, G.; Longo, C.; Moscarella, E.; Specchio, F.; Zalaudek, I. The dermatoscopic universe of basal cell carcinoma. *Dermatol. Pract. Concept.* **2014**, *4*, 11–24. [CrossRef]
9. Pampena, R.; Parisi, G.; Benati, M.; Borsari, S.; Lai, M.; Paolino, G.; Cesinaro, A.M.; Ciardo, S.; Farnetani, F.; Bassoli, S.; et al. Clinical and Dermoscopic Factors for the Identification of Aggressive Histologic Subtypes of Basal Cell Carcinoma. *Front. Oncol.* **2021**, *10*, 630458. [CrossRef]

10. Reiter, O.; Mimouni, I.; Dusza, S.; Halpern, A.C.; Leshem, Y.A.; Marghoob, A.A. Dermoscopic features of basal cell carcinoma and its subtypes: A systematic review. *J. Am. Acad. Dermatol.* **2021**, *85*, 653–664. [CrossRef]

11. Mae, K.; Tsuboi, R.; Irisawa, R.; Sato, T.; Fukushima, N.; Harada, K. Recent reductions in the size of facial pigmented basal cell carcinoma at diagnosis and the surgical margin: A retrospective and comparative study. *J. Dermatol.* **2021**, *5*, 661–666. [CrossRef]

12. Conforti, C.; Giuffrida, R.; Zalaudek, I.; Guarneri, F.; Cannavò, S.P.; Pizzichetta, M.A.; Bonin, S.; Corneli, P.; Bussani, R.; Bazzacco, G.; et al. Dermoscopic Findings in the Presurgical Evaluation of Basal Cell Carcinoma. A Prospective Study. *Dermatol. Surg.* **2021**, *47*, 37–41. [CrossRef]

13. Yélamos, O.; Braun, R.P.; Liopyris, K.; Wolner, Z.J.; Kerl, K.; Gerami, P.; Marghoob, A.A. Usefulness of dermoscopy to improve the clinical and histopathologic diagnosis of skin cancers. *J. Am. Acad. Dermatol.* **2019**, *80*, 365–377. [CrossRef]

14. Van Limbergen, E.; Pötter, R.; Hoskin, P.; Baltas, D. *The GEC ESTRO Handbook of Brachytherapy*, 2nd ed.; European Society for Therapeutic Radiology and Oncology: Brussels, Belgium, 2019.

15. NCCN Guidelines Non Melanoma Skin Cancer, 2021,V.1. Available online: https://www.nccn.org/guidelines/guidelines-process/transparency-process-and-recommendations (accessed on 30 May 2021).

16. Peris, K.; Fargnoli, M.C.; Garbe, C.; Kaufmann, R.; Bastholt, L.; Seguin, N.B.; Bataille, V.; Marmol, V.D.; Dummer, R.; Harwood, C.A.; et al. Diagnosis and treatment of basal cell carcinoma: European consensus-based interdisciplinary guidelines. *Eur. J. Cancer* **2019**, *118*, 10–34. [CrossRef]

17. Rowe, D.E.; Carroll, R.J.; Day, C.L.J. Long-term recurrence rates in previously untreated (primary) basal cell carcinoma: Implications for patient follow-up. *J. Dermatol. Surg. Oncol.* **1989**, *15*, 315–328. [CrossRef]

18. Rowe, D.E.; Carroll, R.J.; Day, C.L.J. Mohs surgery is the treatment of choice for recurrent (previously treated) basal cell carcinoma. *J. Dermatol. Surg. Oncol.* **1989**, *15*, 424–431. [CrossRef]

19. American Joint Committee on Cancer. Cutaneous Squamous Cell Carcinoma of the Head and Neck. In *AJCC Cancer Staging Manual*, 8th ed.; Springer: New York, NY, USA, 2017; pp. 171–181.

20. Messina, J.; Epstein, E.H.J.; Kossard, S.; McKenzie, C.; Patel, R.M.; Patterson, J.W.; Scolyer, R.A. Basal cell carcinoma. In *WHO Classification of Skin Tumours*; IARC Publications: Lyon, France, 2018; pp. 26–34.

21. Cox, J.D.; Stetz, J.; Pajak, T.F. Toxicity criteria of the Radiation Therapy Oncology Group (RTOG) and the European Organization for Research and Treatment of Cancer (EORTC). *Int. J. Radiat. Oncol. Biol. Phys.* **1995**, *31*, 1341–1346. [CrossRef]

22. Guinot, J.L.; Rembielak, A.; Perez-Calatayud, J.; Rodríguez-Villalba, S.; Skowronek, J.; Tagliaferri, L.; Guix, B.; Gonzalez-Perez, V.; Valentini, V.; Kovacs, G.; et al. GEC-ESTRO ACROP recommendations in skin brachytherapy. *Radiother. Oncol.* **2018**, *126*, 377–385. [CrossRef]

23. Tognetti, L.; Cinotti, E.; Fiorani, D.; Couzan, C.; Cavarretta, C.; Chazelle, M.; Labeille, B.; Pianigiani, E.; Cevenini, G.; Perrot, J.L.; et al. Long-term therapy of multiple basal cell carcinomas: Clinicodermoscopic score for monitoring of intermittent vismodegib treatment. *Dermatol. Ther.* **2019**, *32*, e13097. [CrossRef]

24. R Core Team. R: A Language and Environment for Statistical Computing. Available online: http://cran.r-project.org (accessed on 30 May 2021).

25. Dourmishev, L.A.; Rusinova, D.; Botev, I. Clinical variants, stages, and management of basal cell carcinoma. *Indian Dermatol. Online J.* **2013**, *4*, 12–17. [CrossRef] [PubMed]

26. Navarrete-Dechent, C.; Cordova, M.; Liopyris, K.; Aleissa, S.; Rajadhyaksha, M.; Cohen, G.; Marghoob, A.A.; Rossi, A.M.; Barker, C.A. In vivo imaging characterization of basal cell carcinoma and cutaneous response to high-dose ionizing radiation therapy: A prospective study of reflectance confocal microscopy, dermoscopy, and ultrasonography. *J. Am. Acad. Dermatol.* **2021**, *84*, 1575–1584. [CrossRef] [PubMed]

27. Richtig, E.; Arzberger, E.; Hofmann-Wellenhof, R.; Fink-Puches, R. Assessment of changes in lentigo maligna during radiotherapy by in-vivo reflectance confocal microscopy: A pilot study. *Br. J. Dermatol.* **2015**, *172*, 81–87. [CrossRef] [PubMed]

28. Sánchez, R.B.; Llanas, O.P.; Calatayud, J.P.; Estrada, R.B. Dermoscopy margin delineation in radiotherapy planning for superficial or nodular basal cell carcinoma. *Br. J. Dermatol.* **2015**, *172*, 1162–1163. [CrossRef] [PubMed]

29. Diluvio, L.; Bavetta, M.; Di Prete, M.; Orlandi, A.; Bianchi, L.; Campione, E. Dermoscopic monitoring of efficacy of ingenolmebutate in the treatment of pigmented and non-pigmented basal cell carcinomas. *Dermatol. Ther.* **2017**, *30*, e12438. [CrossRef]

30. Aguilar, J.A.; Garcés, M.H.; Bayona, J.I.Y.; Azcona Rodríguez, M.; Martínez de Espronceda Ezquerro, I.; Sarriugarte Aldecoa-Otalora, J. Criterios dermatoscópicos comopredictores de ausencia de respuesta a tratamiento con imiquimod en carcinomas basocelulares superficiales [Dermoscopic signs as predictors of non-response to imiquimod treatment in superficial basal cellcarcinoma]. *An. Sist. Sanit. Navar.* **2019**, *42*, 303–307.

31. Husein-ElAhmed, H.; Fernandez-Pugnaire, M.A. Dermatoscopy-guided therapy of pigmented basal cell carcinoma with imiquimod. *An. Bras. Dermatol.* **2016**, *91*, 764–769. [CrossRef] [PubMed]

32. Dörr, W. Radiobiology of tissue reactions. *Ann. ICRP* **2015**, *44* (Suppl. S1), 58–68. [CrossRef]

Article

Interstitial HDR Brachytherapy in the Treatment of Non-Melanocytic Skin Cancers around the Eye

Paweł Cisek [1,2], Dariusz Kieszko [1], Mateusz Bilski [2,3,*], Radomir Dębicki [1], Ewelina Grywalska [4,5], Rafał Hrynkiewicz [6], Dominika Bębnowska [6], Izabela Kordzińska-Cisek [7], Agnieszka Rolińska [8], Paulina Niedźwiedzka-Rystwej [5,*] and Ludmiła Grzybowska-Szatkowska [2,3]

[1] Department of Brachytherapy, St. John's Cancer Center, 20-090 Lublin, Poland; pcisek@interia.eu (P.C.); dkieszko@cozl.pl (D.K.); rdebicki@cozl.pl (R.D.)
[2] Department of Radiotherapy, Medical University of Lublin, 20-093 Lublin, Poland; ludmila.grzybowska-szatkowska@umlub.pl
[3] Department of Radiotherapy, St. John's Cancer Center, 20-090 Lublin, Poland
[4] Department of Clinical Immunology and Immunotherapy, Medical University of Lublin, 20-093 Lublin, Poland; ewelina.grywalska@umlub.pl
[5] Department of Clinical Immunology, St. John's Cancer Center, 20-090 Lublin, Poland
[6] Institute of Biology, University of Szczecin, Felczaka 3c, 71-412 Szczecin, Poland; rafal.hrynkiewicz@usz.edu.pl (R.H.); dominika.bebnowska@usz.edu.pl (D.B.)
[7] I Department of Clinical Oncology, St. John's Cancer Center, 20-090 Lublin, Poland; ikordzinska@cozl.pl
[8] Department of Applied Psychology, Medical University of Lublin, 20-093 Lublin, Poland; aga.rolinska@wp.pl
* Correspondence: bilskimat@gmail.com (M.B.); paulina.niedzwiedzka-rystwej@usz.edu.pl (P.N.-R.)

Simple Summary: Eyelid tumors account for approximately 3% of all head and neck cancers and 5 to 10% of all skin cancers. Among the basic methods of treating eyelid tumors, apart from surgery, is radiotherapy, but this method carries a high risk of complications within the eye lens and may lead to the development of cataracts. Interstitial HDR brachytherapy is a less invasive method of skin cancer treatment. Unfortunately, the analysis of the literature to date has shown that it is rarely used in the treatment of skin cancer in this location. In our study, we analyzed the results of 28 patients treated with HDR interstitial brachytherapy. We showed that this is a highly effective, short-lived and relatively low burden method of treating patients with neoplasms of the skin of the eyelids, medial and lateral angles, and skin cancer of the cheek, nose and temples with an infiltration of the ocular structures.

Abstract: Background: Eyelid tumors are rare skin cancers, the most common of which is basal cell carcinoma characterized primarily by local growth. In addition to surgery, radiotherapy is among the basic methods of treatment. External beam radiotherapy is associated with the risk of complications within ocular structures, especially the lens. In the case of interstitial brachytherapy, it is possible to administer a high dose to the clinical target volume (CTV), while reducing it in the most sensitive structures. Methods: This paper presents the results of an analysis of 28 patients treated with interstitial high dose rate (HDR) brachytherapy for skin cancers of the upper and lower eyelid; medial and lateral canthus; and the cheek, nose and temples with the infiltration of ocular structures. The patients were treated according to two irradiation schedules: 49 Gy in 14 fractions of 3.5 Gy twice a day for 7 days of treatment, and 45 Gy in 5 Gy fractions twice a day for 5 days. The mean follow-up was 22 months (3–49 months). Results: two patients (6%) had a relapse: a local recurrence within the irradiated area in one of them, and metastases to lymph nodes in the other. The most common early complication was conjunctivitis (74%), and the most common late complication was dry eye syndrome (59%). Conclusions: Interstitial HDR brachytherapy for skin cancers of the upper and lower eyelid; medial and lateral cants; and the cheek, nose and temples with infiltration of ocular structures is a highly effective, short and relatively low burden type of treatment.

Keywords: HDR brachytherapy; non-melanocytic skin cancer; cancer around the eye; radiation therapy

Citation: Cisek, P.; Kieszko, D.; Bilski, M.; Dębicki, R.; Grywalska, E.; Hrynkiewicz, R.; Bębnowska, D.; Kordzińska-Cisek, I.; Rolińska, A.; Niedźwiedzka-Rystwej, P.; et al. Interstitial HDR Brachytherapy in the Treatment of Non-Melanocytic Skin Cancers around the Eye. *Cancers* **2021**, *13*, 1425. https://doi.org/10.3390/cancers13061425

Academic Editor: Girish M. Shah

Received: 21 February 2021
Accepted: 16 March 2021
Published: 20 March 2021

Publisher's Note: MDPI stays neutral with regard to jurisdictional claims in published maps and institutional affiliations.

1. Introduction

Eyelid tumors are rare skin cancers, accounting for about 3% of all head and neck cancers and 5–10% of all skin cancers. Annual incidence is only 1.37 per 100,000 people. Basal cell carcinoma is the most common histopathological type, accounting for about 90% of cases. Squamous cell carcinoma, sebaceous gland carcinoma, Merkel cell carcinoma and melanoma are less common [1,2]. While basal cell carcinoma is primarily characterized by local growth and low metastatic potential (0.028 to 0.55%), squamous cell carcinoma is an aggressive cancer with metastasis percentages between 5 and 12% [1,3,4]. Squamous cell carcinoma also has a worse prognosis, with 5- and 10-year survival rates at 90 and 83–87%, respectively. In basal cell carcinoma, 5- and 10-year survival rates are close to 100% [4]. Assessment by the general practitioners is often a key step towards a patient diagnosis. To help physicians in selecting skin cancer patients, a checklist was developed that provides essential tips to accelerate the patient's progress throughout the diagnosis stage and thus increase the chances of recovery. Unfortunately, the further stages of cancer diagnosis can also be difficult [5]. Despite the fact that these two types of carcinoma are well characterized, they are often misdiagnosed in dermatoscopic diagnosis. In order to avoid basal cell carcinoma being incorrectly defined as squamous cell carcinoma and vice versa, the criteria for its differentiation must be suitably critical [6]. The most common location within the eye is the lower eyelid (50–60%) and medial canthus (25–30%) [1]. The basic method of treatment is surgery; the others are cryotherapy, laser therapy, local chemotherapy, photodynamic therapy, immunotherapy and radiation therapy [7,8]. However, some of these methods may not cause acceptable side effects in patients after treatment. For patients with carcinoma of the eyelids or lips who do not agree to the proposed methods of treatment for aesthetic and functional reasons, radical radiotherapy remains a good solution as the basic therapy [9]. Commonly used radiotherapy techniques include X-ray teleradiotherapy, megavoltage photon teleradiotherapy, electron teleradiotherapy and brachytherapy, but Hedgehog pathway inhibitors may also be an effective and attractive therapy for local basal cell carcinoma located in the eyes and eyelids [10].

The main limitation of the use of radiation therapy is its toxicity to ocular structures, primarily to lenses, which, due to their low radiation tolerance, are exposed to late toxicity consisting of lens fibrosis and, consequently, cataracts [11]. The advantage of interstitial brachytherapy is the possibility of placing applicators inside the target, which makes it possible to administer a high dose within the clinical target volume (CTV). The rapid decrease in the dose with the growing distance from the applicator allows for the protection of adjacent structures, in particular, for a significant dose reduction within the most sensitive structures—the lenses.

2. Materials and Methods

2.1. Group Characteristics

The analysis included 28 patients diagnosed with skin cancers of the upper and lower eyelid; medial and lateral canthus; and the cheek, nose and temples with infiltration of the above-mentioned ocular structures. The patients were treated with HDR brachytherapy at the Brachytherapy Department of the Centre of Oncology of the Lublin Region between 2012 and 2017. All patients received radical treatment. Brachytherapy was used as independent treatment in 24 patients, and as adjuvant treatment after surgery in the other 4 patients. The indication for treatment was the microscopic non-radicality of the procedure. In the case of brachytherapy as an independent treatment, 8 patients had undergone surgery and suffered relapse after the treatment, and for the remaining patients, brachytherapy was the first method of cancer treatment. None of the patients had previously received any other radiation therapy. None of the patients were clinically diagnosed with lymph node metastases (9 patients had previously had a neck ultrasound, and 1 patient had had a CT scan of the neck). No distant metastases were found in any of the patients (all patients had chest X-ray, and 4 of them had abdominal ultrasound as well). The clinical and histopathological characteristics are presented in Table 1.

Table 1. The clinical and histopathological characteristics of patients.

Clinical and Histopathological Factors	Number/Median (Range)
age	82 (64–96)
Sex	
Men	17
Women	11
Histopathological type	
Basal cell carcinoma	24
G1 squamous cell carcinoma	2
G2 squamous cell carcinoma	2
Stage	
T1N0M0	23
T2N0M0	5
Location	
Lower eyelid	14
Upper eyelid	1
Medial canthus	11
Lateral canthus	1
Lower eyelid and medial canthus	1
Number of applicators	5 (3–11)

2.2. Application Procedure

The application procedure in most patients (26) was performed under local infiltration anesthesia with 1% lignocaine. In the remaining patients, the procedure was performed under short intravenous anesthesia with propofol and fentanyl. Flexible applicators of 35 cm long made by Varian Medical System were used. They were inserted under the skin at a depth of 2–5 mm, parallel to each other, according to the Paris system, so that they covered the tumor or the tumor bed with a 1–2 cm healthy skin margin. The applicators were placed densely, every 2–6 mm, which allowed for a rapid decrease in the dose outside the irradiation area (Figure 1). After the application procedure, a tomography for treatment planning was performed using a 32-row Siemens CT scanner. The slice thickness was 1–3 mm. GTV was defined as the area with clinically evident neoplastic infiltration, and CTV comprised the infiltrate with a healthy tissue margin of 0.5–1 cm. In the case that the infiltration border was the eyeball, the CTV did not include the eyeball structures. Based on the location of the applicators, the CTV area was contoured on tomography scans for treatment planning. The eyeball, lens, optic nerve and retina were also drawn, as well as the structure of the opposite eye, if relevant to the location of the tumor.

2.3. Application Procedure

The patients were treated according to 2 irradiation schedules: 49 Gy in 14 fractions of 3.5 Gy twice a day for 7 days of treatment (9 patients) and 45 Gy in 5 Gy fractions twice a day for 5 days (19 patients). The choice of schedule was determined by a number of factors. The shorter one was used in the elderly, patients in worse general condition and patients with less advanced disease. Treatment planning was carried out using the BrachyVision treatment planning system. An Ir 192 source with a diameter of 0.6 mm and an average activity of 10 Ci was used. Treatment was performed using a 24-channel GammaMed or GammaMedplus (Varian) afterloader.

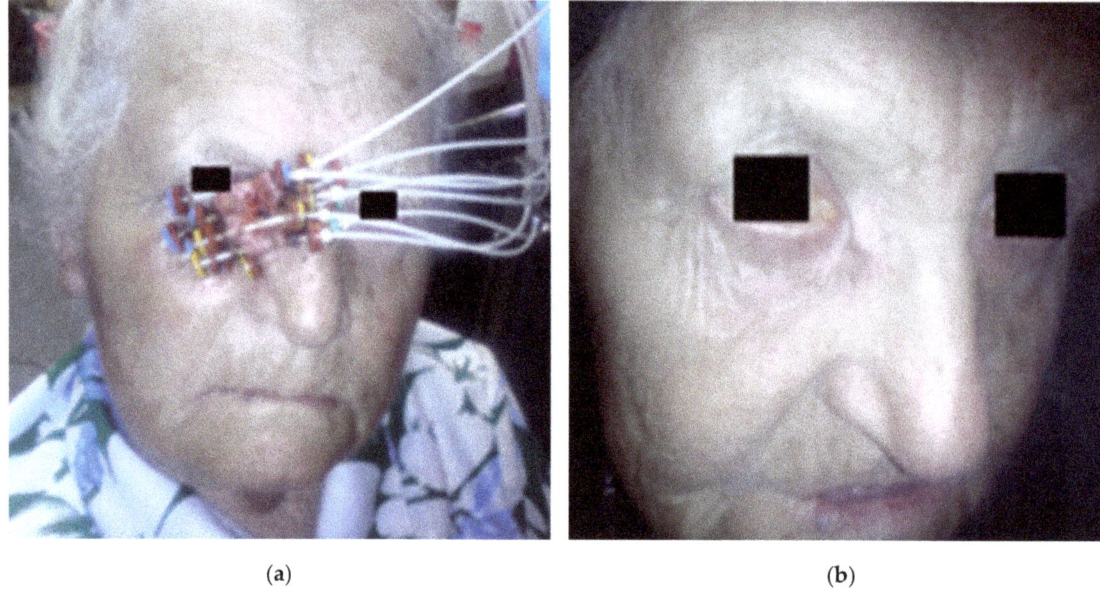

(a)   (b)

Figure 1. (**a**) Patient, age 80, basal cell carcinoma of the lower eyelid and medial canthus of the right eye. Applicators placed in the tumor; (**b**) the same patient, six months after the end of treatment.

2.4. Post-Treatment Surveillance

In the post-treatment period, patients were subjected to periodic clinical evaluation, initially every month for the first 3 months; some of them were also subjected to ophthalmic evaluation. Ultrasound of the neck and abdomen and chest X-ray were also carried out periodically in selected patients. Patients were evaluated for local recurrence of lymph node metastases and distant metastases. The frequency and severity of early and late side-effects of radiation were assessed as well. The Common Terminology Criteria for Adverse Events (CTCAE) v 4.0 scale and the Radiation Therapy Oncology Group (RTOG) scale were used to assess toxicity [12,13].

3. Results

3.1. Dosage and Treatment Planning

In patients with a fraction dose of 3.5 Gy, the average D90 (dose in 90% isodose) was 3.9 ± 0.4 Gy. The median D90 was 3.7 Gy (3.5–4.3 Gy). The mean D100% (dose in 100% isodose) was 2.6 ± 0.6 Gy. The median D100% was 2.7 Gy (2–3.2 Gy). The mean volume of irradiated lesion was 8.0 ± 6.7 cm^3. In patients with a fraction dose of 5 Gy, the mean D90 was 5.4 ± 0.5 Gy. The median D90 was 5.5 Gy (5–6 Gy). The mean D100% (dose in 100% isodose) was 3.7 ± 0.8 Gy. The median D100% was 3.7 Gy (3–4.6 Gy). The mean volume of irradiated lesion was 8.1 ± 6.7 cm^3. The total doses in both fractionation schemes were converted into the biologically effective dose (BED) and the 2 Gy-per-fraction equivalent dose (EQD2) for alpha/beta 10. The data are presented in Table 2.

Table 2. Characteristics of treatment schemes and dosage.

Fractionation Scheme		Median	Mean	Min	Max
3.5 Gy/49 Gy/BID (two times a day)	D90	51.8	54.6	49.0	60.2
	BED D90	71.0	75.9	66.2	86.1
	EQD2 D90	59.1	63.3	55.1	71.7
	D100	37.8	36.4	28.0	44.8
	BED D100	48.0	45.9	33.6	59.1
	EQD2 D100	40.0	38.2	28.0	49.3
5 Gy/45 Gy/BID	D90	48.6	49.5	45.0	54.0
	BED D90	74.8	76.7	67.5	86.4
	EQD2 D90	62.4	63.9	56.3	72.0
	D100	33.3	34.2	27.0	41.4
	BED D100	45.6	47.2	35.1	60.4
	EQD2 D100	38.02	39.3	29.3	50.4

The median volume of irradiated lesion was 8.8 cm^3 (1.4–14.8 cm^3)—in patients treated with a 3.5 Gy dose, it was 10.5 (4.3–13.1) cm^3, and in patients treated with dose of 5 Gy, it was 6.6 (1.4–8.9) cm^3.

Doses in critical organs were reported for the maximum dose in the nearest lens, eyeball, retina and optic nerve. All doses were converted to BED and EQD2 for alpha/beta 3. The data are presented in Table 3.

Table 3. Characteristics of doses achieved in critical organs.

Fractionation Scheme		Median	Mean	Min	Max
3.5 Gy/49 Gy/BID	Dmax lens	18.2	19.1	9.8	32.4
	BED Dmax	25.5	28.8	12.1	55.7
	EQD2 Dmax	15.3	17.3	7.3	33.4
	Dmax eyeball	48.6	48.9	43.4	57.8
	Dmax retina	12.6	11.2	7.0	24.9
	Dmax nerve	10.8	9.8	5.6	22.1
5 Gy/45 Gy/BID	Dmax lens	17.1	17.2	9.9	34.2
	BED Dmax	27.9	28.2	13.5	77.5
	EQD2 Dmax	16.8	16.9	8.2	46.5
	Dmax eyeball	44.1	45.2	36.9	53.1
	Dmax retina	11.7	12.9	9.9	35.1
	Dmax nerve	9.0	9.8	3.6	24.3

During the mean follow-up of 24 months (4–49 months), two patients (7%) had a relapse. In one of them, it was a local recurrence in the irradiated area. This was a patient with basal cell carcinoma of the canthus, and the recurrence, which was confirmed histopathologically, took place during the 8th month of follow-up. The second patient had metastases to submandibular lymph nodes after 3 months. The patient had extensive cutaneous squamous cell carcinoma of the cheek that covered the lower eyelid.

3.2. Toxicity

The early and late toxicity of the radiation therapy and application procedure was analyzed. Complications from the application procedure and acute toxicity were assessed in all 28 patients, and late toxicity was assessed in 20 patients. The data are presented in Table 4.

Table 4. Characteristics of early and late toxicity of radiation therapy and application procedure. CTCAE—Common Terminology Criteria for Adverse Events; RTOG—Toxicity criteria of the Radiation Therapy Oncology Group.

Complications from the Application Procedure	CTCAE/number (Percentage) of Patients	RTOG/number (Percentage) of Patients
oedema	CTCAE 1–28 (100%)	-
hematoma	CTCAE 1–3 (11%)	-
Early complications	**CTCAE/number (percentage) of patients**	**RTOG/number (percentage) of patients**
skin	-	RTOG 1–18 (64%) RTOG 2–9 (32%) RTOG 3–1 (3%)
Conjunctivitis	CTCAE 1–20 (71%) CTCAE 2–1 (3%)	-
Late complications	**CTCA/number (percentage) of patients**	**RTOG/number (percentage) of patients**
Eyelid deformity	CTCAE 1–6 (30%) CTCAE 2–1 (5%)	-
Dry eye syndrome	CTCAE 1–11 (55%)	-
Skin lesions (discoloration, thinning, telangiectasia)	-	RTOG 1–16 (80%)

4. Discussion

An analysis of the available literature indicates the rarity of the use of interstitial HDR brachytherapy in the treatment of skin cancers in this location. Available analyses regarding HDR brachytherapy involve small groups of patients (8–20 patients) [14–18], and LDR brachytherapy is more common (20–160 patients) [19–23]. Studies indicate the high efficiency of brachytherapy, in the range of 94–100%, both among patients treated with HDR brachytherapy and LDR brachytherapy [14–23]. A study by Vavssori et al. included 10 patients with eyelid tumors using contact high dose rate brachytherapy (HDR-BT) with a customized applicator. Except for one patient, all patients were perfectly tolerant to the treatment. After a median follow-up time of 51 months, none of the patients experienced optic neuropathy, retinopathy, lacrimation or visual impairment. Moreover, no relapse or metastasis was observed in patients, and the cosmetic effects were satisfactory. These data indicate that this method is effective and safe in the treatment of eyelid s and may be an effective alternative to other therapies [18]. There are also reports indicating that the therapeutic and cosmetic effect of brachytherapy in this localization may be unfavorable [24]. In the analyzed group of patients, the local control throughout the entire follow-up period was also high and amounted to 97%. These results are similar to those of teleradiotherapy (93–96.5%) [23,25,26] or surgical treatment (80–90%) [27–29]. Meta-analysis covering nearly 10,000 patients comparing EBRT vs. BTR showed a similar efficacy of both treatment methods in non-melanocytic skin tumors with a better cosmetic effect of brachytherapy [30].

Doses in the analyzed group of patients were reported in those ocular structures that are the most sensitive to radiation. Due to the properties of the source and the dense arrangement of applicators, a sharp dose reduction outside the irradiation area was achieved. The median total dose in the organs analyzed for the 3.5 Gy/49 Gy/BID scheme was 18.2, 12.6, 48.6 and 10.8 Gy in the lens, retina, cornea and optic nerve, respectively, and 17.1, 11.7, 44, 1 and 9 Gy. Studies show that the lens is the most sensitive to radiation [11]. Although a late radiation side-effect in the form of a cataract can occur even after a single dose of radiation of 2 Gy, it does not affect the quality of vision [11]. The severity and timing of cataracts are dose dependent. Studies indicate that irradiation up to a dose in the 2.5–6.5 Gy range is associated with a 33% cataract risk within 8 years, and irradiation up to a dose in the 6.5–11.5 Gy range with a cataract risk of 66% within 4 years [31]. Emami et al. [32] showed that there is a 5% risk of cataracts in 5 years (TD5/5) for 10 Gy, and a risk of

50% in 5 years (TD50/5) for 18 Gy. The risk of cataracts is associated with a number of factors, such as age, type of radiation and fraction dose [33–35]. Most of the patients in the analyzed group were elderly. Although the risk of radiation cataracts is lower in older patients than in younger patients irradiated with the same dose, radiation-induced cataracts in the elderly are concurrent with cataracts caused by other factors [33]. Research also indicates that the risk of cataracts increases with a higher fraction dose and is greater for brachytherapy and electron radiation than for photon radiation [35]. In the analyzed group of patients, the use of brachytherapy was associated with a short hospitalization time and a relatively low burden on the patient, although a slightly longer 1.5-week treatment regimen due to a lower fraction dose may be more beneficial in terms of lens protection.

Research indicates that the risk of retinal damage occurs when the 24 Gy dose is exceeded. TD5/5 is 45–50 Gy, and TD50/5 is 55 Gy [32,36]. The tolerance dose was not exceeded in the analyzed group of patients. Similarly, the optic nerve tolerance dose was not exceeded. In the study by Emami et al. [32] and analysis of Mayo et al. [37], the researchers indicate that a dose lower than 50 Gy is associated with a risk of nerve damage of less than 5%. Despite a relatively high dose in the cornea, resulting from the close proximity of applicators, studies indicate that the risk of corneal damage occurs after exceeding the 50 Gy dose, and the risk of corneal ulceration after exceeding 60 Gy [37,38].

5. Conclusions

Interstitial HDR brachytherapy for skin cancers of the upper and lower eyelid; medial and lateral cants; and the cheek, nose and temples with infiltration of ocular structures is a highly effective, short and relatively low burden type of treatment. The profile of expected toxicity is favorable. No severe and late CTCAE $\geq$ 3 or late RTOG $\geq$ 3 toxicity was observed.

Author Contributions: Conceptualization, P.C., D.K., M.B., R.D., I.K.-C. and L.G.-S.; methodology, P.C., D.K., M.B., R.D., E.G., P.N.-R., I.K.-C. and L.G.-S.; software, P.C., D.K., M.B., R.D., I.K.-C. and L.G.-S.; validation, P.C., D.K., M.B., R.D., E.G., R.H., D.B., I.K.-C., A.R., P.N.-R. and L.G.-S.; formal analysis, P.C., D.K., M.B., R.D., E.G., R.H., D.B., I.K.-C., A.R., P.N.-R. and L.G.-S.; investigation, P.C., D.K., M.B., R.D., I.K.-C. and L.G.-S.; resources, P.C., D.K., M.B., R.D., E.G., R.H., D.B., I.K.-C., A.R., P.N.-R. and L.G.-S.; data curation, P.C., D.K., M.B., R.D., I.K.-C. and L.G.-S.; writing—original draft preparation, P.C., D.K., M.B., R.D., E.G., I.K.-C. and L.G.-S.; writing—review and editing, R.H., D.B., I.K.-C., A.R. and P.N.-R.; visualization, P.C., D.K., M.B., R.D., E.G., R.H., D.B., I.K.-C., A.R., P.N.-R. and L.G.-S.; supervision, D.K., M.B., E.G., P.N.-R. and L.G.-S.; project administration, P.C., D.K., M.B., R.D., E.G., R.H., D.B., I.K.-C., A.R., P.N.-R. and L.G.-S.; funding acquisition, D.K., E.G. and L.G.-S. All authors have read and agreed to the published version of the manuscript.

Funding: This research was supported by the Medical University of Lublin (grant no. DS460).

Institutional Review Board Statement: The study was conducted according to the guidelines of the Declaration of Helsinki, and approved by the Ethics Committee of Medical University of Lublin, Poland (approval no. KE-0254/236/2020).

Informed Consent Statement: Informed consent was obtained from all subjects involved in the study. Written informed consent was obtained from the patients to publish this paper.

Data Availability Statement: Due to privacy and ethical concerns, the data that support the findings of this study are available on request from the corresponding author, M.B.

Conflicts of Interest: The authors declare no conflict of interest.

References

1. Frakulli, R.; Galuppi, A.; Cammelli, S.; Macchia, G.; Cima, S.; Gambacorta, M.A.; Cafaro, I.; Tagliaferri, L.; Perrucci, E.; Buwenge, M.; et al. Brachytherapy in Non Melanoma Skin Cancer of Eyelid: A Systematic Review. *J. Contemp. Brachytherapy* **2015**, *7*, 497–502. [CrossRef]
2. Petsuksiri, J.; Frank, S.J.; Garden, A.S.; Ang, K.K.; Morrison, W.H.; Chao, K.S.C.; Rosenthal, D.I.; Schwartz, D.L.; Ahamad, A.; Esmaeli, B. Outcomes after Radiotherapy for Squamous Cell Carcinoma of the Eyelid. *Cancer* **2008**, *112*, 111–118. [CrossRef]
3. Chung, S. Basal Cell Carcinoma. *Arch. Plast. Surg.* **2012**, *39*, 166–170. [CrossRef] [PubMed]

4. Rees, J.R.; Zens, M.S.; Celaya, M.O.; Riddle, B.L.; Karagas, M.R.; Peacock, J.L. Survival after Squamous Cell and Basal Cell Carcinoma of the Skin: A Retrospective Cohort Analysis. *Int. J. Cancer* **2015**, *137*, 878–884. [CrossRef] [PubMed]

5. Moscarella, E.; Lallas, A.; Longo, C.; Alfano, R.; Argenziano, G. Five-Point Checklist for Skin Cancer Detection in Primary Care. *G. Ital. Dermatol. Venereol.* **2019**, *154*, 523–528. [CrossRef]

6. Neagu, N.; Lallas, K.; Maskalane, J.; Salijuma, E.; Papageorgiou, C.; Gkentsidi, T.; Spyridis, I.; Morariu, S.H.; Apalla, Z.; Lallas, A. Minimizing the Dermatoscopic Morphologic Overlap between Basal and Squamous Cell Carcinoma: A Retrospective Analysis of Initially Misclassified Tumours. *J. Eur. Acad. Dermatol. Venereol.* **2020**, *34*, 1999–2003. [CrossRef] [PubMed]

7. Delishaj, D.; Rembielak, A.; Manfredi, B.; Ursino, S.; Pasqualetti, F.; Laliscia, C.; Orlandi, F.; Morganti, R.; Fabrini, M.G.; Paiar, F. Non-Melanoma Skin Cancer Treated with High-Dose-Rate Brachytherapy: A Review of Literature. *J. Contemp. Brachytherapy* **2016**, *8*, 533–540. [CrossRef] [PubMed]

8. Fenton, S.E.; Sosman, J.A.; Chandra, S. Current Therapy for Basal Cell Carcinoma and the Potential Role for Immunotherapy with Checkpoint Inhibitors. *Clin. Ski. Cancer* **2017**, *2*, 59–65. [CrossRef]

9. Fania, L.; Didona, D.; Di Pietro, F.R.; Verkhovskaia, S.; Morese, R.; Paolino, G.; Donati, M.; Ricci, F.; Coco, V.; Ricci, F.; et al. Cutaneous Squamous Cell Carcinoma: From Pathophysiology to Novel Therapeutic Approaches. *Biomedicines* **2021**, *9*, 171. [CrossRef]

10. Villani, A.; Cinelli, E.; Fabbrocini, G.; Lallas, A.; Scalvenzi, M. Hedgehog Inhibitors in the Treatment of Advanced Basal Cell Carcinoma: Risks and Benefits. *Expert Opin. Drug Saf.* **2020**, *19*, 1585–1594. [CrossRef]

11. Jeganathan, V.S.E.; Wirth, A.; MacManus, M.P. Ocular Risks from Orbital and Periorbital Radiation Therapy: A Critical Review. *Int. J. Radiat. Oncol. Biol. Phys.* **2011**, *79*, 650–659. [CrossRef]

12. Common Terminology Criteria for Adverse Events (CTCAE). Available online: https://evs.nci.nih.gov/ftp1/CTCAE/CTCAE_4.03/CTCAE_4.03_2010-06-14_QuickReference_5x7.pdf (accessed on 20 February 2021).

13. RTOG.org. Available online: https://www.rtog.org/ResearchAssociates/AdverseEventReporting/CooperativeGroupCommonToxicityCriteria.aspx (accessed on 20 February 2021).

14. Laskar, S.G.; Basu, T.; Chaudhary, S.; Chaukar, D.; Nadkarni, M.; Gn, M. Postoperative Interstitial Brachytherapy in Eyelid Cancer: Long Term Results and Assessment of Cosmesis after Interstitial Brachytherapy Scale. *J. Contemp. Brachytherapy* **2015**, *6*, 350–355. [CrossRef]

15. Mareco, V.; Bujor, L.; Abrunhosa-Branquinho, A.N.; Ferreira, M.R.; Ribeiro, T.; Vasconcelos, A.L.; Ferreira, C.R.; Jorge, M. Interstitial High-Dose-Rate Brachytherapy in Eyelid Cancer. *Brachytherapy* **2015**, *14*, 554–564. [CrossRef]

16. Azad, S.; Choudhary, V. Treatment Results of High Dose Rate Interstitial Brachytherapy in Carcinoma of Eye Lid. *J. Cancer Res. Ther.* **2011**, *7*, 157–161. [CrossRef] [PubMed]

17. Rio, E.; Bardet, E.; Ferron, C.; Peuvrel, P.; Supiot, S.; Campion, L.; de Montreuil, C.B.; Mahe, M.A.; Dreno, B. Interstitial Brachytherapy of Periorificial Skin Carcinomas of the Face: A Retrospective Study of 97 Cases. *Int. J. Radiat. Oncol. Biol. Phys.* **2005**, *63*, 753–757. [CrossRef] [PubMed]

18. Vavassori, A.; Riva, G.; Durante, S.; Fodor, C.; Comi, S.; Cambria, R.; Cattani, F.; Spadola, G.; Orecchia, R.; Jereczek-Fossa, B.A. Mould-Based Surface High-Dose-Rate Brachytherapy for Eyelid Carcinoma. *J. Contemp. Brachytherapy* **2019**, *11*, 443–448. [CrossRef]

19. Daly, N.J.; De Lafontan, B.; Combes, P.F. Results of the Treatment of 165 Lid Carcinomas by Iridium Wire Implant. *Int. J. Radiat. Oncol. Biol. Phys.* **1984**, *10*, 455–459. [CrossRef]

20. Conill, C.; Sánchez-Reyes, A.; Molla, M.; Vilalta, A. Brachytherapy with 192Ir as Treatment of Carcinoma of the Tarsal Structure of the Eyelid. *Int. J. Radiat. Oncol. Biol. Phys.* **2004**, *59*, 1326–1329. [CrossRef] [PubMed]

21. Ducassou, A.; David, I.; Filleron, T.; Rives, M.; Bonnet, J.; Delannes, M. Retrospective Analysis of Local Control and Cosmetic Outcome of 147 Periorificial Carcinomas of the Face Treated with Low-Dose Rate Interstitial Brachytherapy. *Int. J. Radiat. Oncol. Biol. Phys.* **2011**, *81*, 726–731. [CrossRef]

22. Krengli, M.; Masini, L.; Comoli, A.M.; Negri, E.; Deantonio, L.; Filomeno, A.; Gambaro, G. Interstitial Brachytherapy for Eyelid Carcinoma. *Outcome Analysis in 60 Patients. Strahlenther. Onkol.* **2014**, *190*, 245–249. [CrossRef]

23. Lederman, M. Radiation treatment of cancer of the eyelids. *Br. J. Ophthalmol.* **1976**, *60*, 794–805. [CrossRef]

24. Eftekhari, K.; Anderson, R.L.; Suneja, G.; Bowen, A.; Oberg, T.J.; Bowen, G.M. Local Recurrence and Ocular Adnexal Complications Following Electronic Surface Brachytherapy for Basal Cell Carcinoma of the Lower Eyelid. *JAMA Dermatol.* **2015**, *151*, 1002–1004. [CrossRef]

25. Schlienger, P.; Brunin, F.; Desjardins, L.; Laurent, M.; Haye, C.; Vilcoq, J.R. External radiotherapy for carcinoma of the eyelid: Report of 850 cases treated. *Int. J. Radiat. Oncol. Biol. Phys.* **1996**, *34*, 277–287. [CrossRef]

26. Bertelsen, K.; Gadeberg, C. Carcinoma of the eyelid. *Acta Radiol. Oncol. Radiat. Phys. Biol.* **1978**, *17*, 58–64. [CrossRef] [PubMed]

27. Berking, C.; Hauschild, A.; Kölbl, O.; Mast, G.; Gutzmer, R. Basal cell carcinoma-treatments for the commonest skin cancer. *Dtsch. Ärzteblatt Int.* **2014**, *111*, 389–395. [CrossRef]

28. Gill, H.S.; Moscato, E.E.; Seiff, S.R. Eyelid Margin Basal Cell Carcinoma Managed with Full-Thickness En-Face Frozen Section Histopathology. *Ophthalmic Plast. Reconstr. Surg.* **2014**, *30*, 15–19. [CrossRef]

29. Castley, A.J.; Theile, D.R.; Lambie, D. The Use of Frozen Section in the Excision of Cutaneous Malignancy: A Queensland Experience. *Ann. Plast. Surg.* **2013**, *71*, 386–389. [CrossRef]

30. Zaorsky, N.G.; Lee, C.T.; Zhang, E.; Galloway, T.J. Skin CanceR Brachytherapy vs External Beam Radiation Therapy (SCRiBE) Meta-Analysis. *Radiother. Oncol.* **2018**, *126*, 386–393. [CrossRef] [PubMed]
31. Merriam, G.R.; Szechter, A.; Focht, E.F. *The Effects of Ionizing Radiations on the Eye1*; Columbia University: New York, NY, USA, 2015; pp. 346–385.
32. Emami, B.; Lyman, J.; Brown, A.; Cola, L.; Goitein, M.; Munzenrider, J.E.; Shank, B.; Solin, L.J.; Wesson, M. Tolerance of Normal Tissue to Therapeutic Irradiation. *Int. J. Radiat. Oncol. Biol. Phys.* **1991**, *21*, 109–122. [CrossRef]
33. Eric, J.; Hall, A.J.G. *Radiobiology for the Radiologist*; Lippincott: New York, NY, USA, 1988; Volume 12–13, pp. 439–442.
34. Deeg, H.J.; Flournoy, N.; Sullivan, K.M.; Sheehan, K.; Buckner, C.D.; Sanders, J.E.; Storb, R.; Witherspoon, R.P.; Thomas, E.D. Cataracts after Total Body Irradiation and Marrow Transplantation: A Sparing Effect of Dose Fractionation. *Int. J. Radiat. Oncol. Biol. Phys.* **1984**, *10*, 957–964. [CrossRef]
35. Takeda, A.; Shigematsu, N.; Suzuki, S.; Fujii, M.; Kawata, T.; Kawaguchi, O.; Uno, T.; Takano, H.; Kubo, A.; Ito, H. Late Retinal Complications of Radiation Therapy for Nasal and Paranasal Malignancies: Relationship between Irradiated-Dose Area and Severity. *Int. J. Radiat. Oncol. Biol. Phys.* **1999**, *44*, 599–605. [CrossRef]
36. Evans, J.R.; Sivagnanavel, V.; Chong, V. Radiotherapy for Neovascular Age-Related Macular Degeneration. *Cochrane Database Syst. Rev.* **2010**, *2010*, CD004004. [CrossRef] [PubMed]
37. Gordon, K.B.; Char, D.H.; Sagerman, R.H. Late Effects of Radiation on the Eye and Ocular Adnexa. *Int. J. Radiat. Oncol. Biol. Phys.* **1995**, *31*, 1123–1139. [CrossRef]
38. Barabino, S.; Raghavan, A.; Loeffler, J.; Dana, R. Radiotherapy-Induced Ocular Surface Disease. *Cornea* **2005**, *24*, 909–914. [CrossRef] [PubMed]

Article

Remote Skin Cancer Diagnosis: Adding Images to Electronic Referrals Is More Efficient Than Wait-Listing for a Nurse-Led Imaging Clinic

Leah Jones [1,*], **Michael Jameson** [1,2] and **Amanda Oakley** [1,2]

1   Waikato District Health Board, Hamilton 3204, New Zealand; michael.jameson@waikatodhb.health.nz (M.J.); amanda.oakley@waikatodhb.health.nz (A.O.)

2   Waikato Clinical Campus, University of Auckland, Hamilton 3204, New Zealand

*   Correspondence: leah.jones@waikatodhb.health.nz

Simple Summary: Skin cancer is a significant cause of death and disability, particularly in New Zealand. Expert diagnosis reduces unnecessary excision of benign lesions, reduces patient anxiety, and allows early identification of skin cancer, particularly of melanoma. The study assessed an electronic referral pathway for teledermatology—diagnosing skin lesions remotely using a standardised template with regional, close-up, and dermoscopic images—and compared this to scheduled nurse-led teledermoscopy clinics. A dermatology opinion was reached more rapidly with comparable efficacy when referrals include good quality images, compared to nurse-led imaging clinics.

Abstract: We undertook a retrospective comparison of two teledermatology pathways that provide diagnostic and management advice for suspected skin cancers, to evaluate the time from referral to diagnosis and its concordance with histology. Primary Care doctors could refer patients to either the Virtual Lesion Clinic (VLC), a nurse-led community teledermoscopy clinic or, more recently, to the Suspected Skin Cancer (SSC) pathway, which requires them to attach regional, close-up, and dermoscopic images. The primary objective of this study was to determine the comparative time course between the SSC pathway and VLC. Secondary objectives included comparative diagnostic concordance, skin lesion classification, and evaluation of missed skin lesions during subsequent follow-up. VLC referrals from July to December 2016 and 2020 were compared to SSC referrals from July to December 2020. 408 patients with 682 lesions in the VLC cohort were compared with 480 patients with 548 lesions from the 2020 SSC cohort, matched for age, sex, and ethnicity, including histology where available. Median time (SD) from referral to receipt of teledermatology advice was four (2.8) days and 50 (43.0) days for the SSC and VLC cohorts, respectively ($p < 0.001$). Diagnostic concordance between teledermatologist and histopathologist for benign versus malignant lesions was 70% for 114 lesions in the SSC cohort, comparable to the VLC cohort (71% of 122 lesions). Referrals from primary care, where skin lesions were imaged with variable devices and quality resulted in faster specialist advice with similar diagnostic performance compared to high-quality imaging at nurse-led specialist dermoscopy clinics.

Keywords: teledermatology; teledermoscopy; skin cancer; squamous cell carcinoma; basal cell carcinoma; melanoma; telemedicine; skin neoplasms; referral and consultation

Citation: Jones, L.; Jameson, M.; Oakley, A. Remote Skin Cancer Diagnosis: Adding Images to Electronic Referrals Is More Efficient Than Wait-Listing for a Nurse-Led Imaging Clinic. *Cancers* **2021**, *13*, 5828. https://doi.org/10.3390/cancers13225828

Academic Editor: Aimilios Lallas

Received: 29 September 2021
Accepted: 15 November 2021
Published: 20 November 2021

Publisher's Note: MDPI stays neutral with regard to jurisdictional claims in published maps and institutional affiliations.

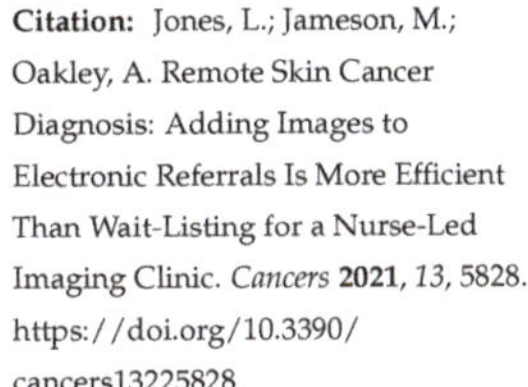

1. Introduction

Skin cancer prevalence is increasing in association with an ageing population with sun-damaged skin [1]. New Zealand has one of the highest rates of melanoma in the world, due to high ultraviolet intensities in the Southern Hemisphere during summer and high proportion of the population with fair skin [2,3]. It has an age-standardised incidence of 30–50 per 100,000 [4,5]. The incidence of basal cell carcinoma (BCC) and squamous cell carcinoma (SCC) is harder to evaluate, given that these are not required to be reported to the cancer registry; however, research has estimated the incidence of cutaneous SCC to be 425–668 per 100,000 and BCC to be 1177 per 10,000, putting it among the highest in the world [6,7]. New Zealand has limited access to publicly-funded dermatology services, with one dermatologist per 274,146 population, far less than the recommended 1 per 100,000 based on referral numbers [8]. In 2016 and 2020 respectively, the Waikato District Health Board, a tertiary dermatology referral centre, had 2.5 and 2.0 full-time equivalent dermatologists for a population of about half a million people [9].

Teledermatology aims to increase access to skin lesion assessment by a dermatologist by using telecommunications technology to deliver healthcare. There are two main methods of teledermatology: video conferencing, in which the patient is reviewed virtually in real-time by a specialist and 'store and forward', where the images are taken to be reviewed at a later time. Advances in technology have increased the opportunities available in teledermatology. It has been gaining in popularity, particularly in light of the COVID-19 pandemic and subsequent restrictions to in-person consults. Recent surveys of Dermatologists in the United States and India showed more than 80% were now offering teledermatology [10]. Our earlier research confirmed that skin tumours could be accurately diagnosed by teledermoscopy [11], which led in January 2010 to the introduction of a store-and-forward method of teledermatology we called the Virtual Lesion Clinic (VLC) [12]. General practitioners (GPs) refer patients with one or more skin lesions of concern to the dermatology service. Patients are allocated to a nurse-led imaging clinic in one of three locations. After reviewing the files, the teledermatologist discharges the patient or refers them to a specialist surgical service for excision, arranges to monitor the lesion at the VLC, or advises the GP on treatment in primary care. By mid-2019, the publicly funded VLC had evaluated more than 11,000 lesions in 6600 patients attending scheduled nurse-specialist clinics in three community locations [13]. The safety and efficacy of the VLC model has been confirmed [14,15].

The introduction of electronic referrals in February 2016 led to a direct Suspected Skin Cancer (SSC) advice service in July 2017. The SSC pathway template expects the referring Primary Care doctor to attach regional, close-up, and dermoscopy images for 1 to 5 lesions. This has become possible mainly due to the ubiquitous use of smartphones with adequate cameras and dermoscopic attachments [16]. The teledermatologist provides advice to the GP directly. Unlike VLC, the SSC pathway requires the GP to arrange skin lesion excision independently. If images are inadequate for diagnosis, this is conveyed to the GP with a request for further images or to request assessment at the VLC. This SSC template is currently used for more than half of the referrals to Dermatology each month (the remainder use the 'General Dermatology' template).

The benefit of increased access to skin cancer screening by teledermoscopy, particularly in screening rural and low-income groups, has been well established [17,18]. A 2018 systematic review that included 16 studies, predominantly cross-sectional and observational, reported diagnostic accuracy in eight studies [19]. This included our local interventional study, which compared dermatologist concordance of teledermoscopy and in-person reviews, where the concordance between dermatologists was 74% [11]. Teledermoscopy diagnosis of 43 skin lesions was concordant with histology in 85% of cases compared to 91% for in-person diagnosis [20].

An interventional study of store-and-forward teledermoscopy compared to previous in-person referrals found the time to diagnose 79 patients reduced from 70 days to 0.5 days ($p < 0.001$) and reduced the time until definitive treatment from 73.5 days to 3.0 days ($p < 0.001$) [21]. The teledermatology images were assessed by two dermatologists independently with at least partial concordance in all cases (complete concordance in 32%) [21]. Of the 29 patients that went on to have in-person consultation, the concordance between teledermatologist and dermatologist was partially concordant in 15 patients (52%) and completely concordant in 11 patients (38%) [21]. A retrospective cohort study of 2385 referrals in a poorly-serviced area in the USA showed an 84% reduction in in-person consultations and reduced wait time from 77 days to 28 days by using teledermatology [22].

The objectives of this retrospective study were to assess the efficacy and efficiency of direct teledermoscopy via GP electronic referral to the SSC pathway in comparison to the VLC pathway. The primary objective was to compare the time from receipt of the referral to return of the Dermatologist advice. Secondary objectives were, firstly, to determine diagnostic concordance between GP and dermatologist (and histology if available) between the SSC 2020 cohort and VLC cohorts from 2020 and 2016, and secondly, the identification of any incidental suspected skin cancers found during follow-up specialist appointments.

2. Materials and Methods

The Health and Disabilities Committee (HDEC) of New Zealand determined the study to be a service review and therefore out of scope for formal ethics approval (22 December 2020).

We undertook a retrospective review to compare outcomes of referrals for possible skin cancer to either the VLC or SSC pathway in 2020 or the VLC in 2016. As there was a reduction in referral numbers during the first four months of 2020, due to healthcare interruptions related to COVID-19, we chose to evaluate referrals received 1 July–31 December in 2016 and 2020 for consistency.

Referrals were identified by searching departmental records. For the 2020 SSC and VLC cohorts, the keywords 'lesion' and 'skin cancer' were used; for the 2016 VLC cohort a unique coding identifier was used. Referrals for reasons other than a suspected skin cancer were excluded.

From the 2020 SSC referrals, 2020 VLC and 2016 VLC referrals, we determined the number of referrals, the time from referral to advice, the number of lesions per referral, and patient demographics—including age, sex, ethnicity, and whether urban, rural, or semi-rural residence—based on the Statistics New Zealand Urban Accessibility Classification [23]. For the SSC pathway cohort, we recorded data on image quality. Propensity score matching was used to create a subset of the 2020 SSC pathway cohort matched for age, gender, and ethnicity to those in the 2016 VLC cohort. This was used to obtain data on time to definitive management, lesion histology, and additional lesions of concern during specialist review.

We classified skin lesions as benign, malignant, or pre-malignant, recorded the specific diagnosis, and determined the diagnostic concordance between GP, dermatologist, and histology report, along with the management advice, the time to definitive treatment, and the number of incidental lesions identified at related in-person specialist appointments (Appendix A).

For the benign/malignant classification, a diagnosis was concordant if the GP or teledermatologist's diagnosis (or in the case of multiple diagnoses, the first diagnosis) was classified in the same group (benign, malignant, pre-malignant) as histology where available, otherwise in the same group as the teledermatologist's diagnosis in the case of GP-dermatologist concordance. For the specific diagnosis, this was deemed completely concordant if the GP or teledermatologist only provided one diagnosis and this was the same as the histology diagnosis. A diagnosis was partially concordant if more than one diagnosis was provided by the GP or teledermatologist and one of those was the same as the histology. If a diagnosis was not provided or was labelled uncertain by the GP or teledermatologist, this was deemed not concordant.

Statistical analysis was performed using Microsoft Excel (2016, version 16.0.4591.1000, Microsoft, Redmond, WA, USA) and IBM SPSS Statistics (version 26, IBM, Armonk, NY, USA) software. The chi-squared test was used for comparison of categorical variables between groups. One-way analysis of variance (ANOVA) was used for comparison of continuous variables. A two-sided p-value of < 0.05 was used to determine statistical significance.

3. Results

Between 1 July and 31 December 2016, 481 patients were referred to the VLC, of whom 400 patients with 682 lesions were eligible for analysis. In the period 1 July to 31 December

2020, 1307 patients of 1495 referred to the SSC pathway, and 108 patients (with 277 lesions) of 134 referred to the VLC, were eligible for analysis (Figure 1). 1307 met the eligibility criteria. Propensity score matching of the SSC cohort to the 2016 VLC cohort identified 481 patients with 548 lesions from the SSC cohort. Patient characteristics are detailed in Table 1.

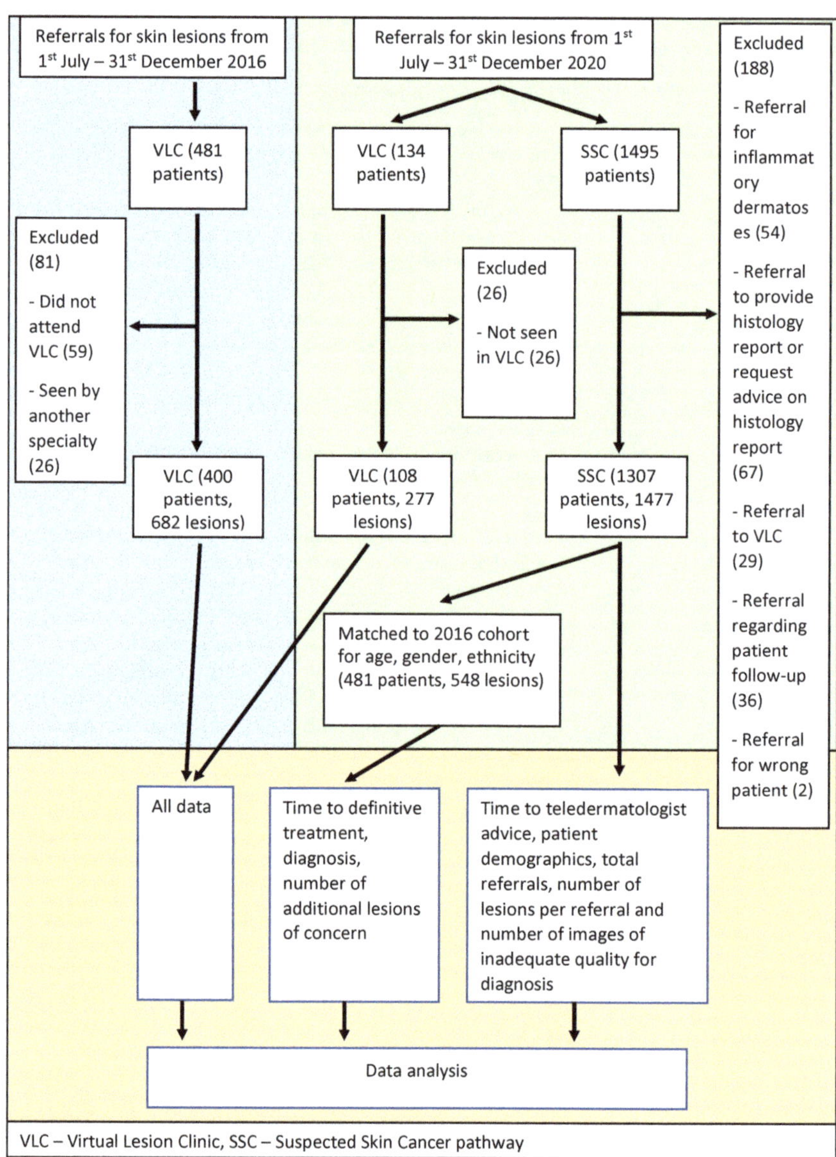

Figure 1. Flow chart.

Table 1. Patient characteristics.

Variable	Total SSC $n = 1307$ (%)	Matched SSC $n = 481$ (%)	2020 VLC $n = 108$ (%)	*p*-Value	2016 VLC $n = 400$ (%)	*p*-Value
			Age:			
Overall mean (SD)	61 yr (19.2)	55 yr (21.0)	59 yr (16.1)	<0.001	55 yr (21.0)	<0.001
0–9 years	25 (2)	11 (2)	1 (1)		11 (3)	
10–19 years	35 (3)	23 (5)	1 (1)		19 (5)	
20–29 years	55 (4)	32 (7)	3 (3)		27 (7)	
30–39 years	73 (6)	45 (9)	7 (6)		33 (8)	
40–49 years	116 (9)	60 (12)	13 (12)		50 (13)	
50–59 years	216 (17)	73 (15)	26 (24)		57 (14)	
60–69 years	306 (23)	100 (21)	30 (28)		89 (22)	
70–79 years	292 (22)	86 (18)	18 (17)		78 (20)	
80–89 years	163 (12)	41 (9)	7 (6)		31 (8)	
90+ years	38 (3)	10 (2)	2 (2)		5 (1)	
			Sex:			
Female	738 (56)	309 (64)	64 (59)		254 (64)	
Male	569 (44)	172 (36)	44 (41)	0.57	146 (37)	0.01
			Ethnicity:			
New Zealand European	1096 (84)	378 (78)	80 (74)		317 (79)	
Maori	73 (6)	30 (6)	12 (11)		26 (7)	
Pasifika	12 (1)	4 (1)	1 (1)		2 (1)	
European, other	82 (6)	50 (10)	11 (10)		42 (11)	
Asian	22 (2)	14 (3)	2 (2)		9 (2)	
Other	22 (2)	5 (1)	2 (2)	0.13	4 (1)	0.04
			Patient Location:			
Urban	770 (59)	290 (60)	37 (35)		211 (53)	
Semi-rural	481 (37)	171 (36)	64 (60)		169 (42)	
Rural	55 (4)	19 (4)	6 (6)	<0.01	20 (5)	0.10
			Referrer location:			
Urban	798 (61)	303 (63)	35 (33)		216 (54)	
Semi-rural	473 (36)	171 (32)	66 (62)		175 (44)	
Rural	20 (2)	8 (2)	5 (5)	<0.01	9 (2)	0.02

SSC—suspected skin cancer pathway; VLC—virtual lesion clinic; yr—years; SD—standard deviation.

Patients in the total SSC group had a mean age of 61 years compared to 59 years for the 2020 VLC group and 55 years for the 2016 VLC group ($p < 0.001$) (Table 1). There was a slight female preponderance (56%, 59%, and 64% respectively). Ethnicity was predominantly Caucasian with 84% New Zealand European in the SSC pathway total cohort, 78% in the SSC matched cohort, 74% in 2020 VLC cohort and 79% in 2016 VLC group. Maori ethnicity was claimed by 6%, 11% and 7% respectively. Analysis of patient and referrer location revealed a similar proportion of rural patients in all groups, however there was a larger portion of patients located semi-rurally in the 2020 VLC (60% for patients and 62% for referrers, $p < 0.01$) and 2016 VLC cohorts (42% for patients and 44% for referrers, $p = 0.1$, $p = 0.02$), compared to 37% for patients and 36% for referrers in the total SSC pathway group.

A group of 45 patients were lost to follow-up after receiving teledermatologist advice (19 (4%) in SSC matched cohort, 13 (12%) in the 2020 VLC cohort and 13 (3%) in the 2016 cohort). In the total SSC group, 71 lesions (5%) were unable to be diagnosed due to poor image quality. Another 115 lesions (9%) were able to be diagnosed but the images were noted by the dermatologist to be of poor quality (Table 2).

Table 2. Image quality for SSC lesions (n = 186 * (%)).

Reason for Poor Quality	Not Able to Diagnose	Able to Diagnose
No dermoscopic image	26 (37)	35 (30)
No macroscopic image	4 (6)	44 (38)
Image out of focus	15 (21)	23 (20)
Other poor quality	22 (31)	15 (13)
Dermoscopy imaging incomplete	2 (3)	3 (3)
No images	10 (14)	0
Unable to open images	2 (3)	0
Different patient's images	1 (1)	0

SSC—suspected skin cancer pathway. * Some lesions had more than one reason for inadequate quality.

Diagnostic and management advice was provided by two teledermatologists in 2020 and by three in 2016 time period; one was common to both time periods. Median time from referral to teledermatologist advice was 4 days (range 0–19) in the SSC pathway total cohort. For the VLC cohorts, the median time to advice was 42 days (range 16–184) in 2020 and 50 days (range 17–313) in 2016 (p < 0.001) (Table 3). Waiting time for a clinic appointment was the rate-limiting factor, with a median time of 26 days (range 0–173) in 2020 and 43 days (range 1–308) in 2016. The time from advice to definitive treatment (e.g., excision, in-person review) was 21.5 days in the SSC pathway cohort compared to 45 days in the 2020 VLC cohort and 60 days in the 2016 cohort. This increases to 94 days and 112 days for the 2020 and 2016 VLC cohorts respectively when calculating time from acceptance of the referral to treatment.

Table 3. Time to advice and treatment.

Variable	Total SSC n = 1307	Matched SSC n = 481	2020 VLC n = 108	p-Value	2016 VLC n = 400	p-Value
Median time from referral to dermatologist advice (SD, range)	4.0 days (2.8, 0–19)	5.0 days (2.6, 0–16)	42.0 days (29.3, 16–184)	<0.001	50.0 days (43.0, 17–313)	<0.001
Median time from referral to triage (SD, range)	N/A	N/A	2.0 days (2.2, 1–15)	<0.001	3.0 days (2.3, 2–25)	<0.001
Median wait time from referral to VLC clinic (SD, range)	N/A	N/A	26.0 days (29.9, 0–173)		43.0 days (40.0, 1–308)	
Median time from advice to definitive treatment (SD, range)	N/A	21.5 days (52.4, 0–236) n = 102	45.0 days (38.7, 4–142) n = 28	0.76	60.0 days (60.8, 2–365) n = 104	<0.001
Median time from referral triage to definitive treatment (SD, range)	N/A	21.5 days (52.4, 0–236)	94.0 days (48.1, 25–194)	<0.001	112.0 days (68.0, 30–378)	<0.001

SSC—suspected skin cancer pathway; VLC—virtual lesion clinic; SD—standard deviation; N/A—not applicable.

Skin lesions were classified as benign or malignant and by specific diagnosis (Table 4, Figures 2 and 3).

In the SSC pathway matched cohort, the teledermatologist recorded benign lesions in 65%, with a benign-to-malignant ratio of 3.0, compared to a benign-to-malignant ratio of 1.3 for GP (Table 4).

In the 2020 and 2016 VLC cohorts, the teledermatologist recorded benign lesions in 64% and 67%, and benign-to-malignant ratio of 3.4 and 3.8 respectively, compared to 1.7 and 0.8 for GP (Table 4).

Analysis of histology of malignant lesions revealed a keratinocytic-to-melanocytic ratio of 1.8 in the SSC matched cohort, 1.2 in the 2020 VLC cohort and 3.4 in the 2016 cohort (Table 5). Analysis of melanomas revealed a melanoma-in-situ-to-invasive melanoma ratio of 3.7 in the SSC matched cohort and 2.0 and 2.3 for the 2020 VLC and 2016 VLC cohorts respectively (Table 5).

Table 4. Lesion classification.

Variable	Matched SSC				
	Dermatologist Diagnosis $n = 528$ (%)	GP Diagnosis $n = 548$ (%)	*p*-Value	Histological Diagnosis $n = 113$ (%)	*p*-Value
Benign	343 (65)	237 (43)		32 (28)	
Pre-malignant	52 (10)	19 (4)		5 (4)	
Malignant	116 (22)	181 (33)		76 (67)	
Uncertain	17 (3)	111 (20)		N/A	
			<0.001		<0.001
Benign:malignant	3.0	1.3		0.4	
	2020 VLC				
	Dermatologist Diagnosis $n = 277$ * (%)	GP diagnosis $n = 172$ (%)	*p*-Value	Histology Diagnosis $n = 33$ (%)	*p*-Value
Benign	177 (64)	30 (17)		11 (33)	
Pre-malignant	40 (14)	11 (6)		2 (6)	
Malignant	52 (19)	18 (11)		20 (61)	
Uncertain	8 (3)	113 (38)		N/A	
			<0.001		0.09
Benign:malignant	3.4	1.7		0.6	
	2016 VLC				
	Dermatologist Diagnosis $n = 680$ * (%)	GP Diagnosis $n = 603$ (%)	*p*-Value	Histology Diagnosis $n = 122$ (%)	*p*-Value
Benign	460 (67)	149 (25)		21 (17)	
Pre-malignant	65 (10)	16 (3)		14 (11)	
Malignant	121 (18)	183 (30)		87 (71)	
Uncertain	34 (5)	255 (42)		N/A	
			<0.001		<0.001
Benign:malignant	3.8	0.8		0.2	

SSC—suspected skin cancer pathway; VLC—virtual lesion clinic; GP—general practitioner; N/A—not applicable; IEC—intraepithelial carcinoma; SCC—squamous cell carcinoma; BCC—basal cell carcinoma. * Includes extra lesions identified by nurse specialist during VLC.

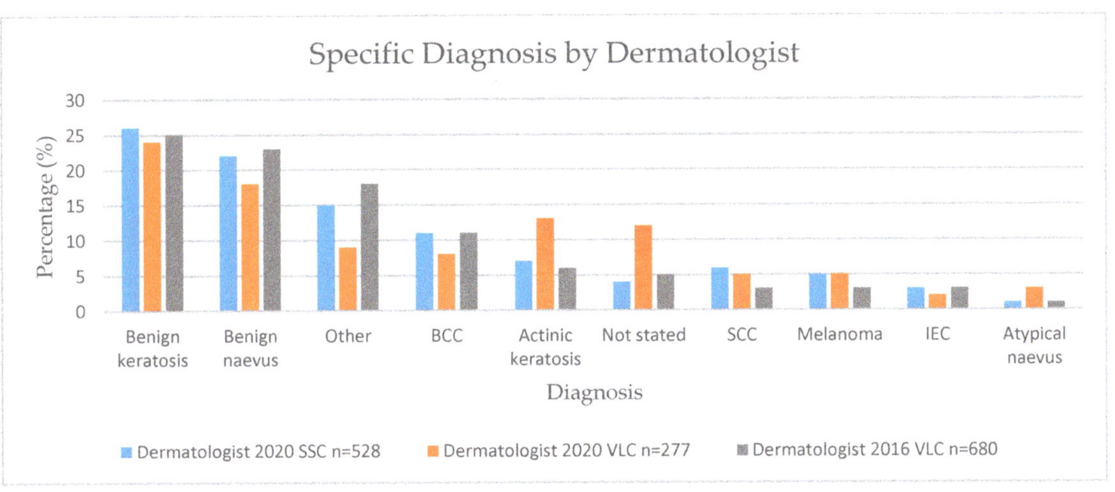

Figure 2. Specific lesion diagnosis by dermatologist. BCC—basal cell carcinoma; SCC—squamous cell carcinoma; IEC—intraepithelial carcinoma.

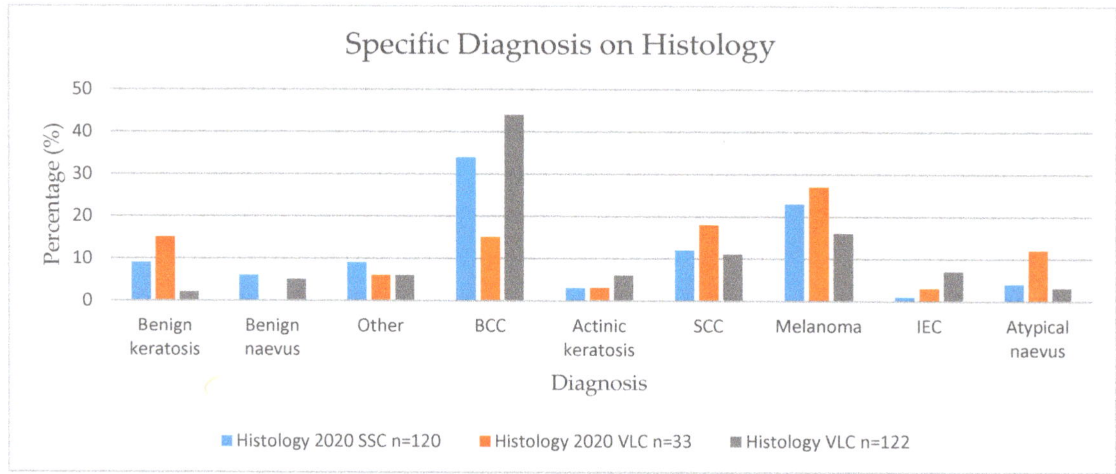

Figure 3. Specific lesion diagnosis on histology. BCC—basal cell carcinoma; SCC—squamous cell carcinoma; IEC—intraepithelial carcinoma.

Table 5. Histology lesion classification.

Variable	Matched SSC $n = 113$	2020 VLC $n = 33$	2016 VLC $n = 122$
Keratinocytic:melanocytic	1.8	1.2	3.4
Total number MIS	22	6	14
Total number melanoma	6	3	6
MIS:melanoma	3.7	2.0	2.3

SSC—suspected skin cancer pathway; VLC—virtual lesion clinic; MIS—melanoma-in-situ.

No further management was recommended by the dermatologist for 56%, 57%, and 69% of matched SSC pathway, 2020 VLC and 2016 VLC cohorts respectively (Table 6). Surgical management with either biopsy or excision was recommended for 26% of lesions in the matched SSC group compared to 16% in the 2020 VLC and 16% in the 2016 VLC groups ($p < 0.001$). In-person review was more common in the VLC cohorts at 7% in 2020 and 5% in 2016, compared to 0% in the SSC pathway.

Table 6. Lesion advice.

Variable	Matched SSC $n = 528$ (%)	2020 VLC $n = 277$ (%)	*p*-Value	2016 VLC $n = 682$ (%)	*p*-Value
No further management	298 (56)	157 (57)		471 (69)	
Monitor	38 (7)	21 (8)		18 (3)	
Topical	55 (10)	37 (13)		50 (7)	
Surgical	136 (26)	43 (16)		112 (16)	
In-person review	1 (0)	19 (7)		31 (5)	
			<0.001		<0.001

SSC—suspected skin cancer pathway; VLC—virtual lesion clinic.

GP–dermatologist diagnostic concordance was available for 528 lesions in the SSC matched cohort, 172 lesions in the 2020 VLC cohort and 601 lesions in the 2016 VLC cohort (Table 7). The concordance for the benign/malignant classification was 58%, 22%, and

32% for the SSC, 2020 VLC and 2016 VLC cohorts, respectively ($p < 0.001$). Concordance for the specific diagnosis was 46% (35% complete, 11% partial), 31% (22% complete, 9% partial), and 24% (17% complete, 7% partial) for the SSC, 2020 VLC and 2016 VLC cohorts, respectively ($p < 0.001$).

Table 7. Diagnostic concordance.

GP-Dermatologist Concordance					
Variable	Matched SSC $n = 528$ (%)	2020 VLC $n = 172$ (%)	p-Value	2016 VLC $n = 601$ (%)	p-Value
	Benign/malignant:				
Concordant	305 (58)	38 (22)		194 (32)	
Not concordant	223 (42)	134 (78)		407 (68)	
			<0.001		<0.001
	Specific diagnosis:				
Completely concordant	183 (35)	37 (22)		101 (17)	
Partially concordant	60 (11)	16 (9)		39 (7)	
Not concordant	285 (54)	119 (69)		461 (77)	
			<0.001		<0.001
Dermatologist-Histology Concordance					
Variable	Matched SSC $n = 114$ (%)	2020 VLC $n = 32$ * (%)	p-Value	2016 VLC $n = 112$ (%)	p-Value
	Benign/malignant:				
Concordant	80 (70)	20 (63)		86 (71)	
Not concordant	34 (30)	12 (38)		36 (30)	
			0.41		0.96
	Specific diagnosis:				
Completely concordant	60 (53)	19 (59)		70 (57)	
Partially concordant	15 (13)	2 (6)		0 (0)	
Not concordant	39 (34)	11 (34)		52 (43)	
			0.54		<0.001
GP-Histology Concordance					
Variable	Matched SSC $n = 114$ (%)	2020 VLC $n = 17$ (%)	p-Value	2016 VLC $n = 98$ (%)	p-Value
	Benign/malignant:				
Concordant	68 (60)	2 (12)		48 (49)	
Not concordant	46 (40)	15 (88)		50 (51)	
			<0.001		<0.001
	Specific diagnosis:				
Completely concordant	46 (40)	1 (6)		40 (41)	
Partially concordant	9 (8)	1 (6)		5 (5)	
Not concordant	59 (52)	15 (88)		53 (54)	
			0.02		0.71

SSC—suspected skin cancer pathway; VLC—virtual lesion clinic. * Includes extra lesions identified by nurse specialist during VLC.

Histology was available for 114 lesions in the SSC pathway matched cohort, 32 lesions in the 2020 VLC cohort, and 122 lesions in the 2016 cohort (Table 7).

For the SSC pathway cohort, dermatologist–histology concordance was 70% for benign/malignant classification, whereas GP–histology concordance was 60% ($p = 0.01$). Dermatologist–histology concordance for specific diagnosis was 66% (53% complete, 13% partial), compared to GP–histology concordance of 48% (40% complete, 8% partial) ($p < 0.001$).

For the 2020 VLC cohort, dermatologist–histology concordance was 63% for benign/malignant classification, compared to 12% for GP–histology concordance ($p = 0.21$). Dermatologist–histology concordance for specific diagnosis was 65% (59% complete, 6% partial), whereas GP–histology concordance was 12% (6% complete, 6% partial) ($p = 0.73$).

For the 2016 VLC cohort, dermatologist–histology concordance was 71% for benign/malignant classification compared to GP-histology concordance of 49% ($p = 0.001$). Dermatologist—histology concordance was 57% for specific diagnosis (57% complete, 0% partial) compared to GP-histology concordance of 46% (41% complete, 5% partial) respectively ($p = 0.03$).

The benign/malignant dermatologist–histology concordance was similar between the SSC pathway and 2016 VLC cohorts at 70% and 71% respectively. Non-concordant results for specific diagnoses were 34% in the SSC cohort and 43% in the 2016 VLC cohort ($p < 0.001$). There were 11 lesions overall (4%) that were identified as being likely benign by dermatologists and malignant by histology with four (4%) in the matched SSC pathway, two (6%) in 2020 VLC and five (4%) in 2016 VLC cohorts. The teledermatology recommendation was 'no further management' for two of these lesions, 'monitoring' for one lesion, 'topical therapy' for one lesion, 'excision/biopsy to remove doubt', and 'in-person review' for two lesions.

A public hospital specialist Dermatology or Plastic Surgery outpatient appointment for a study lesion was made for 156 patients (41 in the matched SSC pathway cohort, 24 in the 2020 VLC cohort and 91 in the 2016 cohort). A mean of 0.2 incidental lesions of concern were identified (0.2, 0.0, and 0.2 in SSC pathway, 2020 VLC and 2016 VLC cohorts, respectively) (Tables S1 and S2). Eight out of 11 lesions (73%) were malignant on histology in the SSC pathway cohort, with a keratinocytic to melanocytic ratio of 3.0. The 2016 VLC cohort was found to have 6 out of 16 lesions (38%) malignant and a keratinocytic to melanocytic ratio of 1.0.

The VLC nurse specialists identified 107 incidental lesions in patients attending imaging clinics (78 in 2016 and 29 in 2020). Histology was available for 26 lesions in the 2016 VLC cohort with a benign-to-malignant ratio of 0.2 and 11 lesions in the 2020 VLC cohort with a benign-to-malignant ratio of 0.0 (Table S3).

4. Discussion

Good quality teledermatology and teledermoscopy allow effective skin lesion diagnosis and management [24]. Electronic referrals from primary to secondary care are increasingly used by district health boards in New Zealand. Availability of outpatient appointments for patients with suspected skin cancer is outstripped by the number of referrals to dermatology; other sites may offer remote advice, hold dedicated clinics, or forward referrals to a surgical specialty. The use of teledermatology worldwide increased during 2020 due to the COVID-19 pandemic compared to few sites using it in 2016. We have compared GP referrals to our SSC pathway, where images of mixed quality were taken using various devices, to our nurse-led VLC, where uniformly high-quality images were taken with standardized cameras.

The study has met our primary objective to demonstrate a statistically significant reduction in time to dermatologist advice using the SSC pathway in 2020 compared to the VLC in 2016 and 2020, with a median time from referral to advice of 4 days compared to 42 days and 50 days, respectively ($p < 0.001$). The longer time to advice included waits due to missed or rescheduled appointments. This is comparable to a US teledermatology service, where the time to advice was reduced from 70 days when using in-person clinic assessments to 0.5 days when using electronic referrals [21]. There was also a reduction in time to definitive treatment when using the SSC pathway compared to the 2016 and 2020 VLC, both for time from referral assessment (or triage for patients referred to the VLC) and for the time from dermatologist advice to treatment. This is likely explained by the differences in how advice is managed in each pathway, with GPs arranging excisions in the SSC pathway. This can lead to excision of lesions of concern within a few days to weeks in the private sector. In contrast, lesions requiring excision identified at VLC are referred to plastic surgery or dermatology at the hospital with longer wait times. The Standards of Service Provision for Melanoma Patients in New Zealand (2014) recommended "patients referred with a high suspicion of melanoma receive their first cancer treatment within 62 days of receipt of referral" [25]. A large retrospective study by Conic et al. showed improved survival in patients treated within 90 days for melanoma stages I–III and improved survival for patients treated within 30 days for stage I melanoma [26]. The reduction in time to treatment by a median of 62.5 days and 90.5 days compared to 2020 VLC and 2016 VLC are likely to significantly improve the proportion of patients meeting these standards.

The higher proportions of melanoma-in-situ found in the SSC pathway matched cohort compared to the 2016 VLC cohort were striking. We suspect that the rapid response to referrals in the SSC pathway allowed GPs to get a second opinion on lesions they would have otherwise managed independently. The total referral numbers for 2020 SSC pathway were significantly higher than for the 2016 VLC and the GP benign-to-malignant ratio in the SSC pathway matched cohort was higher at 1.3 compared to 0.8 in the 2016 VLC. This also hints at a lower threshold for referral.

Interpretation is complicated by differing rates of diagnostic uncertainty in the referrals: the SSC pathway template required the GP to select a suspected diagnosis (including an option, 'uncertain' in 20%), but this was not required for referrals to the VLC in 2016 where any referral without a diagnosis was deemed uncertain (42%).

The histology benign-to-malignant ratio was higher in the SSC pathway matched cohort (0.6) than the 2016 VLC cohort (0.2). This could indicate that dermatologists were conservative in diagnosing SSC pathway lesions due to some images of inferior quality. It is important to note that both ratios are very low. A recent systematic review estimated the number-needed to biopsy to find one melanoma for global dermatologists to be 7.5 [27]. In our study, the number needed to biopsy for melanoma was 4.2 in the SSC pathway cohort and 6.1 in the 2016 VLC cohort, despite including keratinocytic skin cancers. This raises the concern of missed melanoma. The high ratio of melanoma-in-situ to melanoma indicating enhanced early diagnosis of melanoma through the SSC pathway, reduces this possibility. Conversely, given these higher rates of melanoma-in-situ in the 2020 SSC pathway cohort, we need to be mindful of the possibility of overdiagnosis. The natural history of melanoma-in-situ is still not completely established [28]. Studies with high melanoma-in-situ excision rates or those assessing skin cancer screening, have frequently failed to show a mortality benefit [28,29]. In addition, the considerable variability in histopathologist classification of melanoma-in-situ versus atypical naevus further complicates this assessment [30]. There is currently no consensus on the best way to manage the risk of overdiagnosis, whilst still ensuring early treatment of invasive melanoma. Perhaps better education in skin lesion assessment will lead to improvements in this area.

Dermatologists recommended 'no further action' for fewer lesions in the SSC pathway cohort (56%) and 2020 VLC cohort (57%), compared to the 2016 VLC cohort (69%). They more frequently recommended surgery (by 10%), monitoring (4%), and topical treatment (3%) in the SSC pathway compared to 2016 VLC. This may reflect reduced confidence due to the inferior quality of the teledermatology images or the dermatologists' individual practices, with only one out of three diagnosing the 2016 cohort also diagnosing in 2020. In the 2020 VLC group, the main increases were 5% increased monitoring (due to more immunosuppressed and transplant patients requiring surveillance), 6% increased topical treatment, and 2% more in-person review. The rates of surgical treatment were the same as in 2016 at 16%.

The SSC pathway dermatologist–histology diagnostic concordance for benign/malignant of 70% and the 2016 VLC of 71% are consistent with a systematic review of 21 studies in which teledermatologist-histology concordance was 51–85% [31]. There was a slight improvement in overall concordance for specific diagnoses in the SSC pathway cohort at 66% (53% complete, 13% partial), compared to 2016 VLC at 57% (57% complete, 0% partial). Partial SSC pathway concordance reflects a lack of confidence in lesions of poor quality.

One potential problem encountered when calculating concordance is the ambiguity among histopathologists, dermatologists, and GPs regarding classification of some lesions, particularly atypical naevus and melanoma-in-situ (one is benign and the other malignant) [32]. SCC-in-situ was categorised as pre-malignant and SCC as malignant, so clinician classification as the former and pathologist classification as the latter would be discordant; in practice, both diagnoses would be managed similarly. These classification differences may have increased non-concordance between dermatologist and histology.

The average age of patients in the total SSC pathway cohort was 6 years older than in the 2016 VLC cohort; this is only partially explained by an aging population in the Waikato area. Statistical data from NZ stats show the median age of the Waikato region to be 37.6 years in 2013 and 38.2 years in 2018 [33]. The convenience of the SSC referral method may encourage GPs to refer a greater proportion of older patients, as has been found in other studies [18,34–36]. A higher proportion of 2020 VLC referrals were from semi-rural locations compared to 2020 SSC pathway and 2016 VLC. This suggests that GPs in these locations have less access or familiarity with dermoscopy.

New Zealand Europeans made up the majority of patients referred to the SSC pathway and 2016 VLC. Only 6% and 7% of patients referred to the SSC pathway and to the 2016 VLC respectively were Maori, whereas the total proportion of Maori in the Waikato region was 21% in 2018, likely reflecting lower rates of skin cancer associated with darker skin types [33].

The dermatologists indicated that the large majority of referrals to the SSC pathway had images of adequate quality (86%); poor quality was usually associated with lack of dermoscopic images, images out of focus, and, less often, lack of macroscopic images. Other reasons for poor quality included excessive blood/hair or insufficient gel used for dermoscopy images. Melanoma diagnosis requires high quality macroscopic and dermoscopic images to diagnose confidently. A systematic review assessing the use of smartphones to detect melanoma reported that up to 20% of images were of insufficient quality for diagnosis [37]. A randomised controlled trial assessing smartphone images with dermoscopy in participants without healthcare training found 5% were inadequate for diagnosis [38].

Not all patients referred to a dermatologist with a skin cancer have had a full skin check by a competent health professional. Patients with one skin cancer are likely to have others [39–41]. We reviewed the small subset of patients who had a related in-person assessment by a hospital plastic surgeon or dermatologist: incidental lesions were identified in 11 patients (27%) in the matched SSC pathway cohort, 2 patients (8%) in the 2020 VLC cohort, and 22 patients (24%) in the 2016 VLC cohort with histology available for few of these lesions (11, 2, and 16 respectively). We can extrapolate that many of the patients diagnosed with skin cancer through the SSC pathway will have had other significant lesions.

Limitations: The majority of skin lesions in 2020 were assessed by the SSC pathway. Only those in which the GP did not provide images or images were deemed inadequate were seen in VLC. This cohort has a potential selection bias for lesions that were more difficult to diagnose or were more likely to be malignant that may explain a lower diagnostic concordance in 2020 VLC compared to the SSC pathway. This could potentially be an issue, to a lesser extent in the 2016 VLC, if referrers had a higher threshold for referral. Low numbers and non-significant p-value may also reflect the small sample size. Histology reviews were limited to biopsies requested by the dermatologists; we did not look for histology of any lesion that the teledermatologist had not indicated should be excised and thus we did not calculate a false negative rate. Some SSC referrals were missed due to misleading referral description on manual searching. Follow-up time was limited to 4 months after the 2020 cohorts, excluding 19% of the 2020 VLC cohort as they had not yet received a planned outpatient appointment or surgical procedure. This compares to 17% of the 2016 VLC cohort that did not attend. For the matched SSC pathway cohort, histology was unavailable for 27 lesions (19% of those planned for excision), compared to the 2016 VLC cohort where histology was not found for 8 lesions (6%).

5. Conclusions

We found that referrals for skin lesions imaged in primary care with variable devices and image quality results in faster dermatologist advice with similar diagnostic performance, compared to referrals to nurse-led specialist clinics where there is consistently high-quality imaging. Use of the SSC pathway encourages early referral and diagnosis of atypical melanocytic lesions. Detection of incidental lesions in some of the skin cancer patients seen in-person is a reminder to encourage full skin examination and a low threshold for referral of suspicious lesions. Improved diagnostic performance through the SSC pathway could be achieved by more consistent adherence to the imaging requirements of the SSC referral pathway. Our results support increased uptake of teledermatological solutions in skin lesion assessment. This will be important as support tools offering augmented intelligence are integrated into clinical practice in the future. The lack of in-person lesion evaluation by dermatologists creates a burden on primary care to undertake skin checks and identify suspicious lesions requiring a second opinion.

Supplementary Materials: The following are available online at https://www.mdpi.com/article/10.3390/cancers13225828/s1, Table S1: Patient demographics of incidental lesions during specialist follow-up, Table S2: Incidental lesions during specialist follow-up, Table S3: Incidental lesions during VLC.

Author Contributions: Conceptualization, A.O.; Methodology, A.O., L.J. and M.J.; Software, L.J.; Validation, L.J., A.O. and M.J.; Formal analysis, L.J.; Investigation, L.J.; Resources, A.O. and L.J.; Data curation, L.J.; Writing—original draft preparation, L.J.; Writing—review and editing, A.O. and M.J.; Visualization, L.J.; Supervision, A.O. and M.J.; Project administration, L.J. and A.O.; Funding acquisition, L.J. and A.O. All authors have read and agreed to the published version of the manuscript.

Funding: This research was funded by the Waikato Medical Research Foundation (grant number 324) and the APC was funded by the Waikato Medical Research Foundation.

Institutional Review Board Statement: Ethical review and approval were waived for this study by the Health and Disability Ethics Committees with formal acknowledgement 22 December 2020, due to the study being an audit or related activity and not involving the use of human tissue without consent.

Informed Consent Statement: All patients attending the Virtual Lesion Clinic consented to the use of their data. The majority of patients referred to the eTriage system consented for their images to be used for research and audit. However, the data used for this study did not evaluate images. Patient consent was waived due to the use of retrospective, de-identified information from the healthcare database.

Data Availability Statement: Original data for this study is retained on a protected District Hospital Server. The authors have not prepared a dataset suitable for a public archive.

Acknowledgments: The Waikato District Health Board dermatologists and registrars for their contribution to the research and skin triage pathways.

Conflicts of Interest: Jones and Jameson have no conflict of interest to declare. Oakley has two conflicts to declare. Amanda Oakley is paid a consultation fee for diagnosing private patients for MoleMap New Zealand, the provider of the VLC service. She was not paid a consultation fee for diagnosing cases for the VLC, as this was part of her hospital commitment. Until December 2019, Amanda Oakley was a board member of Waikato Medical Research Foundation (WMRF) and on the Grants Committee. She was no longer in these positions when an application for funding was submitted. The funder had no role in the design of the study; in the collection, analyses, or interpretation of data; in the writing of the manuscript, or in the decision to publish the results.

Appendix A

Table A1. Diagnosis and Outcome Classification.

Specific Diagnoses	Includes:
Melanocytic naevus	Junctional naevus, dermal naevus, compound naevus, blue naevus, acral naevus, Myerson naevus, cockade naevus, other benign naevus
Atypical naevus	Atypical naevus, Spitz naevus, Reed naevus
Benign keratosis	Seborrhoeic keratosis, solar lentigo, lichen planus-like keratosis, porokeratosis, viral wart
Angioma	Angioma
Dermatofibroma	Dermatofibroma
Actinic keratosis	Actinic keratosis, actinic cheilitis
IEC	Intraepithelial carcinoma
SCC	Squamous cell carcinoma, keratoacanthoma, cutaneous horn
BCC	Nodular basal cell carcinoma, morphoeic basal cell carcinoma, pigmented basal cell carcinoma, recurrent basal cell carcinoma, superficial basal cell carcinoma
Melanoma	Superficial spreading melanoma, nodular melanoma, acral melanoma, lentigo maligna melanoma, melanoma-in-situ, lentigo maligna.
Other	All other lesions not covered in the above diagnoses
Benign–Malignant Classification	**Includes:**
Benign	Melanocytic naevus, atypical naevus, benign keratosis, angioma, dermatofibroma, benign lesions in the other category
Malignant	Melanoma, BCC, SCC
Pre-malignant	IEC, actinic keratosis
Uncertain	Lesions labelled uncertain or diagnosis not provided
Outcome Advice	**Includes:**
No action	No further lesion assessment or treatment required
Monitor	GP or VLC skin lesion monitoring
Topical	Cryotherapy, 5-fluorouracil, or imiquimod treatment
Surgical	Excision or biopsy
Review	Face-to-face review

IEC—intraepidermal carcinoma; SCC—squamous cell carcinoma; BCC—basal cell carcinoma; VLC—virtual lesion clinic.

References

1. Population Pyramids of the World from 1950 to 2100. Available online: https://www.populationpyramid.net/new-zealand/ (accessed on 1 October 2020).
2. Selected Cancers 2015, 2016, 2017. Available online: https://www.health.govt.nz/publication/selected-cancers-2015-2016-2017 (accessed on 9 September 2021).
3. Salmon, P.J.; Chan, W.C.; Griffin, J.; McKenzie, R.; Rademaker, M. Extremely high levels of melanoma in Tauranga, New Zealand: Possible causes and comparisons with Australia and the northern hemisphere. *Australas. J. Dermatol.* **2007**, *48*, 208–216. [CrossRef] [PubMed]
4. Whiteman, D.C.; Green, A.C.; Olsen, C.M. The growing burden of invasive melanoma: Projections of incidence rates and numbers of new cases in six susceptible populations through 2031. *J. Investig. Dermatol.* **2016**, *136*, 1161–1171. [CrossRef] [PubMed]
5. Sneyd, M.J.; Cox, B. Melanoma in Maori, Asian, and Pacific peoples in New Zealand. *Cancer Epidemiol. Biomarkers Prev.* **2009**, *18*, 1706–1713. [CrossRef] [PubMed]
6. Elliott, B.M.; Douglass, B.R.; McConnell, D.; Johnson, B.; Harmston, C. Incidence, demographics and surgical outcomes of cutaneous squamous cell carcinoma diagnosed in Northland, New Zealand. *N. Z. Med. J.* **2018**, *131*, 61–68.
7. Pondicherry, A.; Martin, R.; Meredith, I.; Rolfe, J.; Emanuel, P.; Elwood, M. The burden of non-melanoma skin cancers in Auckland, New Zealand. *Australas. J. Dermatol.* **2018**, *59*, 210–213. [CrossRef] [PubMed]
8. Rowan, D.; Lamb, S.; Greig, D.; Keefe, M.; Chan, W.C.; Koo, W.; Agnew, K.; Giles, A.; Graves, B. Health Workforce New Zealand Dermatology Workforce Service. Ministry of Health NZ. Available online: https://www.health.govt.nz/system/files/documents/pages/dermatology-workforce-service-forecast-nov14.docx (accessed on 16 August 2021).

9.   Ministry of Health. Population of Waikato DHB. Available online: https://www.health.govt.nz/new-zealand-health-system/my-dhb/waikato-dhb/population-waikato-dhb (accessed on 28 September 2021).

10.   Elsner, P. Teledermatology in the times of COVID-19—A systematic review. *J. Dtsch. Dermatol. Ges.* **2020**, *18*, 841–845. [CrossRef] [PubMed]

11.   Tan, E.; Yung, A.; Jameson, M.; Oakley, A.; Rademaker, M. Successful triage of patients referred to a skin lesion clinic using teledermoscopy (IMAGE IT trial). *Br. J. Dermatol.* **2010**, *162*, 803–811. [CrossRef] [PubMed]

12.   Lim, D.; Oakley, A.; Rademaker, M. Better, sooner, more convenient: A successful teledermoscopy service. *Australas. J. Dermatol.* **2012**, *53*, 22–25. [CrossRef] [PubMed]

13.   Teoh, N.; Oakley, A. 9-year review of Waikato Teledermatoscopy Service (abstract). Proceedings of the Waikato Clinical Campus Biannual Research Seminar. *N. Z. Med. J.* **2019**, *132*, 112.

14.   Sunderland, M.; Teague, R.; Gale, K.; Rademaker, M.; Oakley, A.; Martin, R.C. E-referrals and teledermatoscopy grading for melanoma: A successful model of care. *Australas. J. Dermatol.* **2020**, *61*, 147–151. [CrossRef] [PubMed]

15.   Congalton, A.T.; Oakley, A.M.; Rademaker, M.; Bramley, D.; Martin, R.C.W. Successful melanoma triage by a virtual lesion clinic (teledermatoscopy). *J. Eur. Acad. Dermatol. Venereol.* **2015**, *29*, 2423–2428. [CrossRef] [PubMed]

16.   Oakley, A. Mobile teledermatology is here to stay. *Br. J. Dermatol.* **2015**, *172*, 856–857. [CrossRef] [PubMed]

17.   Rustad, A.M.; Lio, P.A. Pandemic pressure: Teledermatology and health care disparities. *J. Patient Exp.* **2021**, *8*, 2374373521996982. [CrossRef]

18.   Maddukuri, S.; Patel, J.; Lipoff, J.B. Teledermatology addressing disparities in health care access: A review. *Curr. Dermatol. Rep.* **2021**, 1–8. [CrossRef] [PubMed]

19.   Bruce, A.; Mallow, J.; Theeke, L. The use of teledermoscopy in the accurate identification of cancerous skin lesions in the adult population: A systematic review. *J. Telemed. Telecare* **2018**, *24*, 75–83. [CrossRef]

20.   Piccolo, D.; Smolle, J.; Argenziano, G.; Wolf, I.H.; Braun, R.; Cerroni, L.; Ferrari, A.; Hofmann-Wellenhof, R.; Kenet, R.O.; Magrini, F.; et al. Teledermoscopy: Results of a multicentre study on 43 pigmented skin lesions. *J. Telemed. Telecare* **2000**, *6*, 132–137. [CrossRef] [PubMed]

21.   Carter, Z.A.; Goldman, S.; Anderson, K.; Li, X.; Hynan, L.S.; Chong, B.F.; Dominguez, A.R. Creation of an internal teledermatology store-and-forward system in an existing electronic health record: A pilot study in a safety-net public health and hospital system. *JAMA Dermatol.* **2017**, *153*, 644–650. [CrossRef]

22.   Naka, F.; Lu, J.; Porto, A.; Villagra, J.; Wu, Z.H.; Anderson, D. Impact of dermatology eConsults on access to care and skin cancer screening in underserved populations: A model for teledermatology services in community health centers. *J. Am. Acad. Dermatol.* **2018**, *78*, 293–302. [CrossRef] [PubMed]

23.   Urban Rural Profile Categories. Available online: https://www.otago.ac.nz/healthsciences/otago706060.pdf (accessed on 24 August 2021).

24.   Chuchu, N.; Dinnes, J.; Takwoingi, Y.; Matin, R.N.; Bayliss, S.E.; Davenport, C.; Moreau, J.F.; Bassett, O.; Godfrey, K.; O'Sullivan, C.; et al. Teledermatology for diagnosing skin cancer in adults. *Cochrane Database Syst. Rev.* **2018**, *12*, CD013193. [CrossRef]

25.   National Melanoma Tumour Standards Working Group. Standards of Service Provision for Melanoma Patients in New Zealand—Provisional. Available online: https://www.melnet.org.nz/uploads/standards-of-service-provision-melanoma-patients-jan1 4.pdf (accessed on 3 November 2021).

26.   Conic, R.Z.; Cabrera, C.I.; Khorana, A.A.; Gastman, B.R. Determination of the impact of melanoma surgical timing on survival using the National Cancer Database. *J. Am. Acad. Dermatol.* **2017**, *78*, 40–46. [CrossRef] [PubMed]

27.   Nelson, K.C.; Swetter, S.M.; Saboda, K.; Chen, S.C.; Curiel-Lewandrowski, C. Evaluation of the Number-Needed-to-Biopsy Metric for the Diagnosis of Cutaneous Melanoma. A Systematic Review and Meta-analysis. *JAMA Dermatol.* **2019**, *155*, 1167–1174. [CrossRef] [PubMed]

28.   Kutzner, H.; Jutzi, T.B.; Krahl, D.; Krieghoff-Henning, E.I.; Heppt, M.V.; Hekler, A.; Schmitt, M.; Maron, R.C.; Fröhling, S.; von Kalle, C.; et al. Overdiagnosis of melanoma—Causes, consequences and solutions. *J. Dtsch. Dermatol. Ges.* **2020**, *18*, 36–1243. [CrossRef] [PubMed]

29.   Nufer, K.L.; Raphael, A.P.; Soyer, H.P. Dermoscopy and overdiagnosis of melanoma in situ. *JAMA Dermatol.* **2018**, *154*, 398–399. [CrossRef]

30.   Elmore, J.G.; Barnhill, R.L.; Elder, D.E.; Longton, G.M.; Pepe, M.S.; Reisch, L.M.; Carney, P.A.; Titus, L.J.; Nelson, H.D.; Onega, T.; et al. Pathologists' diagnosis of invasive melanoma and melanocytic proliferations: Observer accuracy and reproducibility study. *BMJ* **2017**, *358*, j3798. [CrossRef]

31.   Finnane, A.; Dallest, K.; Janda, M.; Soyer, H.P. Teledermatology for the diagnosis and management of skin cancer: A systematic review. *JAMA Dermatol.* **2017**, *153*, 319–327. [CrossRef] [PubMed]

32.   Stefanato, C.M. The "dysplastic nevus" conundrum: A look back, a peek forward. *Dermatopathology* **2018**, *5*, 53–57. [CrossRef] [PubMed]

33.   Age and Sex by Ethnic Group (Grouped Total Responses), for Census Usually Resident Population Counts 2006, 2013, and 2018 Censused (RC, TA, SA2, DHB). Available online: http://nzdotstat.stats.govt.nz/wbos/Index.aspx (accessed on 24 September 2021).

34.   Coustasse, A.; Sarkar, R.; Abodunde, B.; Metzger, B.J.; Slater, C.M. Use of teledermatology to improve dermatological access in rural areas. *Telemed. e-Health* **2019**, *25*, 1022–1032. [CrossRef] [PubMed]

35. Garcia-Romero, M.T.; Prado, F.; Dominguez-Cherit, J.; Hojyo-Tomomka, M.T.; Arenas, R. Teledermatology via a social networking web site: A pilot study between a general hospital and a rural clinic. *Telemed. e-Health* **2011**, *17*, 652–655. [CrossRef]
36. Greisman, L.; Nguyen, T.M.; Mann, R.E.; Baganizi, M.; Jacobson, M.; Paccione, G.A.; Friedman, A.J.; Lipoff, J.B. Feasibility and cost of a medical student proxy-based mobile teledermatology consult service with Kisoro, Uganda, and Lake Atitlán, Guatemala. *Int. J. Dermatol.* **2015**, *54*, 685–692. [CrossRef] [PubMed]
37. Rat, C.; Hild, S.; Sérandour, J.R.; Gaultier, A.; Quereux, G.; Dreno, B.; Nguyen, J.M. Use of Smartphones for Early Detection of Melanoma: Systematic Review. *J. Med. Internet Res.* **2018**, *20*, e9392. [CrossRef] [PubMed]
38. Janda, M.; Horsham, C.; Vagenas, D.; Loescher, L.J.; Gillespie, N.; Koh, U.; Curiel-Lewandrowski, C.; Hofmann-Wellenhof, R.; Halpern, A.; Whiteman, D.C.; et al. Accuracy of mobile digital teledermoscopy for skin self-examinations in adults at high risk of skin cancer: An open-label, randomised controlled trial. *Lancet Digit. Health* **2020**, *2*, e129–e137. [CrossRef]
39. Bradford, P.T.; Freedman, D.M.; Goldstein, A.M.; Tucker, M.A. Increased risk of second primary cancers after a diagnosis of melanoma. *Arch. Dermatol.* **2010**, *146*, 265–272. [CrossRef] [PubMed]
40. Bartos, V. Development of multiple-lesion basal cell carcinoma of the skin: A comprehensive review. *Sisli Etfal Hastan. Tip Bul.* **2019**, *53*, 323–328. [PubMed]
41. Ciążyńska, M.; Kamińska-Winciorek, G.; Lange, D.; Lewandowski, B.; Reich, A.; Sławińska, M.; Pabianek, M.; Szczepaniak, K.; Hankiewicz, A.; Ułańska, M. The incidence and clinical analysis of non-melanoma skin cancer. *Sci. Rep.* **2021**, *11*, 4337. [CrossRef] [PubMed]

Review

Merkel Cell Carcinoma: New Trends

Ellen M. Zwijnenburg [1], Satish F.K. Lubeek [2], Johanna E.M. Werner [3], Avital L. Amir [4], Willem L.J. Weijs [5], Robert P. Takes [6], Sjoert A.H. Pegge [7], Carla M.L. van Herpen [8], Gosse J. Adema [1] and Johannes H. A. M. Kaanders [1,*]

[1] Department of Radiation Oncology, Radboudumc, 6525 GA Nijmegen, The Netherlands; e.zwijnenburg@radboudumc.nl (E.M.Z.); gosse.adema@radboudumc.nl (G.J.A.)
[2] Department of Dermatology, Radboudumc, 6525 GA Nijmegen, The Netherlands; satish.lubeek@radboudumc.nl
[3] Department of Surgery, Radboudumc, 6525 GA Nijmegen, The Netherlands; annelies.werner@radboudumc.nl
[4] Department of Pathology, Radboudumc, 6525 GA Nijmegen, The Netherlands; avital.amir@radboudumc.nl
[5] Department of Maxillofacial Surgery, Radboudumc 6525 GA Nijmegen, The Netherlands; willem.weijs@radboudumc.nl
[6] Department of Head and Neck Surgery, Radboudumc, 6525 GA Nijmegen, The Netherlands; robert.takes@radboudumc.nl
[7] Department of Radiology and Nuclear Medicine, Radboudumc, 6525 GA Nijmegen, The Netherlands; sjoert.pegge@radboudumc.nl
[8] Department of Medical Oncology, Radboudumc, 6525 GA Nijmegen, The Netherlands; carla.vanherpen@radboudumc.nl
* Correspondence: j.kaanders@radboudumc.nl; Tel.: +31-629-501-943

Citation: Zwijnenburg, E.M.; Lubeek, S.F.K.; Werner, J.E.M.; Amir, A.L.; Weijs, W.L.J.; Takes, R.P.; Pegge, S.A.H.; van Herpen, C.M.L.; Adema, G.J.; Kaanders, J.H.A.M. Merkel Cell Carcinoma: New Trends. *Cancers* **2021**, *13*, 1614. https://doi.org/10.3390/cancers13071614

Academic Editor: Aimilios Lallas

Received: 10 March 2021
Accepted: 29 March 2021
Published: 31 March 2021

Publisher's Note: MDPI stays neutral with regard to jurisdictional claims in published maps and institutional affiliations.

Simple Summary: In this review, we discuss a rare skin cancer that occurs mostly in elderly people called "Merkel cell carcinoma" (MCC). The incidence is increasing due to ageing of the population, increased sun exposure, and the use of medication that inhibits the immune system. Unlike most other skin cancers, MCC grows rapidly and forms metastases easily. We discuss the biology and treatment of MCC. Management should be by an experienced and multidisciplinary team, and treatment must start quickly. The standard practice of MCC treatment is surgery followed by radiotherapy. However, because it concerns an elderly and often frail population, (extensive) surgery may not always be feasible due to the associated morbidity. In those situations, radiotherapy alone is a good alternative. An important new development is immunotherapy that can cause long-lasting responses in a significant proportion of the patients with recurrent or metastatic MCC.

Abstract: Merkel cell carcinoma (MCC) is a rare neuroendocrine tumor of the skin mainly seen in the elderly. Its incidence is rising due to ageing of the population, increased sun exposure, and the use of immunosuppressive medication. Additionally, with the availability of specific immunohistochemical markers, MCC is easier to recognize. Typically, these tumors are rapidly progressive and behave aggressively, emphasizing the need for early detection and prompt diagnostic work-up and start of treatment. In this review, the tumor biology and immunology, current diagnostic and treatment modalities, as well as new and combined therapies for MCC, are discussed. MCC is a very immunogenic tumor which offers good prospects for immunotherapy. Given its rarity, the aggressiveness, and the frail patient population it concerns, MCC should be managed in close collaboration with an experienced multidisciplinary team.

Keywords: Merkel cell carcinoma; surgery; radiotherapy; immunotherapy; biomarkers

1. Introduction

Merkel cell carcinoma (MCC) is a rare and aggressive malignancy of neuroendocrine origin that originates in the skin. MCC is mostly seen on the facial skin and extremities of elderly Caucasians and is associated with UV-exposure and infection with the Merkel cell polyomavirus (MCPyV). More than 70% of patients are above 70 years of age at diagnosis [1,2]. Clinically negative prognostic indicators for survival are tumor burden, regional metastases, gender (male), location in face or trunk as compared to upper extremities, and immunosuppression [1–5]. Histopathological negative prognostic factors are depth of invasion, lymph vascular invasion, and T-cell infiltration [6,7]. Association with MCPyV is reported in more than 80% of MCC patients in the Western world [8].

The clinical presentation of MCC is a fast growing, painless, reddish-purple cutaneous nodule. The incidence of MCC is low, but increasing in most Western countries [9]. In the US, there was a 95% increase in the absolute number of cases reported to SEER-18, from 334 in 2000 to 652 in 2013 [10]. In Australia, the incidence among men increases at an annual rate of 4.2%. Remarkably, while the incidence of MCC in men is rising, the incidence in women has been decreasing since 2002 [11]. The reason for this is unclear. Most likely causes for the overall increase are ageing of the population, increased UV-exposure, both voluntary (sun bathing) and involuntary (outdoor jobs, depletion of the ozone layer) [12], and increased use of immunosuppressive medication. Additionally, better recognition by pathologists with the availability of more specific immunohistochemical markers for this tumor may have contributed. Tumors that were previously classified as "unspecified small cell carcinoma" may now be identified as MCC.

Patients with MCC are at a high risk for loco-regional recurrence and distant metastases. Like melanoma, MCC shows a strong tendency to form satellite lesions in the skin. At presentation, 50–65% of patients have localized disease, 25–50% have regional metastases, and about 10% present with distant metastases [13,14]. The variation in reported incidence of lymph node metastases can be explained by variations in diagnostic procedures, i.e., use of ultrasound with or without fine needle cytology, FDG-PET-scan, and/or sentinel node procedure. The 5-year survival, independent of age is 50–60% for localized disease, and for lymphogenic and hematogenic metastatic disease this is 30–35% and 14%, respectively [1,13–15]. It must be noted that in this elderly population, a significant number of MCC patients die due to other causes. In a Dutch cohort of 351 patients, the 5-year overall survival was 58%, but the MCC-related survival was 78% [16]. Farley et al. reported a high overall 5-year survival rate of 70% and a 5-year disease-specific survival of 84% [17].

2. Etiology, Pathology, and Tumor Biology

On routine hematoxylin and eosin staining MCC is typically characterized as a monomorphous small round blue cell tumor with round or oval nuclei, finely dispersed chromatin, indistinct nucleoli, and scant cytoplasm. There is a high mitotic rate. Variations are seen, in particular in MCPyV-negative MCC. Histopathological confirmation of MCC requires immunohistochemistry to differentiate from other small cell neoplasms such as metastatic small-cell lung cancer, melanoma, or lymphoma. Molecular markers diagnostic for MCC include neuroendocrine markers (chromogranin A, synaptophysin, CD56), cytokeratin 20 (dot-like pattern), neurofilament, and MCPyV large T antigen (LT). Negative TTF1, S-100, and leukocyte common antigen (LCA) can be used to differentiate MCC from small-cell lung cancer, melanoma, and lymphoma, respectively.

The small round blue cells of MCC share ultrastructural and immunohistochemical characteristics with benign Merkel cells. However, the suggestion that the Merkel cell is the cell of origin of MCC is debated. Benign Merkel cells are typically located in the basal layer of the epidermis whereas MCC is reported to originate from any layer of the skin. Furthermore, it is suggested that MCPyV-positive and negative MCC may arise from distinct cells of origin [18]. Additionally, morphological and immunophenotypical features differ between MCPyV-positive and negative MCC [19,20]. Virus-negative tumors have more heterogeneous cytological features and frequently display elongated nuclei, resembling the spindle-shape variant of small cell lung cancer, larger cell size, more abundant cytoplasm, and prominent nuclei. There is evidence that UV-associated MCC derives from an epidermal progenitor cell, whereas MCPyV-associated tumors are suggested to be of non-epithelial origin or, alternatively, from cutaneous appendage precursor cells [21].

MCPyV seroprevalence in the general population is very high and increases with age from 50% in childhood to 80% in adults above 50 years of age [22,23]. MCPyV remains latent in most immunocompetent hosts, but when the immune system weakens, this can lead to viral reactivation. MCPyV is a small circular double-stranded DNA virus with a genome of about 5400 base pairs that is divided into early and late regions [24,25]. The early region encodes large and small tumor (T) antigens and a 57 KDa protein of uncertain function. The late gene region encodes structural proteins, the major capsid protein VP1, and the minor capsid proteins VP2 and VP3. Despite the high MCPyV seropositivity in the population, the incidence of MCC is very low. This is explained by increasing evidence for the hypothesis that oncogenesis by MCPyV requires the rare combination of two essential events. First, clonal integration of the viral genome in the host genome must take place. A crucial next event is a mutation with loss of expression of the C-terminus of the "large T antigen", by which viral replication is inhibited with a subsequent increase in the synthesis of the viral oncoproteins large T antigen and small T antigen that promote cell cycle progression and survival [18]. MCPyV-positive MCC typically carry a low mutational load whereas UV-induced oncogenesis of virus-negative MCC is characterized by a cascade of oncogenic mutations as a result of accumulating DNA-damage [18].

There is equivocal data on the prognostic significance of MCPyV-status [18]. One retrospective analysis of 282 cases used large T antigen immunohistochemistry with two distinct antibodies as well as quantitative MCPyV PCR to assess MCPyV-status [26]. Fifty-three of 282 MCC (19%) were identified as virus-negative. In multi-variate analysis including stage, age, gender, and immune status, virus-negative patients were 1.5 times more likely to die from MCC albeit that this was not a statistically significant difference ($p = 0.14$). The favorable outcome of MCPyV-positive MCC is linked to increased immune-cell infiltration, in particular of CD8+ T cells, suggesting an antitumor immune response as the underlying mechanism [18,27].

Ongoing expression of viral oncoproteins is necessary for survival and progression of MCPyV-associated MCC [28]. These persistently expressed viral antigens may at some point trigger a host immune response [29,30]. Spontaneous regressions of MCC have been reported, either or not after discontinuation of immunosuppressive medication. In patients that do not use immunosuppressives, spontaneous regressions have been reported shortly after biopsy. Likely, the biopsy can generate an inflammatory environment stimulating antitumor immune responses [31–34].

3. Diagnostics and Staging

Physical examination includes assessment of the primary tumor and documentation by light photography, palpation of regional lymph node basins, and inspection of the entire skin surface by a dermatologist. The latter is important because MCC patients often develop other skin cancers as well and the dermatologist is the best qualified professional to do this. In case of advanced primary tumors a CT- or MR-scan can be considered to assess invasion of deeper structures. There is no general consensus on the role of imaging in the work-up of MCC patients with clinically localized disease and the current NCCN practice guideline does not recommend routine baseline imaging [35]. However, a recent retrospective analysis revealed that of 492 patients with no signs or symptoms of regional or distant spread, 65 (13%) were upstaged by diagnostic imaging (CT, MRI, or FDG-PET-CT) with consequences for treatment [36]. Confirmative data are provided by a literature review on the role of FDG-PET-CT [37]. In addition, or as an alternative to PET-CT, ultrasound with fine needle cytology can be considered for the evaluation of regional lymph nodes [38,39].

However, guidelines and reviews concur that the preferred diagnostic method for assessment of lymph node status is sentinel lymph node biopsy (SLNB) [18,35,37–40]. In 25–45% of patients clinically staged N0, SLNB demonstrated lymph node metastases with the majority of studies indicating a number close to 30% [41–49]. Various predictors for sentinel lymph node positivity have been identified including tumor size, but even for tumors <1 cm and <0.5 cm, the risk is still 20–31% [42,46,47] and 14% [44], respectively. Based on clinical and pathological characteristics, no subgroup of patients could be identified to have a likelihood of a positive sentinel lymph node lower than 15% [42]. An analysis of 1174 patients undergoing SLNB yielded a hazard ratio of death of 3.15 (95% CI 1.98–5.04, $p < 0.001$) for patients with positive sentinel lymph nodes versus those with negative sentinel lymph nodes [46]. These data strongly support the recommendation of SLNB for assessment of regional lymph nodes in MCC. However, in none of the patient cohorts discussed above, PET-CT or ultrasound with or without cytology was routinely performed in the diagnostic work-up. What the added value of SLNB is after state-of-the-art imaging is an issue that needs to be further addressed. Of patients with clinically uninvolved regional nodes 17% were upstaged by PET-CT, because of detection of regional and/or distant metastases, indicating that futile SLNB can be avoided in these patients [36,37].

Staging is according to AJCC, 8th edition which is based on an analysis of prognostic factors from 9387 MCC cases in the US [13].

4. Treatment

For various diseases with high prevalence at old age such as rheumatoid arthritis, psoriasis, and chronic lymphatic leukemia, immunosuppressive medication is frequently prescribed. It has been suggested that patients affected by rheumatologic diseases and treated with biologic immunosuppressives, including anti-TNF, are at an increased risk of MCC development [50]. Due to this possible cause-effect relationship, after diagnosis of MCC, immediate discontinuation of this medication should be considered, at least temporarily [5]. If discontinuation is not possible, dose reduction or replacement medication with less immunosuppressive effect may be an alternative. Burden and risk of progression of the underlying disease must be weighed against the potential detrimental effects on the tumor. For solid-organ transplant recipients, this is a particular problem because there is the risk of losing the transplanted organ. It is not known if restart of immunosuppressives after an adequate disease-free interval is safe or if drugs with another mechanism can be an alternative option.

4.1. Surgery and Adjuvant Radiotherapy

Radical excision is generally considered the treatment of choice [35,38,39]. Wide resection margins, varying between 1 cm and 3 cm depending on localization, are recommended. The reason for this is not only to obtain free resection margins, but also to include potential small satellite lesions close to the primary tumor. Wide excision must be balanced against the functional and cosmetic consequences, especially for tumors arising in the facial skin. The recommendation for wide excision is mostly based on experience from the 20th century when patients were often treated with surgery alone and local recurrence rates were in the order of 25% to 45% [51–53]. A more recent retrospective analysis of a cohort of 240 patients of whom 70% received postoperative radiotherapy reported much lower local recurrence rates of 2.9%, 2.8%, and 5.2% for margins of 1 cm, 1–2 cm, and >2 cm, respectively [54]. This suggests that if postoperative radiotherapy is given routinely, margins of 1 cm should be sufficient. More recent studies demonstrate that patients with localized disease that undergo surgery and postoperative radiotherapy not only have a significantly better locoregional tumor control, but also a better survival compared to those treated with surgery alone. Data were extracted from two US databases: SEER database (National Cancer Institute) and National Cancer Database (American College of Surgeons). Large cohorts of 1665, 4815, and 6908 cases were analyzed for the role of adjuvant radiotherapy in MCC [15,55,56]. Adjuvant radiotherapy improved 5-year overall survival rates by 10–15% depending on the size and stage of the tumor. Even for small primary tumors, there was a survival advantage [55]. A recent meta-analysis (29 studies, 17,179 patients) confirmed that adjuvant radiotherapy improves survival significantly (HR 0.81, $p < 0.001$) and reduces the risk of local and regional recurrence by 80% and 70%, respectively [57]. Apart from the fact that MCC is very radiosensitive, with radiotherapy, much larger skin surfaces and wider lymph drainage areas can be treated than with surgery. Whether small node-negative tumors should be routinely treated with postoperative radiotherapy remains a matter of debate. Even in a subgroup analysis of 168 patients with tumors <1 cm, Mojica et al. reported an improved median survival from 48 to 93 months with adjuvant radiotherapy [55]. Frohm et al., on the other hand, reported acceptable loco-regional control rates with surgery alone in tumors with largest diameter <2 cm [58]. Of 104 patients, 18 (17%) developed a local, in-transit, and/or regional recurrence. Two comments are to be placed with this study: First, the 17% is an absolute rate, the actuarial rate (corrected for duration of follow-up) most likely is higher, and second, the majority (13) of the recurrences were in the regional lymph nodes, which is a poor prognostic sign. Based on these and other data, the Danish guideline suggests that adjuvant radiotherapy may be omitted in selected low-risk cases, i.e., primary tumor <1 cm, negative margin status, no lymph vascular invasion, negative sentinel node biopsy, and no chronic immunosuppression [36].

After a negative SLNB, the risk of regional recurrence is relatively low (9–16%) [40,41,45] and treatment of the lymph node regions is generally not recommended. A positive SLNB must be followed by a complete lymph node dissection or regional radiotherapy. Despite subsequent treatment, the cumulative regional recurrence rate in SLNB positive patients is 11–28% [43,48]. If no SLNB is performed, elective treatment of at least the first draining lymph node level either by surgery or by radiotherapy is recommended. In case of clinically manifest regional metastases, a therapeutic lymph node dissection is performed. In virtually all cases, postoperative radiotherapy is indicated because it reduces the regional recurrence risk and improves the 3-year disease-specific survival from 48% to 76% [59].

The target volume for radiotherapy includes the primary tumor bed after excision with a margin for microscopic spread. The Danish guideline recommends margins of 1–2 cm, but this is not evidence-based [39]. Expert opinion is that margins should be generous, up to 3 cm, but adjusted to critical structures and sensitive organs, especially in the face. In case of clinically node negative disease, the regional nodal stations should be irradiated electively. This is not indicated after a negative sentinel node procedure. If lymph node metastases are present, the positive nodal level is treated as well as the next draining level.

A dose of 50–56 Gy in 2-Gy fractions is recommended in case of negative resection margins, 56–60 Gy for microscopically positive margins, and 60–66 Gy for grossly positive resection margins [35]. Depending on the condition of the patient, the size of the target volume and vulnerability of the tissues to be irradiated, hypofractionated schedules with biologically equivalent tumor dose can be used.

Side effects during and shortly after radiotherapy include dry or moist desquamation of skin. Other acute side effects depend on the area treated. Treatment of regional nodes in the head and neck area can cause mucositis with dysphagia. Skin and mucosal reactions generally heal within a few weeks. Hair loss, fibrosis, lymphedema, and xerostomia are potential long-term effects. These are generally mild with modern radiation techniques such as intensity modulated radiotherapy (IMRT) and volumetric modulated arc therapy (VMAT). The risk and severity of lymphedema increase if postoperative radiotherapy is applied after lymph node dissection [60].

4.2. Definitive Radiotherapy

MCC is notorious for its rapid growth and metastatic potential. After excision, re-excision, or lymph node dissection, it is not unusual that there is a delay before adjuvant radiotherapy is started [61]. The reason for this is multifactorial. Patients are mostly elderly and frail and have multiple comorbidities. Prolonged postoperative recovery and wound healing disturbances are common. Furthermore, because of its rarity, unfamiliarity with the disease still often causes unwanted delays in referrals and treatment. Two recent studies analyzed the time elapsed between surgery and radiotherapy, and concurred that the risk of loco-regional recurrence increased if the delay was greater than 8 weeks (25% vs. 10% and 37% vs. 0%, $p < 0.01$) [62,63]. These findings are supported by an earlier Australian publication [64].

Definitive radiation monotherapy is used as an alternative to surgery for patients who are poor surgical candidates or for those in whom surgery would result in significant functional compromise [65]. Given this selection bias, it is difficult to compare the results of primary radiotherapy versus surgery in retrospective cohorts. An attempt was made by a propensity score matched analysis using patient data from the National Cancer Database [66]. MCC patients treated with definitive radiotherapy were identified and matched with another patient treated with surgery (with or without adjuvant radiotherapy) accounting for age, co-morbidity score, stage, and grade. There were 1227 patients treated between 2004 and 2014 in each group. For stage I-II disease, 5-year overall survival was 61% in the surgery group and 42% in the radiotherapy group. For stage III, this was 34% and 21%, respectively. However, it is noteworthy that, despite the matching, there were significant differences between the groups. Patients in the radiotherapy group had larger and more advanced tumors, were less often treated in high volume academic centers, and had longer delays from diagnosis to start of treatment. The authors acknowledge these limitations and appreciate that "long-term survival can be obtained in patients with locally advanced and regionally metastatic disease with definitive radiotherapy". Single center studies report good loco-regional control rates with radiotherapy alone, ranging from 75% to 95%, similar to results of single center studies with surgery plus adjuvant radiotherapy [16,59,67–73]. The studies with radiotherapy alone had smaller patient numbers, but also included more advanced stages. A systematic review including 23 studies encompassing 264 patients reported a cumulative post-radiotherapy in-field control rate of 88% [74]. Figure 1 shows a patient with a large MCC on the cheek that was treated with radiotherapy alone. There was a durable complete regression until the last follow-up one-and-a-half years later.

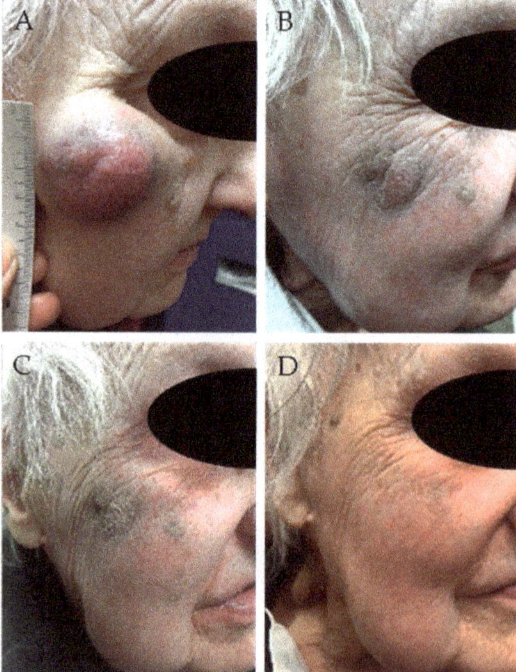

Figure 1. Ninety-four-year-old lady with Merkel cell carcinoma (MCC) of the right cheek, treated with radiotherapy (48 Gy). (**A**) Before treatment; (**B**) after 8 fractions (32 Gy); (**C**) after 12 fractions (48 Gy); (**D**) 6 weeks after completion of radiotherapy.

Although surgery is generally considered the primary treatment for MCC, this has developed empirically. There are no prospective clinical studies comparing surgery with or without adjuvant radiotherapy with radiotherapy alone. MCC is very radiosensitive. In vitro data confirm that it is even more sensitive than small-cell lung cancer, a tumor where radiotherapy and chemotherapy are the main treatment modalities and surgery is of minor relevance [75]. Therefore, primary radiotherapy is an excellent alternative for surgery with postoperative radiotherapy, also for operable patients with resectable disease [76]. A single modality treatment can spare these elderly patients the burden of additional morbidity and reduce health care costs.

4.3. Systemic Treatment

There is no role for chemotherapy in the primary treatment of MCC. Adjuvant chemotherapy for patients with stage I-III disease does not improve survival [15,35,77]. In the palliative setting, chemotherapy can be given for metastatic disease. Cytostatic drugs mostly used are carboplatin (or cisplatin) and etoposide or a combination of cyclophosphamide, doxorubicine (or epirubicine), and vincristine. There is often a rapid response (53–76%), but rarely is this long-lasting [77]. Additionally, these drugs are toxic for the elderly patient and often not tolerated. The progression-free survival varies from 3 to 8 months [77].

Chemotherapy will fade further into the background with the upsurge of immunotherapy. MCC is a very immunogenic tumor, indicating that there is great potential for immunotherapy. Avelumab is a fully human monoclonal antibody that targets the programmed death-ligand 1 (PD-L1). Expression of programmed cell death protein 1 (PD-1) ligands, in particular PD-L1, is upregulated in a variety of tumors including MCC, and blockade of PD-L1 signal can sensitize tumors to cytotoxic T lymphocyte killing [78]. In a phase II study that included 88 patients with metastatic MCC previously treated with chemotherapy, the immune checkpoint inhibitor avelumab produced a response in 33% of patients of which 11% had a complete response [79]. The median time to response was 6.1 weeks and was not associated with MCPyV or PD-L1 status. In 71% of responders, there was a durable effect of more than one year. Experience from daily practice showed higher response rates of 47–57% and complete response in almost 25% of the patients, albeit that the duration of response was shorter (median 8 months) [80,81].

Immunotherapy with the PD-1 inhibitor pembrolizumab as first-line treatment for irresectable recurrence or metastatic disease was studied in a cohort of 50 patients [82]. A complete response was observed in 24% of patients and a partial response in 32%. Additionally, in this study, responses were prolonged with a 2-year progression-free survival of 48%.

Recently, a study was published on nivolumab, another PD-1 inhibitor, in the neoadjuvant setting [83]. Patients with resectable MCC received one or two courses of nivolumab, once per two weeks, starting 4 weeks before tumor resection. Of 39 included patients, 36 were operated, and in 17 (47%) tumors, a pathological complete response was observed. Three patients did not undergo surgery because of tumor progression or side effects of the treatment. Four other patients had progressive disease under nivolumab. The observation that 7 of 39 patients (18%) experienced detrimental effects under the neoadjuvant treatment is not trivial. It means that in these cases, while the tumor is progressing, valuable time is lost with potential deleterious postponement of local treatment. This may adversely affect the prognosis and shows that it is important that patients are closely observed during neoadjuvant treatment. Additionally, this emphasizes the importance of developing biomarkers predictive for response to immunotherapy.

Avelumab is currently considered first-line treatment for metastatic MCC, but it is expected that soon immunotherapy will also take a role in the primary treatment of localized disease, either in the (neo)adjuvant setting or concurrently with local treatment, be it surgery or radiotherapy. Currently, 9 clinical trials that study the role of immunotherapy specifically in MCC are registered in clinicaltrials.gov (Table 1). Five of these are trials that combine immunotherapy with concurrent radiotherapy. The hypothesis is that the immune activating properties of radiotherapy can potentiate immunotherapy.

Table 1. Clinical trials registered in ClinicalTrials.gov that study the role of immunotherapy specifically in MCC (date of search: 9 March 2021).

ClinicalTrials. Gov Identifier	Type of Study	Investigational Drug	Mode of Action	Eligibility	Recruitment Status
NCT02584829	Phase I/II	Avelumab *	PD-L1 inhibition	Stage IV	active, not recruiting
NCT04160065	Phase I	IFx-Hu2.0 (intratumoral)	Emm55 protein expression	Advanced	recruiting
NCT04291885	Phase II, randomized	Avelumab	PD-L1 inhibition	Stage I-III	recruiting
NCT03271372	Phase III, randomized	Avelumab	PD-L1 inhibition	Stage III	recruiting
NCT03988647	Phase II	Pembrolizumab *	PD-1 inhibition	Stage IV	recruiting
NCT03798639	Phase I, randomized	Nivolumab * ipilimumab	PD-1 inhibition CTLA-4 inhibition	pathol. Stage IIIA-B	recruiting
NCT03712605	Phase III, randomized	Pembrolizumab	PD-1 inhibition	Stage I-III	recruiting
NCT03304639	Phase II, randomized	Pembrolizumab *	PD-1 inhibition	Stage III-IV	active, not recruiting
NCT04261855	Phase Ib/II	Avelumab *	PD-L1 inhibition	Stage IV	recruiting

* Trials that combine immunotherapy with concurrent radiotherapy.

Figure 2 shows a patient with extensive lymphogenic metastases from a MCC progressive under avelumab. Avelumab was discontinued, and palliative radiotherapy was initiated. In the last week of radiotherapy, avelumab was restarted. After two months, there was a complete remission and the patient is still free of disease at last follow-up two years later. Of interest in this context is that out-of-field abscopal effects have been reported following short-course radiotherapy in patients with MCC progressive on PD-1 checkpoint blockade [84]. These observations suggest that the combination of radiotherapy and immunotherapy may be a potent therapeutic strategy, not only for advanced metastatic disease, but also for earlier stages. Better understanding of the mechanisms behind the interactions between the two modalities will be obtained by current and future research [85].

Although immunotherapy is often better tolerated than many chemotherapy regimens, it is not without side-effects. The list of potential side-effects is long, but many are rare. The toxicity profiles of avelumab and pembrolizumab largely overlap and most frequent are infusion-related (allergic) reactions, fatigue, diarrhea, nausea, and fever. Toxicities more specific for immunotherapy include autoimmune endocrine dysfunctions, pneumonitis, colitis, and hepatitis [86,87]. Reported overall incidence of adverse effects for avelumab was 28–46%, of which 8–9% were grade 3–4 adverse reactions [80,83]. In the study with pembrolizumab as first line for advanced and metastatic MCC, treatment-related adverse events of any grade occurred in 48 of 50 (96%) patients, of which 14 (28%) were grade 3 or higher. Seven patients discontinued pembrolizumab as a result of treatment toxicity [82].

A strategy to avoid adverse events is intratumoral immunotherapy for accessible lesions. In a pilot study, 15 MCC patients were subjected to intratumoral delivery of plasmid interleukin-12 via electroporation [88]. All patients completed at least one cycle without noteworthy systemic toxicity.

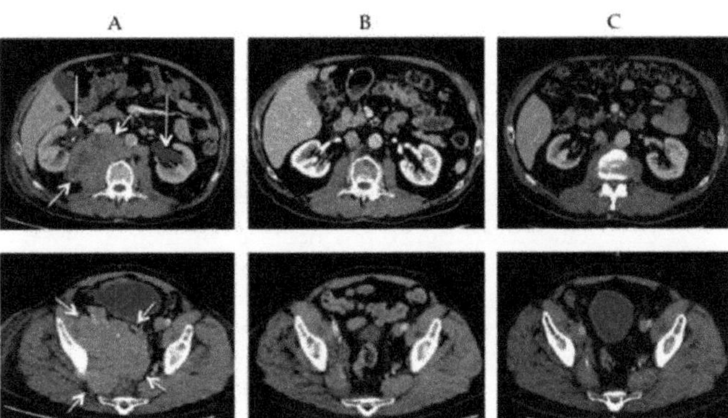

Figure 2. CT-scans of a 66-year-old man with massive lymphogenic metastases in abdomen and pelvis (short arrows) progressive under avelumab. Subsequent treatment with radiotherapy and restart of avelumab in last week of radiotherapy. (**A**) Before radiotherapy, note bilateral hydronephrosis (long arrows); (**B**) two months after radiotherapy (13 × 3 Gy), complete regression and recovery of hydronephrosis; (**C**) two years after radiotherapy; persistent complete regression.

For patients that do not respond to immunotherapy, alternative targeted therapies are needed. The mutational profile of MCPyV-positive tumors is different from that of MCV-negative tumors. It may be important to consider this biologic distinction when selecting a targeted therapy because driver mutations are more likely to be present in the MCPyV negative tumors that have a high mutational burden. About 80% of MCC's are driven by integration of MCPyV and ongoing T-antigen oncoprotein expression is needed for tumor progression. T-cell responses to these antigens are reported [89], which might explain the rare cases of spontaneous regression [33,90] but in the vast majority of cases, this immune response is suboptimal and ineffective for various reasons. It has been suggested to enhance this immune response by therapeutic vaccination to T-antigen under the assumption that the "nonself" viral antigen can trigger a stronger and more tumor-specific response compared to less cancer specific overexpressed oncoproteins [91,92]. Therapeutic MCPyV vaccination has been explored in a murine melanoma tumor line [93]. It could be demonstrated that a vaccine encoding the amino terminus of MCPyV large T antigen generated an antitumor effect mediated by CD4+ T-cells. Another group exploited dendritic cells loaded with large T-antigen and showed induction of antigen-specific T-cell responses in blood from healthy donors and MCC patients [94]. MCC has been shown to be an excellent model for further exploration of therapeutic anti-tumor vaccination.

One recent mechanistic insight relates to the activation of lysine-specific histone demethylase 1 (LSD1)-mediated dysregulation of gene expression by MCPyV small T antigen [95,96]. It was observed that all of six tested MCPyV-positive MCC cell lines responded to LSD1 inhibition, whereas three MCPyV-negative MCC cell lines did not [95]. Another study showed that LSD1 is a potent inhibitor of anti-tumor immunity and responsiveness to immunotherapy [97]. This suggests that a combination of LSD1 depletion and PLD1 blockade is a strategy that needs to be explored in clinical trials.

A drug currently under clinical investigation in MCC is called domatinostat (4SC-202). Domatinostat inhibits both class I histone deacetylases (HDAC's) and LSD1, and enhances the expression of major histocompatibility (MHC) class I and -II genes [98,99]. As a result, the immunogenicity of tumor cells is increased with improved recognition by cytotoxic T-cells. The combination with a checkpoint inhibitor is expected to improve the effect of immunotherapy, especially in patients that do not respond to anti-PD-(L)1 treatment alone. A clinical study combining domatinostat and avelumab in patients with advanced or metastatic MCC that have progressed on previous avelumab or pembrolizumab monotherapy is currently recruiting (NCT04393753).

Other therapeutic opportunities include somatostatin analogues, tyrosine kinase inhibitors, electrochemotherapy, talimogene laherparepvec (TVEC), and many others. However, most of these strategies have not been investigated systematically in properly designed prospective clinical trials. Reports are mostly retrospective and involve small patient numbers.

5. Biomarkers

Biomarkers can be prognostic or predictive. "Predictive assays" need to be distinguished conceptually from "prognostic factors". The latter are determined empirically and, although useful, they merely indicate favorable or unfavorable outcomes, but offer no basis for selection of more effective treatment strategies. A predictive assay provides a mechanistic basis and identifies a biological target for personalized treatment.

As prognostic molecular biomarkers for MCC, p63, p53, survivin, CD34, hedgehog proteins, and several others have been suggested, but none of these emerge as very strong and robust prognosticators [18]. The prognostic relevance of MCPyV remains equivocal [100]. Serum neuron-specific enolase (NSE) has become of interest recently. NSE is found in neuroendocrine tissues and is expressed in the cytoplasm of MCC cells [101]. NSE was determined in serum samples of 84 MCC patients at baseline and during follow-up [102]. Baseline NSE levels correlated with extent of disease, but not with relapse-free survival or overall survival. Interestingly, NSE was particularly useful in detecting progression of the disease with a negative predictive value of 98%. Another, earlier study in 60 patients, however, did not find associations of NSE blood levels with recurrence or survival [103]. The clinical relevance of NSE as a prognostic biomarker needs to be further validated.

Immune-response related tumor characteristics are the most likely predictive biomarker candidates for immunotherapy. In the three previously discussed phase II trials with immunotherapy MCPyV-status, total mutational burden (TMB), CD8+ T-cell density, and PD-L1 expression have been investigated as putative biomarkers [79,82,83]. For TMB and CD8+ T-cell density, non-significant trends for associations with tumor response were found. PD-L1 expression ($\geq 1\%$ tumor cells positive) varied from 26% to 82% in the three studies, and no associations were found with response to immune checkpoint blockade. In two studies, however, there was a trend for better overall survival for cases with positive PD-L1 expression [79,82]. The large variation in PD-L1 expression between the studies is remarkable. This might be explained by tissue sampling errors, differences in immunohistochemistry protocols, inter-observer differences, and/or differences between untreated and recurrent cases. In a retrospective analysis of a small cohort of 27 MCC patients, PD-1, but not PD-L1, expression was associated with immunotherapy response [104]. Response rate was 77% in PD-1 positive tumors vs. 21% in PD-1 negative tumors (p <0.01). The value of PD-L1 and PD-1 as predictive biomarkers for immunotherapy in MCC needs to be further explored in larger clinical trials.

6. Multi-Disciplinary and Expertise

Given the rarity and the aggressiveness of MCC and the rapid evolution of new treatment opportunities, management of MCC requires a multidisciplinary team, preferably in a high-volume center. An analysis of 5304 cases from the US National Cancer Database with stage I-III MCC demonstrated that 5-year overall survival was 62.3% at high volume facilities vs. 56.8% at lower-volume facilities (p < 0.001) [105]. That being said, it will not always be feasible or desirable to refer these elderly and frail patients to centers located further away. A solution can be to set up multidisciplinary consultation networks to provide the best attainable care for these patients near their own living environment.

7. Conclusions

MCC is a rare skin cancer, albeit with rapidly increasing incidence. Management requires a multidisciplinary team, preferably within a network of an expertise center. Tumor progression and metastasis formation is often fast, and early recognition of MCC and expeditious diagnostic workup and treatment initiation are vital. Effective treatments are available that have improved prognosis significantly. Surgery with adjuvant radiotherapy is the standard for localized disease, but radiotherapy alone is a good alternative. Immune checkpoint inhibitors offer durable responses in a significant proportion of the patients with metastatic or recurrent disease. However, treatment morbidity is not negligible in this elderly and frail patient population. The challenge is to accomplish as high as possible cure rates with limited and acceptable toxicity and morbidity.

Author Contributions: E.M.Z. and J.H.A.M.K. had the idea for the article and performed the literature search and data analysis. E.M.Z. and J.H.A.M.K. drafted and revised the manuscript. S.F.K.L., J.E.M.W., A.L.A., W.L.J.W., R.P.T., S.A.H.P., C.M.L.v.H., and G.J.A. critically reviewed and edited the manuscript. All authors have read and agreed to the published version of the manuscript.

Funding: This research received no external funding.

Institutional Review Board Statement: Not applicable.

Informed Consent Statement: Informed consent for publication was obtained from the patient pictured in Figure 1.

Data Availability Statement: Not applicable.

Conflicts of Interest: All authors declare no conflict of interest.

References

1. Albores-Saavedra, J.; Batich, K.; Chable-Montero, F.; Sagy, N.; Schwartz, A.M.; Henson, D.A. Merkel cell carcinoma demographics, morphology, and survival based on 3870 cases: A population based study. *J. Cutan. Pathol.* **2010**, *37*, 20–27. [CrossRef] [PubMed]
2. Reichgelt, B.A.; Visser, O. Epidemiology and survival of Merkel cell carcinoma in the Netherlands. A population-based study of 808 cases in 1993–2007. *Eur. J. Cancer* **2011**, *47*, 579–585. [CrossRef] [PubMed]
3. Ma, J.E.; Brewer, J.D. Merkel cell carcinoma in immunosuppressed patients. *Cancers* **2014**, *6*, 1328–1350. [CrossRef] [PubMed]
4. Brewer, J.D.; Shanafelt, T.D.; Otley, C.C.; Roenigk, R.K.; Cerhan, J.R.; Kay, N.E.; Weaver, A.L.; Call, T.G. Chronic lymphocytic leukemia is associated with decreased survival of patients with malignant melanoma and Merkel cell carcinoma in a SEER population-based study. *J Clin. Oncol.* **2012**, *30*, 843–849. [CrossRef] [PubMed]
5. Bryant, M.K.; Ward, C.; Gaber, C.E.; Strassle, P.D.; Ollila, D.W.; Laks, S. Decreased survival and increased recurrence in Merkel cell carcinoma significantly linked with immunosuppression. *J. Surg. Oncol.* **2020**, *122*, 653–659. [CrossRef] [PubMed]
6. Andea, A.A.; Coit, D.G.; Amin, B.; Busam, K.J. Merkel cell carcinoma: Histologic features and prognosis. *Cancer* **2008**, *113*, 2549–2558. [CrossRef]
7. Farah, M.; Reuben, A.; Spassova, I.; Yang, R.K.; Kubat, L.; Nagarajan, P.; Ning, J.; Li, W.; Aung, P.P.; Curry, J.L.; et al. T-cell repertoire in combination with T-cell density predicts clinical outcomes in patients with Merkel cell carcinoma. *J. Investig. Dermatol.* **2020**, *140*, 2146–2156. [CrossRef]
8. Becker, J.; Houben, R.; Ugurel, S.; Trefzer, U.; Pföhler, C.; Schrama, D. Polyomavirus is frequently present in Merkel cell carcinoma of European patients. *J. Investig. Dermatol.* **2009**, *129*, 248–250. [CrossRef] [PubMed]
9. Stang, A.; Becker, J.C.; Nghiem, P.; Ferlay, J. The association between geographic location and incidence of Merkel cell carcinoma in comparison to melanoma: An international assessment. *Eur. J. Cancer* **2018**, *94*, 47–60. [CrossRef]
10. Paulson, K.G.; Park, S.Y.; Vandeven, N.A.; Lachance, K.; Thomas, H.; Chapuis, A.G.; Harms, K.L.; Thompson, J.A.; Bhatia, S.; Stang, A.; et al. Merkel cell carcinoma: Current US incidence and projected increases based on changing demographics. *J. Am. Acad. Dermatol.* **2018**, *78*, 457–463. [CrossRef]
11. Garbutcheon-Singh, K.B.; Curchin, D.J.; McCormack, C.J.; Smith, S.D. Trends in the incidence of Merkel cell carcinoma in Victoria, Australia, between 1986 and 2016. *Australas. J. Dermatol.* **2020**, *61*, e34–e38. [CrossRef]
12. Pinault, L.; Bushnik, T.; Fioletov, V.; Peters, C.E.; King, W.D.; Tjepkema, M. The risk of melanoma associated with ambient summer ultraviolet radiation. *Health Rep.* **2017**, *28*, 3–11.
13. Harms, K.L.; Healy, M.A.; Nghiem, P.; Sober, A.J.; Johnson, T.M.; Bichakjian, C.K.; Wong, S.L. Analysis of prognostic factors from 9387 Merkel cell carcinoma cases forms the basis for the new 8th edition AJCC staging system. *Ann. Surg. Oncol.* **2016**, *23*, 3564–3571. [CrossRef] [PubMed]
14. Iyer, J.G.; Storer, B.E.; Paulson, K.G.; Lemos, B.; Philips, J.L.; Bichakjian, C.K.; Zeitouni, N.; Gershenwald, J.E.; Sondak, V.; Otley, C.C.; et al. Relationship among primary tumor size, number of involved nodes, and survival for 8044 cases of Merkel cell carcinoma. *J. Am. Acad. Dermatol.* **2014**, *70*, 637–643. [CrossRef] [PubMed]
15. Bhatia, S.; Storer, B.E.; Iyer, J.G.; Moshiri, A.; Parvathaneni, U.; Byrd, D.; Sober, A.J.; Sondak, V.K.; Gershenwald, J.E.; Nghiem, P. Adjuvant radiation therapy and chemotherapy in Merkel cell carcinoma: Survival analyses of 6908 cases from the National Cancer Data Base. *J. Natl. Cancer Inst.* **2016**, *108*, djw042. [CrossRef] [PubMed]

16. Van Veenendaal, L.M.; van Akkooi, A.C.J.; Verhoef, C.; Grünhagen, D.J.; Klop, W.M.C.; Valk, G.D.; Tesselaar, M.E.T. Merkel cell carcinoma: Clinical outcome and prognostic factors in 351 patients. *J. Surg. Oncol.* **2018**, *117*, 1768–1775. [CrossRef] [PubMed]
17. Farley, C.R.; Perez, M.C.; Soelling, S.J.; Delman, K.A.; Harit, A.; Wuthrick, E.J.; Messina, J.L.; Sondak, V.K.; Zager, J.S.; Lowe, M.C. Merkel cell carcinoma outcomes: Does AJCC8 underestimate survival? *Ann. Surg. Oncol.* **2020**, *27*, 1978–1985. [CrossRef] [PubMed]
18. Harms, P.W.; Harms, K.L.; Moore, P.S.; DeCaprio, J.A.; Nghiem, P.; Wong, M.K.K.; Brownell, I.; on behalf of the International Workshop on Merkel Cell Carcinoma Research (IWMCC) Working Group. The biology and treatment of Merkel cell carcinoma: Current understanding and research priorities. *Nat. Rev. Clin. Oncol.* **2018**, *15*, 763–776. [CrossRef]
19. Kervarrec, T.; Tallet, A.; Miquelestorena-Standley, E.; Houben, R.; Schrama, D.; Gambichler, T.; Berthon, P.; Le Corre, Y.; Hainaut-Wierzbicka, E.; Aubin, F.; et al. Morphologic and immunophenotypical features distinguishing Merkel cell polyomavirus-positive and negative Merkel cell carcinoma. *Mod. Pathol.* **2019**, *32*, 1605–1616. [CrossRef]
20. Moran, J.M.T.; Biecek, P.; Donizy, P.; Wu, C.-L.; Kopczynski, M.D.; Pieniazek, M.; Rys, J.; Hoang, M.P. Large nuclear size correlated with better overall survival, Merkel cell polyomavirus positivity, and terminal deoxynucleotidyl transferase expression in Merkel cell carcinoma. *J. Am. Acad. Dermatol.* **2021**, *84*, 550–552. [CrossRef]
21. Kervarrec, T.; Samimi, M.; Guyétant, S.; Sarma, B.; Chéret, J.; Blanchard, E.; Berthon, P.; Schrama, D.; Houben, R.; Touzé, A. Histogenesis of Merkel cell carcinoma: A comprehensive review. *Front. Oncol.* **2019**, *9*, 451. [CrossRef] [PubMed]
22. Tolstov, Y.L.; Pastrana, D.V.; Feng, H.; Becker, J.C.; Jenkins, F.J.; Moschos, S.; Chang, Y.; Buck, C.B.; Moore, P.S. Human Merkel cell polyomavirus infection II. MCV is a common human infection that can be detected by conformational capsid epitope immunoassays. *Int. J. Cancer* **2009**, *125*, 1250–1256. [CrossRef] [PubMed]
23. Touzé, A.; Gaitan, J.; Arnold, F.; Cazal, R.; Fleury, M.J.; Combelas, N.; Sizaret, P.-Y.; Guyetant, S.; Maruani, A.; Baay, M.; et al. Generation of Merkel Cell Polyomavirus (MCV)-like particles and their application to detection of MCV antibodies. *J. Clin. Microbiol.* **2010**, *48*, 1767–1770. [CrossRef] [PubMed]
24. Feng, H.; Shuda, M.; Chang, Y.; Moore, P.S. Clonal integration of a polyomavirus in human Merkel cell carcinoma. *Science* **2008**, *319*, 1096–1100. [CrossRef] [PubMed]
25. Coursaget, P.; Samimi, M.; Nicol, J.T.J.; Gardair, C.; Touzé, A. Human Merkel cell polyomavirus: Virological background and clinical implications. *APMIS* **2013**, *121*, 755–769. [CrossRef]
26. Moshiri, A.S.; Doumani, R.; Yelistratova, L.; Blom, A.; Lachance, K.; Shinohara, M.M.; Delaney, M.; Chang, O.; McArdle, S.; Thomas, H.; et al. Polyomavirus-negative Merkel cell carcinoma: A more aggressive subtype based on analysis of 282 cases using multimodal tumor virus detection. *J. Investig. Dermatol.* **2017**, *137*, 819–827. [CrossRef] [PubMed]
27. Ricci, C.; Righi, A.; Ambrosi, F.; Gibertoni, D.; Maletta, F.; Uccella, S.; Sessa, F.; Asioli, S.; Pellilli, M.; Maragliano, R.; et al. Prognostic impact of MCPyV and TIL subtyping in Merkel cell carcinoma: Evidence from a large European cohort of 95 patients. *Endocr. Pathol.* **2020**, *31*, 21–32. [CrossRef]
28. Schrama, D.; Sarosi, E.M.; Adam, C.; Ritter, C.; Kaemmerer, U.; Klopocki, E.; König, E.M.; Utikal, J.; Becker, J.C.; Houben, R. Characterization of six Merkel cell polyomavirus-positive Merkel cell carcinoma cell lines: Integration pattern suggest that large T antigen truncating events occur before or during integration. *Int. J. Cancer* **2019**, *145*, 1020–1032. [CrossRef]
29. Afanasiev, O.K.; Yelistratova, L.; Miller, N.; Nagase, K.; Paulson, K.; Iyer, J.G.; Ibrani, D.; Koelle, D.M.; Nghiem, P. Merkel polyomavirus-specific T cells fluctuate with Merkel cell carcinoma burden and express therapeutically targetable PD-1 and Tim-3 exhaustion markers. *Clin. Cancer Res.* **2013**, *19*, 5351–5360. [CrossRef]
30. Iyer, J.G.; Afanasiev, O.K.; McClurkan, C.; Paulson, K.; Nagase, K.; Jing, L.; Marshak, J.O.; Dong, L.; Carter, J.; Lai, I.; et al. Merkel cell polyomavirus-specific CD8+ and CD4+ T-cell responses identified in Merkel cell carcinomas and blood. *Clin. Cancer Res.* **2011**, *17*, 6671–6680. [CrossRef]
31. Friedlaender, M.M.; Rubinger, D.; Rosenbaum, B.; Amir, G.; Siguencia, E. Temporary regression of Merkel cell carcinoma metastases after cessation of cyclosporine. *Transplantation* **2002**, *73*, 1849–1850. [CrossRef]
32. Sais, G.; Admella, C.; Soler, T. Spontaneous regression in primary cutaneous neuroendocrine (Merkel cell) carcinoma: A rare immune phenomenon? *J. Eur. Acad. Derm. Venereol.* **2002**, *16*, 82–83. [CrossRef]
33. Connelly, T.J.; Cribier, B.; Brown, T.J.; Yanguas, I. Complete spontaneous regression of Merkel cell carcinoma: A review of 10 reported cases. *Dermatol. Surg.* **2000**, *26*, 853–856. [CrossRef]
34. Bertolotti, A.; Conte, H.; Francois, L.; Dutriaux, C.; Ezzedine, K.; Mélard, P.; Vergier, B.; Taieb, A.; Jouary, T. Merkel cell carcinoma: Complete clinical remission associated with disease progression. *JAMA Dermatol.* **2013**, *149*, 501–502. [CrossRef]
35. Bichakjian, C.K.; Olencki, T.; Aasi, S.Z.; Alam, M.; Andersen, J.S.; Blitzblau, R.; Bowen, G.M.; Contreras, C.M.; Daniels, G.A.; Decker, R.; et al. Merkel cell carcinoma, Version 1.2018. NCCN Clinical Practice Guidelines in Oncology. *J. Natl. Compr. Cancer Netw.* **2018**, *16*, 742–774. [CrossRef] [PubMed]
36. Singh, N.; Alexander, N.A.; Lachance, K.; Lewis, C.W.; McEvoy, A.; Akaike, G.; Byrd, D.; Behnia, S.; Bhatia, S.; Paulson, K.G.; et al. Clinical benefit of baseline imaging in Merkel cell carcinoma: Analysis of 584 patients. *J. Am. Acad. Dermatol.* **2021**, *84*, 330–339. [CrossRef]
37. Sachpekidis, C.; Sidiropoulou, P.; Hassel, J.C.; Drakoulis, N.; Dimitrakopoulou-Strauss, A. Positron emission tomography in Merkel cell carcinoma. *Cancers* **2020**, *12*, 2897. [CrossRef] [PubMed]

38. Lebbe, C.; Becker, J.C.; Grob, J.-J.; Malvehy, J.; del Marmol, V.; Pehamberger, H.; Peris, K.; Saiag, P.; Middleton, M.R.; Bastholt, L.; et al. Diagnosis and treatment of Merkel cell carcinoma. European consensus-based interdisciplinary guideline. *Eur. J. Cancer.* **2015**, *51*, 2396–2403. [CrossRef] [PubMed]
39. Naseri, S.; Steiniche, T.; Ladekarl, M.; Bønnelykke-Behrndtz, M.L.; Hölmich, L.R.; Langer, S.W.; Venzo, A.; Tabaksblat, E.; Klausen, S.; Skaarup Larsen, M.; et al. Management recommendations for Merkel cell carcinoma-a Danish perspective. *Cancers* **2020**, *12*, 554. [CrossRef] [PubMed]
40. Cassler, N.M.; Merrill, D.; Bichakjian, C.K.; Brownell, I. Merkel cell carcinoma therapeutic update. *Curr. Treat. Options Oncol.* **2016**, *17*, 36. [CrossRef]
41. Allen, P.J.; Browne, W.B.; Jaques, D.P.; Brennan, M.F.; Busam, K.; Coit, D.G. Merkel cell carcinoma: Prognosis and treatment of patients from a single institution. *J. Clin. Oncol.* **2005**, *23*, 2300–2309. [CrossRef] [PubMed]
42. Schwartz, J.L.; Griffith, K.A.; Lowe, L.; Wong, S.L.; McLean, S.A.; Fullen, D.R.; Lao, C.D.; Hayman, J.A.; Bradford, C.R.; Rees, R.S.; et al. Features predicting sentinel lymph node positivity in Merkel cell carcinoma. *J. Clin. Oncol.* **2011**, *29*, 1036–1041. [CrossRef] [PubMed]
43. Sims, J.R.; Grotz, T.E.; Pockaj, B.A.; Joseph, R.W.; Foote, R.L.; Otley, C.C.; Weaver, A.L.; Jakub, J.W.; Price, D.L. Sentinel lymph node biopsy in Merkel cell carcinoma: The Mayo Clinic experience of 150 patients. *Surg. Oncol.* **2018**, *27*, 11–17. [CrossRef]
44. Smith, F.O.; Yue, B.; Marzban, S.S.; Walls, B.L.; Carr, M.; Jackson, R.S.; Puleo, C.A.; Padhya, T.; Cruse, C.W.; Gonzalez, R.J.; et al. Both tumor depth and diameter are predictive of sentinel node status and survival in Merkel cell carcinoma. *Cancer* **2015**, *121*, 3252–3260. [CrossRef] [PubMed]
45. Fritsch, V.A.; Camp, E.R.; Lentsch, E.J. Sentinel lymph node status in Merkel cell carcinoma of the head and neck: Not a predictor of survival. *Head Neck.* **2014**, *36*, 571–579. [CrossRef]
46. Conic, R.R.Z.; Ko, J.; Saridakis, S.; Damiani, G.; Funchain, P.; Vidimos, A.; Gastman, B.R. Sentinel lymph node biopsy in Merkel cell carcinoma: Predictors of sentinel lymph node positivity and association with overall survival. *J. Am. Acad. Dermatol.* **2019**, *81*, 364–372. [CrossRef]
47. Harounian, J.A.; Molin, N.; Galloway, T.J.; Ridge, D.; Bauman, J.; Farma, J.; Reddy, S.; Lango, M.N. Effect of sentinel lymph node biopsy and LVI on Merkel cell carcinoma prognosis and treatment. *Laryngoscope* **2021**, *131*, E828–E835. [CrossRef]
48. Gunaratne, D.A.; Howle, J.A.A.R.; Veness, M.J. Sentinel lymph node biopsy in Merkel cell carcinoma; a 15 year institutional experience and statistical analysis of 721 reported cases. *Br. J. Dermatol.* **2016**, *174*, 273–281. [CrossRef]
49. Karunaratne, Y.G.; Gunaratne, D.A.; Veness, M.J. Systematic review of sentinel lymph node biopsy in Merkel cell carcinoma of the head and neck. *Head Neck.* **2018**, *40*, 2704–2713. [CrossRef]
50. Rotondo, J.C.; Bononi, I.; Puozzo, A.; Govoni, M.; Foschi, V.; Lanza, G.; Gafà, R.; Gaboriaud, P.; Touzé, F.A.; Selvatici, R.; et al. Merkel cell carcinomas arising in autoimmune disease affected patients treated with biologic drugs, including anti-TNF. *Clin. Cancer Res.* **2017**, *23*, 3929–3934. [CrossRef] [PubMed]
51. O'Connor, W.J.; Brodland, D.G. Merkel cell carcinoma. *Dermatol. Surg.* **1996**, *22*, 262–267. [CrossRef]
52. Yiengpruksawan, A.; Coit, D.G.; Thaler, H.T.; Urmacher, C.; Knapper, W.K. Merkel cell carcinoma. Prognosis and management. *Arch. Surg.* **1991**, *126*, 1514–1519. [CrossRef] [PubMed]
53. Haag, M.L.; Glass, L.F.; Fenske, N.A. Merkel cell carcinoma. Diagnosis and treatment. *Dermatol. Surg.* **1995**, *21*, 669–683. [CrossRef]
54. Perez, M.C.; de Pinho, F.R.; Holstein, A.; Oliver, D.E.; Naqvi, S.M.H.; Kim, Y.; Messina, J.L.; Burke, E.; Gonzalez, R.J.; Sarnaik, A.A.; et al. Resection margins in Merkel cell carcinoma: Is a 1-cm margin wide enough? *Ann. Surg. Oncol.* **2018**, *25*, 3334–3340. [CrossRef] [PubMed]
55. Mojica, P.; Smith, D.; Ellenhorn, J.D.I. Adjuvant radiation therapy is associated with improved survival in Merkel cell carcinoma of the skin. *J. Clin. Oncol.* **2007**, *25*, 1043–1047. [CrossRef] [PubMed]
56. Chen, M.M.; Roman, S.A.; Sosa, J.A.; Judson, B.L. The role of adjuvant therapy in the management of head and neck Merkel cell carcinoma: An analysis of 4815 patients. *JAMA Otolaryngol. Head Neck Surg.* **2015**, *141*, 137–141. [CrossRef]
57. Petrelli, F.; Ghidini, A.; Torchio, M.; Prinzi, N.; Trevisan, F.; Dallera, P.; De Stefani, A.; Russo, A.; Vitali, E.; Bruschieri, L.; et al. Adjuvant radiotherapy for Merkel cell carcinoma: A systematic review and meta-analysis. *Radiother. Oncol.* **2019**, *134*, 211–219. [CrossRef]
58. Frohm, M.L.; Griffith, K.A.; Harms, K.L.; Hayman, J.A.; Fullen, D.R.; Nelson, C.C.; Wong, S.L.; Schwartz, J.L.; Bichakjian, C.K. Recurrence and survival in patients with Merkel cell carcinoma undergoing surgery without adjuvant radiation therapy to the primary site. *JAMA Dermatol.* **2016**, *152*, 1001–1007. [CrossRef]
59. Strom, T.; Carr, M.; Zager, J.S.; Naghavi, A.; Smith, F.O.; Cruse, C.W.; Messina, J.L.; Russell, J.; Rao, N.G.; Fulp, W.; et al. Radiation therapy is associated with improved outcomes in Merkel cell carcinoma. *Ann. Surg. Oncol.* **2016**, *23*, 3572–3578. [CrossRef]
60. Henderson, M.A.; Burmeister, B.H.; Ainslie, J.; Fisher, R.; Di Iulio, J.; Smithers, B.M.; Hong, A.; Shannon, K.; Scolyer, R.A.; Carruthers, S.; et al. Adjuvant lymph-node field radiotherapy versus observation only in patients with melanoma at high risk of further lymph-node field relapse after lymphadenectomy (ANZMTG 01.02/TROG 02.01): 6-year follow-up of a phase 3, randomised controlled trial. *Lancet Oncol.* **2015**, *16*, 1049–1060. [CrossRef]
61. Poulsen, M. Radiation therapy rather than surgery for merkel cell carcinoma: The advantages of radiation therapy. *Int. J. Radiat. Oncol. Biol. Phys.* **2017**, *100*, 14–15. [CrossRef]

62. Alexander, N.; Schaub, S.; Hippe, D.; Lachance, K.; Liao, J.J.; Apisarnthanarax, S.; Bhatia, S.; Tseng, Y.D.; Nghiem, P.T.; Parvathaneni, U. Merkel cell carcinoma recurrence risk increases with the delay op post-operative radiation therapy. In Proceedings of the First International Symposium on Merkel Cell Carcinoma, Tampa, FL, USA, 21–22 October 2019.

63. Zwijnenburg, E.M.; Lubeek, S.F.; Werner, J.E.; Torres-Rivas, J.; Span, P.N.; Takes, R.J.; Weijs, W.L.; Kaanders, J.H. Merkel cell carcinoma is an oncologic emergency. In Proceedings of the Proceedings First International Symposium on Merkel Cell Carcinoma, Tampa, FL, USA, 21–22 October 2019.

64. Tsang, G.; O'Brien, P.; Robertson, R.; Hamilton, C.; Wratten, C.; Denham, J. All delays before radiotherapy risk progression of Merkel cell carcinoma. *Australas. Radiol.* **2004**, *48*, 371–375. [CrossRef]

65. Hong, A.M.; Stretch, J.R.; Thompson, J.F. Treatment of primary Merkel cell carcinoma: Radiotherapy can be an effective, less morbid alternative to surgery. *Eur. J. Surg. Oncol.* **2021**, *47*, 483–485. [CrossRef]

66. Wright, G.P.; Holtzman, M.P. Surgical resection improves median overall survival with marginal improvement in long-term survival when compared with definitive radiotherapy in Merkel cell carcinoma: A propensity score matched analysis of the National Cancer Database. *Am. J. Surg.* **2018**, *215*, 384–387. [CrossRef]

67. Hui, A.C.; Stillie, A.L.; Seel, M.; Ainslie, J. Merkel cell carcinoma: 27-year experience at the Peter MacCallum Cancer Centre. *Int. J. Radiat. Oncol. Biol. Phys.* **2011**, *80*, 1430–1435. [CrossRef]

68. Santamaria-Barria, J.A.; Boland, G.M.; Yeap, B.Y.; Nardi, V.; Dias-Santagata, D.; Cusack, J.C., Jr. Merkel cell carcinoma: 30-year experience from a single institution. *Ann. Surg. Oncol.* **2013**, *20*, 1365–1373. [CrossRef]

69. Bishop, A.J.; Garden, A.S.; Gunn, G.B.; Rosenthal, D.I.; Beadle, B.M.; Fuller, C.D.; Levy, L.B.; Gillenwater, A.M.; Kies, M.S.; Esmaeli, B.; et al. Merkel cell carcinoma of the head and neck: Favorable outcomes with radiotherapy. *Head Neck* **2016**, *38* (Suppl. 1), E452–E458. [CrossRef] [PubMed]

70. Mendenhall, W.M.; Morris, C.G.; Kirwan, J.M.; Amdur, R.J.; Shaw, C.; Dziegielewski, P.T. Management of cutaneous Merkel cell carcinoma. *Acta Oncol.* **2018**, *57*, 320–323. [CrossRef] [PubMed]

71. Pape, E.; Rezvoy, N.; Penel, N.; Salleron, J.; Martinot, V.; Guerresschi, P.; Dziwniel, V.; Darras, S.; Mirabel, X.; Mortier, L. Radiotherapy alone for Merkel cell carcinoma: A comparative and retrospective study of 25 patients. *J. Am. Acad. Dermatol.* **2011**, *65*, 983–990. [CrossRef]

72. Veness, M.; Foote, M.; Gebski, V.; Poulsen, M. The role of radiotherapy alone in patients with Merkel cell carcinoma: Reporting the Australian experience of 43 patients. *Int. J. Radiat. Oncol.* **2010**, *78*, 703–709. [CrossRef] [PubMed]

73. Harrington, C.; Kwan, W. Outcomes of Merkel cell carcinoma treated with radiotherapy without radical surgical excision. *Ann. Surg. Oncol.* **2014**, *21*, 3401–3405. [CrossRef] [PubMed]

74. Gunaratne, D.A.; Howle, J.R.; Veness, M.J. Definitive radiotherapy for Merkel cell carcinoma confers clinically meaningful in-field locoregional control: A review and analysis of the literature. *J. Am. Acad. Dermatol.* **2017**, *77*, 142–148.e1. [CrossRef] [PubMed]

75. Leonard, J.H.; Ramsey, J.R.; Kearsley, J.H.; Birrell, G.W. Radiation sensitivity of Merkel cell carcinoma cells. *Int. J. Radiat. Oncol. Biol. Phys.* **1995**, *32*, 1401–1407. [PubMed]

76. Kok, D.L.; Wang, A.; Xu, W.; Chua, M.S.T.; Guminski, A.; Veness, M.; Howle, J.; Tothill, R.; Kichendasse, G.; Poulsen, M.; et al. The changing paradigm of managing Merkel cell carcinoma in Australia: An expert commentary. *Asia Pac. J. Clin. Oncol.* **2020**, *16*, 312–319. [CrossRef]

77. Tello, T.L.; Coggshall, K.; Yom, S.S.; Yu, S.S. Merkel cell carcinoma: An update and review: Current and future therapy. *J. Am. Acad. Dermatol.* **2018**, *78*, 445–454. [CrossRef]

78. Boyerinas, B.; Jochems, C.; Fantini, M.; Heery, C.R.; Gulley, J.L.; Tsang, K.Y.; Schlom, J. Antibody-dependent cellular cytotoxicity activity of a novel anti-PD-L1 antibody avelumab (MSB0010718C) on human tumor cells. *Cancer Immunol. Res.* **2015**, *3*, 1148–1157.

79. D'Angelo, S.P.; Bhatia, S.; Brohl, A.S.; Hamid, O.; Mehnert, J.M.; Terheyden, P.; Shih, K.C.; Brownell, I.; Lebbé, C.; Lewis, K.D.; et al. Avelumab in patients with previously treated metastatic Merkel cell carcinoma: Long-term data and biomarker analyses from the single-arm phase 2 JAVELIN Merkel 200 trial. *J. Immunother. Cancer* **2020**, *8*, e000674. [CrossRef]

80. Levy, S.; Aarts, M.J.B.; Eskens, F.A.L.M.; Keymeulen, K.B.M.I.; Been, L.B.; Grünhagen, D.; van Akkooi, A.; Jalving, M.; Tesselaar, M.E.T. Avelumab for advanced Merkel cell carcinoma in the Netherlands: A real-world cohort. *J. Immunother. Cancer* **2020**, *8*, e001076. [CrossRef]

81. Walker, J.W.; Lebbé, C.; Grignani, G.; Nathan, P.; Dirix, L.; Fenig, E.; Ascierto, P.A.; Sandhu, S.; Munhoz, R.; Benincasa, E.; et al. Efficacy and safety of avelumab treatment in patients with metastatic Merkel cell carcinoma: Experience from a global expanded access program. *J. Immunother. Cancer* **2020**, *8*, e000313. [CrossRef] [PubMed]

82. Nghiem, P.T.; Bhatia, S.; Lipson, E.J.; Sharfman, W.H.; Kudchadkar, R.R.; Brohl, A.S.; Friedlander, P.A.; Daud, A.; Kluger, H.M.; Reddy, S.A.; et al. Durable tumor regression and overall survival in patients with advanced Merkel cell carcinoma receiving pembrolizumab as first-line therapy. *J. Clin. Oncol.* **2019**, *37*, 693–702. [CrossRef]

83. Topalian, S.L.; Bhatia, S.; Amin, A.; Kudchadkar, R.R.; Sharfman, W.H.; Lebbé, C.; Delord, J.-P.; Dunn, L.A.; Shinohara, M.M.; Kulikauskas, R.; et al. Neoadjuvant nivolumab for patients with resectable Merkel cell carcinoma in the Checkmate 538 Trial. *J. Clin. Oncol.* **2020**, *38*, 2476–2487. [CrossRef] [PubMed]

84. Xu, M.J.; Wu, S.; Daud, A.I.; Yu, S.S.; Yom, S.S. In-field and abscopal response after short-course radiation therapy in patients with metastatic Merkel cell carcinoma progressing on PD-1 checkpoint blockade: A case series. *J. Immunother. Cancer* **2018**, *6*, 43. [CrossRef] [PubMed]

85. Arina, A.; Gutiontov, S.I.; Weichselbaum, R.R. Radiotherapy and immunotherapy for cancer: From "systemic" to "multisite". *Clin. Cancer Res.* **2020**, *26*, 2777–2782. [CrossRef] [PubMed]
86. Ferrari, S.M.; Fallahi, P.; Elia, G.; Ragusa, F.; Ruffilli, I.; Patrizio, A.; Galdiero, M.R.; Baldini, E.; Ulisse, S.; Marone, G.; et al. Autoimmune endocrine dysfunctions associated with cancer immunotherapies. *Int. J. Mol. Sci.* **2019**, *20*, 2560. [CrossRef] [PubMed]
87. Wang, Y.; Kong, D.; Wang, C.; Chen, J.; Li, J.; Liu, Z.; Li, X.; Wang, Z.; Yao, G.; Wang, X. A systematic review and meta-analysis of immune-related adverse events of anti-PD-1 drugs in randomized controlled trials. *Technol. Cancer Res. Treat.* **2020**, *19*, 1533033820967454. [CrossRef] [PubMed]
88. Bhatia, S.; Longino, N.V.; Miller, N.J.; Kulikauskas, R.; Iyer, J.G.; Ibrani, D.; Blom, A.; Byrd, D.R.; Parvathaneni, U.; Twitty, C.G.; et al. Intratumoral delivery of plasmid IL12 via electroporation leads to regression of injected and noninjected tumors in Merkel cell carcinoma. *Clin. Cancer Res.* **2020**, *26*, 598–607. [CrossRef]
89. Lyngaa, R.; Pedersen, N.W.; Schrama, D.; Thrue, C.A.; Ibrani, D.; Met, O.; Thor Straten, P.; Nghiem, P.; Becker, J.C.; Hadrup, S.R. T-cell responses to oncogenic merkel cell polyomavirus proteins distinguish patients with Merkel cell carcinoma from healthy donors. *Clin. Cancer Res.* **2014**, *20*, 1768–1778. [CrossRef]
90. Ciudad, C.; Avilés, J.A.; Alfageme, F.; Lecona, M.; Suárez, R.; Lázaro, P. Spontaneous regression in Merkel cell carcinoma: Report of two cases with a description of dermoscopic features and review of the literature. *Dermatol. Surg.* **2010**, *36*, 687–693. [CrossRef]
91. Tabachnick-Cherny, S.; Pulliam, T.; Church, C.; Koelle, D.M.; Nghiem, P. Polyomavirus-driven Merkel cell carcinoma: Prospects for therapeutic vaccine development. *Mol. Carcinog.* **2020**, *59*, 807–821. [CrossRef]
92. Longino, N.V.; Yang, J.; Iyer, J.G.; Ibrani, D.; Chow, I.T.; Laing, K.J.; Campbell, V.L.; Paulson, K.G.; Kulikauskas, R.M.; Church, C.D.; et al. Human CD4(+) T cells specific for Merkel cell polyomavirus localize to Merkel cell carcinomas and target a required oncogenic domain. *Cancer Immunol. Res.* **2019**, *7*, 1727–1739. [CrossRef]
93. Zeng, Q.; Gomez, B.P.; Viscidi, R.P.; Peng, S.; He, L.; Ma, B.; Wu, T.C.; Hung, C.F. Development of a DNA vaccine targeting Merkel cell polyomavirus. *Vaccine* **2012**, *30*, 1322–1329. [CrossRef]
94. Gerer, K.F.; Erdmann, M.; Hadrup, S.R.; Lyngaa, R.; Martin, L.M.; Voll, R.E.; Schuler-Thurner, B.; Schuler, G.; Schaft, N.; Hoyer, S.; et al. Preclinical evaluation of NF-κB-triggered dendritic cells expressing the viral oncogenic driver of Merkel cell carcinoma for therapeutic vaccination. *Ther. Adv. Med. Oncol.* **2017**, *9*, 451–464. [CrossRef] [PubMed]
95. Park, D.E.; Cheng, J.; McGrath, J.P.; Lim, M.Y.; Cushman, C.; Swanson, S.K.; Tillgren, M.L.; Paulo, J.A.; Gokhale, P.C.; Florens, L.; et al. Merkel cell polyomavirus activates LSD1-mediated blockade of non-canonical BAF to regulate transformation and tumorigenesis. *Nat. Cell. Biol.* **2020**, *22*, 603–615. [CrossRef]
96. Mauri, F.; Blanpain, C. Targeting the epigenetic addiction of Merkel cell carcinoma. *EMBO Mol. Med.* **2020**, *12*, e13347. [CrossRef] [PubMed]
97. Sheng, W.; LaFleur, M.W.; Nguyen, T.H.; Chen, S.; Chakravarthy, A.; Conway, J.R.; Li, Y.; Chen, H.; Yang, H.; Hsu, P.H.; et al. LSD1 ablation stimulates anti-tumor immunity and enables checkpoint blockade. *Cell* **2018**, *174*, 549–563.e19. [CrossRef]
98. Bretz, A.C.; Parnitzke, U.; Kronthaler, K.; Dreker, T.; Bartz, R.; Hermann, F.; Ammendola, A.; Wulff, T.; Hamm, S. Domatinostat favors the immunotherapy response by modulating the tumor immune microenvironment (TIME). *J. Immunother. Cancer* **2019**, *7*, 294. [CrossRef]
99. Song, L.; Bretz, A.C.; Gravemeyer, J.; Spassova, I.; Muminova, S.; Gambichler, T.; Sriram, A.; Ferrone, S.; Becker, J.C. The HDAC Inhibitor domatinostat promotes cell-cycle arrest, induces apoptosis, and increases immunogenicity of Merkel cell carcinoma cells. *J. Invest. Dermatol.* **2020**, *141*, 903–912.e4. [CrossRef]
100. Samimi, M.; Gardair, C.; Nicol, J.T.; Arnold, F.; Touzé, A.; Coursaget, P. Merkel cell polyomavirus in Merkel cell carcinoma: Clinical and therapeutic perspectives. *Semin. Oncol.* **2015**, *42*, 347–358. [CrossRef] [PubMed]
101. Gu, J.; Polak, J.M.; Van Noorden, S.; Pearse, A.G.; Marangos, P.J.; Azzopardi, J.G. Immunostaining of neuron-specific enolase as a diagnostic tool for Merkel cell tumors. *Cancer* **1983**, *52*, 1039–1043. [CrossRef]
102. Van Veenendaal, L.M.; Bertolli, E.; Korse, C.M.; Klop, W.M.C.; Tesselaar, M.E.T.; van Akkooi, A.C.J. The clinical utility of Neuron-Specific Enolase (NSE) serum levels as a biomarker for Merkel cell carcinoma (MCC). *Ann. Surg. Oncol.* **2021**, *28*, 1019–1028. [CrossRef] [PubMed]
103. Gaiser, M.R.; Daily, K.; Hoffmann, J.; Brune, M.; Enk, A.; Brownell, I. Evaluating blood levels of neuron specific enolase, chromogranin A, and circulating tumor cells as Merkel cell carcinoma biomarkers. *Oncotarget* **2015**, *6*, 26472–26482. [CrossRef] [PubMed]
104. Knepper, T.C.; Montesion, M.; Russell, J.S.; Sokol, E.S.; Frampton, G.M.; Miller, V.A.; Albacker, L.A.; McLeod, H.L.; Eroglu, Z.; Khushalani, N.I.; et al. The genomic landscape of Merkel cell carcinoma and clinicogenomic biomarkers of response to immune checkpoint inhibitor therapy. *Clin. Cancer Res.* **2019**, *25*, 5961–5971. [CrossRef] [PubMed]
105. Yoshida, E.J.; Luu, M.; Freeman, M.; Essner, R.; Gharavi, N.M.; Shiao, S.L.; Mallen-St. Clair, J.; Hamid, O.; Ho, A.S.; Zumsteg, Z.S. The association between facility volume and overall survival in patients with Merkel cell carcinoma. *J. Surg. Oncol.* **2020**, *122*, 254–262. [CrossRef] [PubMed]

Review

Regenerative Wound Dressings for Skin Cancer

Teodor Iulian Pavel [1], Cristina Chircov [1], Marius Rădulescu [2,*] and
Alexandru Mihai Grumezescu [1]

1   Department of Science and Engineering of Oxide Materials and Nanomaterials, Faculty of Applied
    Chemistry and Materials Science, University Politehnica of Bucharest, RO-060042 Bucharest, Romania;
    teodor_iulian.pavel@stud.fils.upb.ro (T.I.P.); cristina.chircov@upb.ro (C.C.); agrumezescu@upb.ro (A.M.G.)
2   Department of Inorganic Chemistry, Physical Chemistry and Electrochemistry, Faculty of Applied Chemistry
    and Materials Science, University Politehnica of Bucharest, 1–7 Polizu Street, 011061 Bucharest, Romania
*   Correspondence: marius.radulescu@upb.ro; Tel.: +40-21-402-3997

Received: 26 August 2020; Accepted: 11 October 2020; Published: 13 October 2020

Simple Summary: As the currently and commonly applied treatment strategies for skin cancer are highly invasive and possibly disfiguring, new approaches should focus on developing wound dressings that could promote both tumor eradication and skin regeneration. In this context, we aim to provide a complete overview on the limitations of currently available topical field treatments and to emphasize on the potential of natural biocompounds with anti-cancer and anti-microbial effects that could be introduced into wound dressings consisting of biopolymers with regenerative capacities. This paper could represent the first step towards the scientific advancement of regenerative wound dressings for skin cancer therapy.

Abstract: Skin cancer is considered the most prevalent cancer type globally, with a continuously increasing prevalence and mortality growth rate. Additionally, the high risk of recurrence makes skin cancer treatment among the most expensive of all cancers, with average costs estimated to double within 5 years. Although tumor excision is the most effective approach among the available strategies, surgical interventions could be disfiguring, requiring additional skin grafts for covering the defects. In this context, post-surgery management should involve the application of wound dressings for promoting skin regeneration and preventing tumor recurrence and microbial infections, which still represents a considerable clinical challenge. Therefore, this paper aims to provide an up-to-date overview regarding the current status of regenerative wound dressings for skin cancer therapy. Specifically, the recent discoveries in natural biocompounds as anti-cancer agents for skin cancer treatment and the most intensively studied biomaterials for bioactive wound dressing development will be described.

Keywords: wound dressings; skin cancer; tumor excision; tumor recurrence; natural biocompounds; anti-cancer agents; bioactive wound dressing

---

1. Introduction

Among the noncommunicable diseases, which account for 71% of global deaths, cancer is the second leading one, with 18.1 million cases and 9.6 million deaths worldwide in 2018. According to the World Health Organization (WHO), the numbers will double by 2040, with the highest increase in low- and middle-income countries [1–3]. Cancer is characterized by an abnormal growth of cells, which further invade and spread throughout different body organs through metastasis [4–6]. Its metabolism is highly complex, depending on a series of factors, including genetic and epigenetic alterations, the surrounding environment, the tissue of origin, and the systemic host metabolism [7].

Consequently, cancer cells possess a remarkable ability of surviving and adapting to various stress conditions, such as oxidative and metabolic stress, hypoxia, and nutrient deprivation [8].

Skin cancer is considered the most prevalent cancer type globally and in the United States, with a continuously increasing prevalence and mortality growth rate [9–12]. Skin cancer is characterized by an imbalance in cell homeostasis and excessive cell proliferation as a result of cancer-associated gene mutations, such as skin proto-oncogenes and tumor suppressors within skin cells [12]. Depending on the type of cells affected, there are two major types of skin cancer, namely non-melanoma and cutaneous melanoma [12]. On one hand, non-melanoma skin cancers, predominantly comprising basal cell carcinoma and cutaneous squamous cell carcinoma, originate from the keratinocytes within the epidermis and account for approximately five million new cases and 65,000 associated deaths yearly [10–13]. Other types of non-melanoma skin cancers include Merkel cell carcinoma, Kaposi sarcoma, dermatofibrosarcoma protuberans, primary cutaneous B-cell lymphoma, sebaceous carcinoma, and atypical fibroxanthoma, which are significantly rarer [13,14]. On the other hand, melanoma originates from melanocytes within the deepest layer of the epidermis and, although its prevalence is considerably lower, it has the worst prognosis, with 280,000 new cases and 60,000 associated deaths reported yearly [10–13]. Moreover, the incidence of skin cancers is continuously increasing, which could be associated with higher UV radiation exposure [11,15].

Reducing cancer-related mortality rates has become a major challenge faced by societies, governments, and medical and scientific communities [16]. However, conventional treatment options, including chemotherapy, radiotherapy, immunotherapy, and gene and hormone therapy, are associated with various drawbacks that limit their efficiency [4]. In this regard, cancer treatment generally involves a combination of therapies in order to control the evolution of the disease [17]. Nonetheless, chemotherapy is still considered the most efficient strategy and it is widely used in most cases, with more than 200 anti-cancer drugs developed that include cytostatics, anti-hormonal drugs, recombinant proteins and antibodies for molecular targeted therapy, and supportive care drugs [16–18]. In the case of non-melanoma skin cancer, radical tumor excision remains the most effective approach among the available strategies [14,19,20]. However, radical excision may not be possible due to patient co-morbidities or unfavorable cosmetic defects, and non-surgical approaches, such as cryotherapy, curettage, electrodessication, topical therapy, photodynamic therapy, or radiotherapy become the only option [13,14,19,20]. By contrast, treatment of melanoma involves surgery followed by radiotherapy, immunotherapy, and chemotherapy [13,20]. Nevertheless, there is a high recurrence risk in skin cancer therapy, which makes its treatment among the most expensive of all cancers, with average costs estimated to double within 5 years [9,12,20]. Additionally, surgical interventions could be disfiguring, requiring additional skin grafts for covering the defects [20].

Generally, the skin plays four fundamental roles, namely body protection against physical, chemical, and bacteriological damages, thermoregulation through skin vasculature and eccrine sweat glands, prevention of dehydration, and conduction of neurosensory information which further contributes to endocrine function and immune surveillance regulations [21,22]. Hence, maintaining its integrity is fundamental. In this context, post-surgery management should involve the application of wound dressings for promoting skin regeneration and preventing tumor recurrence and microbial infections, which still represents a considerable clinical challenge [23]. Such wound dressings should maintain a moist environment and allow for fluid exchange, which would promote wound healing and regeneration [24], and provide a controlled release of bioactive compounds for anti-cancer and anti-microbial purposes [25].

Therefore, the aim of this paper is to provide an up-to-date overview regarding the current status of regenerative wound dressings for skin cancer therapy. Specifically, the recent discoveries in natural biocompounds as anti-cancer agents for skin cancer treatment and the most intensively studied biomaterials for bioactive wound dressing development will be described.

2. Topical Field Treatments

Generally, topical field treatments used for skin cancer therapies involve the application of creams or gels containing imiquimod, 5-fluorouracil, ingenol mebutate, or diclofenac [26,27], with regimens lasting from 2 to 90 days, depending on the tumor complexity and dosage. However, such prolonged and complex treatments are usually associated with non-adherence and non-persistence to the prescribed treatment [28]. Additionally, these agents are chemically and pharmacologically different, with various related side effects (Table 1) [29].

Table 1. Characteristics of the main topical field treatment options.

Agent	Chemical Formula	Commercial Formulation	Treatment Schedule	Side Effects
Imiquimod	1-Isobutyl-1*H*-imidazo(4,5-c)-quinolin-4-amine	3.75% cream (Zyclara) 5% cream (Aldara)	5–7 times per week once a day 6–12 weeks	Erosion, ulceration, healing with scarring or hyperpigmentation, erythema, vesiculation, edema, weeping, pruritus, scaling, crusting, burning, and pain
5-Fluorouracil	5-Fluoro-1*H*-pyrimidine-2,4-dione	2% solution 5% solution 5% cream (Efudix)	Twice daily 2 or more weeks	Allergic contact and irritant dermatitis, erythema, pruritus, erosions, hyper- or hypopigmentation, and pain
Ingenol mebutate	Ingenol-3-angelate	0.015% gel 0.05% gels	2 to 7 consecutive days	Erythema, edema, pruritus, and pain
Diclofenac	2-(2-(2,6-Dichloro-phenylamino) phenyl)acetic acid	3% gel	Twice daily for 8 weeks	Erythema, erosion, allergic contact dermatitis, photoallergy, and pruritus

2.1. Imiquimod

Imiquimod (1-isobutyl-1*H*-imidazo(4,5-c)quinolin-4-amine, Figure 1), is a low molecular weight nucleoside analog of the imidazoquinoline family with immunomodulating properties and indirect anti-viral and anti-tumor effects [30–36]. Additionally, it acts as a potent antagonist for the toll-like receptors 7 and 8 [30,31,35,37]. Initially, imiquimod was used for the treatment of human papilloma virus-associated genital and perianal warts. Presently, it is commonly applied for treating actinic keratosis and superficial basal cell carcinoma, with no sufficient information to prove its efficiency for nodular basal cell carcinoma or squamous cell carcinoma [30–32,34,36].

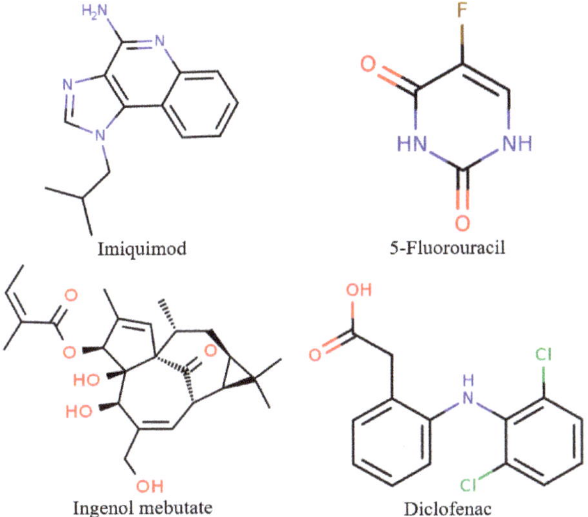

Figure 1. The chemical structures of the four agents commonly used for topical field treatment of skin cancers.

The anti-tumor effects of imiquimod, although not completely understood, could be attributed to two major underlying mechanisms. On one hand, it activates the dendritic cells and monocytes where the toll-like receptors are predominantly expressed, which leads to the NF-κB-dependent secretion of pro-inflammatory cytokines and chemokines, such as interferon-α, tumor necrosis factor-α, and interleukin-6 and -8. Consequently, antigen-presenting cells and innate immunity components are activated, which will generate profound T-helper 1-mediated anti-tumor immune responses [30–35]. On the other hand, imiquimod facilitates pro-inflammatory activities by interfering with adenosine receptor signaling pathways [30].

Imiquimod for topical administration is used as 3.75% or 5% creams and treatment involves 5 to 7 times per week applications, once a day, for 6 to 12 weeks [31]. In this manner, it provides a non-surgical, non-invasive, self-administered therapeutic option with relatively low costs [35,38]. For 6-week treatments using 5% creams, clearance rates are between 52% and 81% [31,32]. However, imiquimod treatments have been associated with a variety of side effects, including erosion, ulceration, healing with scarring or hyperpigmentation, erythema, vesiculation, edema, weeping, pruritus, scaling, crusting, burning, and pain [31–34]. Additionally, it was observed that higher clearance rates lead to higher inflammatory reactions [31,32]. Moreover, imiquimod is characterized by an inability to permeate through the dermis layer due to its low water solubility and the interactions between the amine groups and the aninonic components of the skin [36,38].

2.2. 5-Fluorouracil

5-Fluorouracil (5-fluoro-1*H*-pyrimidine-2,4-dione, Figure 1), is a pyrimidine analogue [31,39–41] which belongs to the class of anti-metabolite drugs [31,35,41–43]. As a chemotherapeutic agent, it has been widely applied in the treatment of malignant diseases [42]. Moreover, it has been used for the topical treatment of superficial basal cell carcinoma and, while it is not recommended for nodular or infiltrative basal cell carcinoma, it is used in very old patients with no other therapeutic options [27,31,42,43].

The main underlying mechanism for its anti-tumor character involves blocking the conversion of deoxyuridine into thymidine as a consequence to the irreversible binding to thymidylate synthase through the cofactor 5,10-methylenetetrahydrofolate. In this manner, DNA synthesis in cancer cells is inhibited, which results in reduced cell proliferation and increased apoptosis [35,41,43]. 5-Fluorouracil metabolism depends on its degradation by the dihydropyrimidine dehydrogenase enzyme which reduces the molecule to its inactive form, dihydrofluorouracil [44].

For topical applications, 5-fluorouracil is available as 2% and 5% solutions or 5% cream and it is used twice a day for 2 or more weeks. For superficial basal cell carcinoma treatment, clearance rate is 90% and for squamous cell carcinoma in situ, between 48% to 85% [31,43]. Its advantage relies on the possibility of the patient to apply it at home, but it can be challenging for elderly individuals who might need assistance [31,43]. Common side effects include allergic contact and irritant dermatitis, erythema, pruritus, erosions, hyper- or hypopigmentation, and pain [31,43,44]. In this case, there is no correlation between clearance rates and adverse reactions [31].

2.3. Ingenol Mebutate

Ingenol mebutate, also known as ingenol-3-angelate (Figure 1), is a diterpene ester derived from the *Euphorbia peplus* plant species [45]. Known for its anti-cancer properties, ingenol mebutate is a novel therapeutic agent used for the treatment of skin conditions, including actinic keratosis, verrucae, and superficial basal cell carcinoma [31,45,46].

There is a dual underlying mechanism that could be attributed to its anti-cancer potential, consisting of mitochondrial destruction induced by increased intracellular calcium levels and subsequent epidermal cell death through necrosis and apoptosis. The following stage involves protein kinase C activation, which further stimulates infiltration of neutrophils, production of pro-inflammatory cytokines, expression of endothelial adhesion molecules, and formation of tumor-specific antibody, leading to a neutrophil-mediated antibody-dependent cellular cytotoxicity [31,35,45–47]. In this manner, inflammation manifests within hours of application through erythema and edema and subsequent pustules, epidermal flaking, and crusting, which subside in less than 2 weeks [47].

Ingenol mebutate is available in 0.015% and 0.05% gels and it is administered for 2 to 7 consecutive days, with clearance occurring in approximately 63% of patients [31,46,47]. While its application is advantageous due to self-administration and short application periods, it is usually associated with adverse events, including erythema, edema, pruritus, and pain [31,47,48].

2.4. Diclofenac

Diclofenac (2-(2-(2,6-dichlorophenylamino)phenyl)acetic acid, Figure 1), also known for its sodium salt sold as Voltaren, is a non-steroidal anti-inflammatory drug widely used for actinic keratosis treatment and can be an adjuvant for basal cell carcinoma, squamous cell carcinoma, and melanoma skin metastases therapy [35,49,50].

Its mechanism of action, although not completely elucidated, could involve the inhibition of cyclooxygenase-1 and -2, enzymes involved in reducing prostaglandin formation from arachidonic acid, which reduces PGE2 synthesis and dysplastic keratinocytes in cancerous lesions [31,35,50,51]. Additionally, it might interfere with the SHH and Wnt signaling and lead to cancer cell apoptosis and inhibit angiogenesis and proliferation [31,35,50].

The current formulation contains 3% diclofenac in 2.5% hyaluronic acid which is applied twice daily for 8 weeks. Clearance rates range between 38% to 47% for actinic keratosis and 64.3% for superficial basal cell carcinoma [31,35,51]. Although it can be self-applied, there are frequent side effects associated, including erythema, erosion, allergic contact dermatitis, photoallergy, and pruritus [31].

3. Natural Anti-Cancer Agents for Skin Cancer

Oxidation by solar radiation has become a principal cause for the development of skin diseases through the excessive production of reactive oxygen species which consequently leads to inflammation and DNA and protein damages [52,53]. Recent studies have reported a direct causal connection between inflammation and cancer development. In this context, cancer-related inflammation occurs through two main pathways, namely the intrinsic pathway produced by genetic events as a causing factor of inflammation and neoplastic transformation and the extrinsic pathway through which carcinogenesis is promoted by inflammatory conditions [54].

In this regard, plants are important sources which produce secondary metabolites for protection purposes. Specifically, such compounds exhibit DNA protection, antioxidant, anti-inflammatory, chemopreventive, and chemotherapeutic activities. Among them, flavonoids, phenolic acids, lignins, stilbenes, and retinoids, are the most commonly studied for their anti-cancer potential, especially for skin cancers [52,53].

3.1. Flavonoids

Flavonoids are a class of antioxidants biosynthesized via the shikimic acid pathway from acetic acids or phenylalanine derivatives [55]. Although they can be divided into flavones, flavonols, isoflavonoids, isoflavones, flavanones, flavanols, and anthocyanidins, most of the bioactive compounds belong to the former three groups, which have received great scientific interest in the past years [56–58]. There is an increasing number of studies reporting the anti-cancer or cancer preventive effects of flavonoids against prostate, colorectal, breast, thyroid, lung, ovarian, and skin cancers [59,60].

The chemopreventive character of flavonoids relies on their potential to inhibit new cancer cell development, prevent carcinogens from reaching activation sites, decrease compound toxicity by inhibiting their metabolism [59]. The molecular mechanisms responsible involve apoptosis induction, cell cycle arrest by inhibiting key regulators, metabolizing enzymes inhibition and subsequent inactivation of carcinogenic compounds, reactive oxygen species scavenging, angiogenesis inhibition, DNA repair mechanism initiation, and cancer cell proliferation and invasiveness suppression [56,59–61].

Additionally, some flavonoid compounds have shown to prevent cancer relapse and chemotherapy failure by considerably inhibiting multidrug resistance [59].

In skin cancers, flavonoids have shown anti-inflammatory, anti-proliferative, anti-angiogenic, and apoptotic activities. Specifically, topical administration of flavonoids leads to skin absorption and consequent activation of a cascade of protective signaling pathways and cell cycle arrest in G0-G1 and G2-M phases [60]. Additionally, dietary intake of flavonoid-rich products has shown to ensure DNA protection of skin cells exposed to carcinogenic factors, such as UV radiation [56]. Among flavonoid compounds, apigenin, quercetin, silymarin, diosmetin, genistein, fisetin, and luteolin, have been identified as potential anti-cancer agents in skin cancers (Table 2) [57,59].

Table 2. The main flavonoid compounds with anti-cancer potential for skin cancer treatment—chemical formula and structure, sources, and study references.

Compound	Chemical Formula	Chemical Structure	Sources	References
Apigenin	5,7-Dihydroxy-2-(4-hydroxyphenyl)-4*H*-chromen-4-one		Parsley, celery, chamomile, oranges, onions, honey, thyme, oregano, rosemary, basil, coriander, tea, beer, and wine	[62–65]
Quercetin	2-(3,4-Dihydroxy-phenyl)-3,5,7-trihydroxy-4*H*-chromen-4-one		Apples, citrus, red onion, and the roots and leaves of many vegetables	[66–71]
Silymarin	3,5,7-Trihydroxy-2-[3-(4-hydroxy-3-methoxyphenyl)-2-(hydroxymethyl)-2,3-dihydro-1,4-benzodioxin-6-yl]-3,4-dihydro-2*H*-1-benzopyran-4-one		Milk thistle seeds	[72–74]
Diosmetin	5,7-Dihydroxy-2-(3-hydroxy-4-methoxyphenyl)-4*H*-chromen-4-one		Rosemary, bergamot juice, and citrus juice	[75–78]
Genistein	5,7-Dihydroxy-3-(4-hydroxyphenyl)-4*H*-chromen-4-one		Soy	[79–82]
Fisetin	2-(3,4-Dihydroxy-phenyl)-3,7-dihydroxy-4*H*-chromen-4-one		Strawberries, apples, persimmons, grapes, onions, kiwi, and kale	[82–84]
Luteolin	2-(3,4-Dihydroxy-phenyl)-5,7-dihydroxy-4*H*-chromen-4-one		Carrots, peppers, celery, olive oil, peppermint, thyme, rosemary, and oregano	[85,86]

3.2. Phenolic Acids

Phenolic acids are a class of antioxidant compounds which are formed through the substitution of hydrogen atoms present on the benzene rings by a carboxylic and at least one hydroxyl group. They are ubiquitously present in plants and human metabolites and, by contrast to flavonoids, have suitable water solubility and high bioavailability [87]. Phenolic acids exhibit their anti-cancer properties by inducing apoptosis through the ASK-1, caspase-3, JNK-p38, and pRb pathways, suppressing cell cycle by p21, bcl2, and bcl-x upregulation and bim, bax, puma ans noxa downregulation, and reducing proliferation and angiogenesis by EGFR, MAPK, mTOR, PI3K/Akt, FAK/PTK2, and JAK/STAT upregulation. Additionally, they are involved in modulation of inflammatory and cytokine genes expression, which are further implicated in cellular proliferation, invasion, and metastasis inhibition [12]. Furthermore, as they are highly potent antioxidants owing to the presence of the hydroxyl substituent on the benzene ring, phenolic acids can act as chemopreventive agents for UV-induced skin cancer [88].

Among them, syringic acid, ferulic acid, and caffeic acid have received a considerable interest in the recent years (Table 3). Specifically, studies have shown that treatment with syringic acid led to suppressed UV-induced cyclooxygenase-2, matrix metalloproteinase-1, and prostaglandin E2 expression and activator protein-1 activity [89]. Additionally, it inhibited the Nox/PTP-κ/EGFR pathway and subsequently reactive oxygen species formation. Its anti-cancer character might be attributed to the presence of the methoxy groups on the benzene ring [90,91]. Similarly, ferulic acid has shown to exhibit photoprotective character which could prevent UV-induced carcinogenesis. The antioxidant properties are generally based on its structural features, namely the hydroxyl group which neutralizes the reactive oxygen species as it acts as an electron donor, the vinyl chain and the methoxyl group which increase molecule stability, and the carboxylic group which prevents lipid peroxidation [92]. Moreover, caffeic acid acts as a chemopreventive agent by modulating inflammatory signaling [93], suppressing the rapamycin cascade signaling, inducing apoptosis [94], and altering cell cycle and caspase gene expression [95].

Table 3. The main phenolic acids with anti-cancer potential for skin cancer treatment – chemical formula and structure, sources, and study references.

Compound	Chemical Formula	Chemical Structure	Sources	References
Syringic acid	4-Hydroxy-3,5-dimethoxybenzoic acid	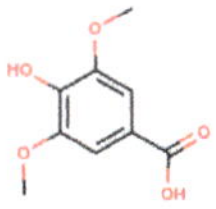	Olives, dates, spices, pumpkin, grapes, acai palm, honey, and red wine	[90,91,96]
Ferulic acid	3-(3-Hydroxy-4-methoxy-phenyl)prop-2-enoic acid	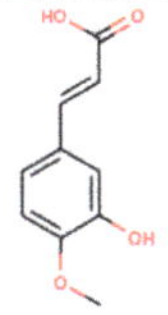	Whole grains, spinach, parsley, grapes, rhubarb, and cereal seeds, mainly wheat, oats, rye, and barley	[92,97]
Caffeic acid	3-(3,4-Dihydroxy-phenyl)prop-2-enoic acid	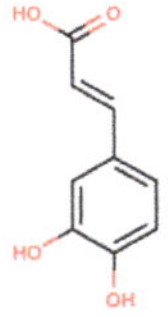	Coffee, wine, tea, and propolis	[93–95,98]

3.3. Lignin

Lignin is the second most abundant renewable resource on earth, comprising a three-dimensional heterogenous and phenolic polymer network synthesized in the cell wall of higher plants. The term involves a variety of natural aromatic compounds obtained through the oxidative coupling of monomeric precursors [99,100]. The main monomers implicated in lignin structure, also known as monolignols, are coniferyl alcohol, sinapyl alcohol, and *p*-coumaryl alcohol (Figure 2) [100,101]. Moreover, lignin structures is highly influenced by the extraction processes and the presence of different functional moieties, such as hydroxyl, methoxyl, carbonyl, and carboxyl groups. Through hydroxyalkylation reaction processes, such as phenolation, demethylation, or methylolation, lignin is transformed to phenolic compounds [101].

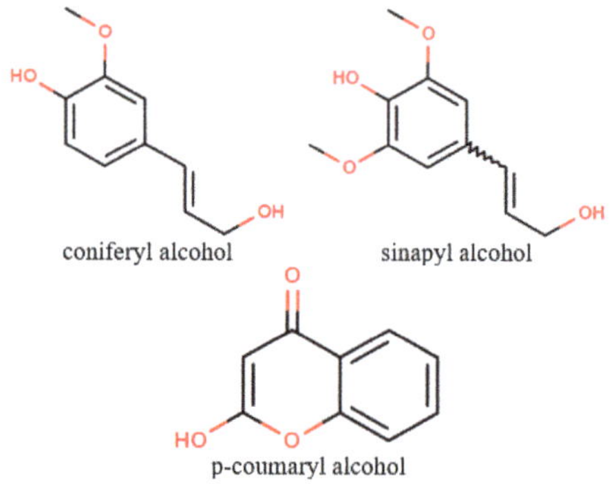

coniferyl alcohol

sinapyl alcohol

p-coumaryl alcohol

Figure 2. The chemical structures of the common monolignol precursors of lignin.

Although it has not yet been converted into high-value products at large scales, lignin has shown to exhibit promising functions, including antioxidant, anti-microbial, and UV blocking. Additionally, as they do not cause cytotoxicity, lignin-based products could be applied for biomedical purposes [99]. In this context, lignin is a widely used bio-based UV-blocking material owing to its UV-absorbing phydroxyphenyl, guaiacyl, and syringyl phenylpropanoid units [99,102].

3.4. Stilbenes

Stilbenes, a class of non-flavonoid phenolic compounds, generally consist of C_6-C_2-C_6 structures with two hydroxyl groups on the A ring and one on the B ring and are regarded as 1,2-diphenylethylenes [103–106]. Most stilbenes are found in plants as aglycones or glycosides, thus offering anti-fungal and anti-bacterial protection [105,106]. Additionally, studies have demonstrated strong antioxidant and chemopreventive properties of the stilbenoid group [103]. Most common stilbene examples are the phytoalexins resveratrol and the resveratrol metabolites, pterostilbene and piceatannol (Table 4) [103,105].

Owing to its promising anti-inflammatory, antioxidant, anti-cancer, anti-mutagenic, anti-aging, and anti-allergenic properties, resveratrol is one of the most intensively studied stilbenes. Initially identified as an SIRT1 activator which regulates energy homeostasis and mitochondrial biogenesis within cells, it is now studied for its apoptosis promoting capacity by enhancing sensitivity to tumor necrosis factor-α and suppressing NF-κB activation [104,106,107]. In skin cancer applications, studies have shown that topical administration of resveratrol not only improved skin elasticity, hydration, and luminosity [108], but also provided protection against UV radiations and UV-induced carcinogenesis by regulating protein activity regarding apoptosis, decreasing reactive oxygen species production, inhibiting tumor incidence and tumorigenesis, and modulating cell cycle molecules and cell signaling pathways [109]. Additionally, it has proved its potential to initiate senescence in squamous cells carcinoma cells through its autolysosome form blockade and Rictor protein expression downregulation, which altered cancer cell skeleton and suppressed cancer progression [110].

Table 4. The main stilbenes with anti-cancer potential for skin cancer treatment—chemical formula and structure, sources, and study references.

Compound	Chemical Formula	Chemical Structure	Sources	References
Resveratrol	5-[(*E*)-2-(4-Hydroxy-phenyl)ethenyl] benzene-1,3-diol	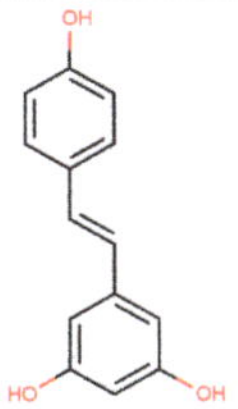	Grapes, wine, nuts, and berries	[108–110]
Pterostilbene	4-[(*E*)-2-(3,5-Dimethoxy-phenyl)ethenyl] phenol	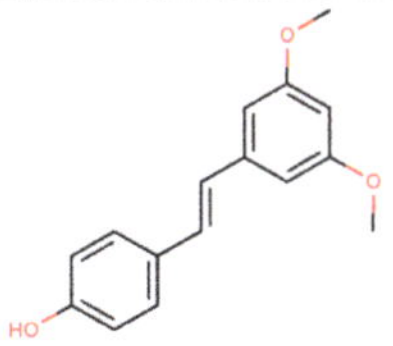	Grapes, wine, nuts, and berries	[111,112]
Piceatannol	5-[(*E*)-2-(3,4-Dihydroxy-phenyl)ethenyl] benzene-1,3-diol	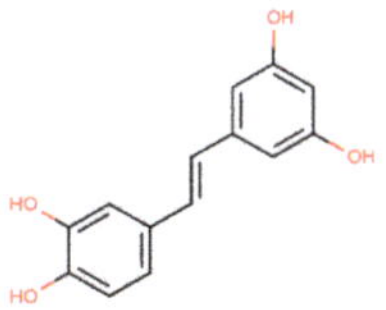	Grapes, berries, and red wine	[113,114]

Pterostilbene and piceatannol, two resveratrol analogs, have also proved anti-cancer activities similar or superior to resveratrol levels [104]. On one hand, pterostilbene, which has a higher lipophilicity and, consequently higher bioavailability and membrane permeability, possesses intrinsic antioxidant properties by activating the nuclear factor erythroid 2-related factor 2 and anti-inflammatory character by targeting inducible nitric oxide synthase, cyclooxygenases, leukotrienes, NF-κB, tumor necrosis factor-α, and interleukins [111]. Moreover, it can also increase lysosome size and induce membrane destabilization and caspase-independent cell death [112]. On the other hand, piceatannol leads to Bax upregulation, Bcl-2 downregulation, and caspase-3 activation, which induced melanoma cell apoptosis [114].

3.5. Retinoids

Retinoids are polyphenolic compounds derived from vitamin A. Among its derivatives, retinol and retinoic acid (Figure 3), also known as tretinoin, are the most commonly used for oral or topical administration to prevent skin cancer [21]. Mechanisms that could be attributed to its anti-tumor effects are based on inhibiting UV-induced phosphorylation of ERK1/2, JNK, and p38 proteins of the MAPK family [12]. Furthermore, oral administration has also proved efficient against basal cell carcinoma, squamous cell carcinoma, and actinic keratosis [115].

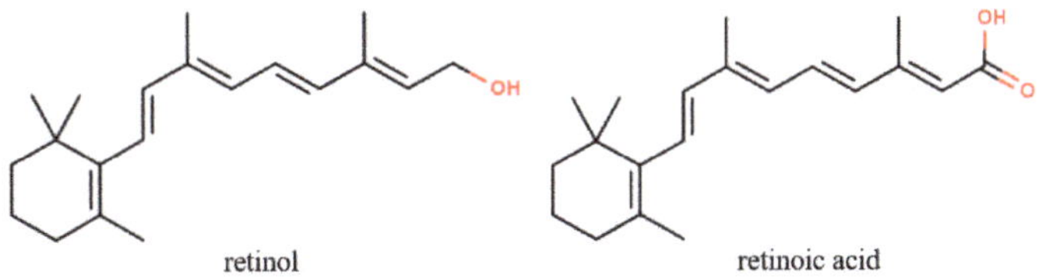

Figure 3. The chemical structures of retinol and retinoic acid, the most common retinoids topically applied for skin cancer.

4. Wound Dressings in Skin Cancer

The skin has the ability to repair itself following injuries due to surgery, trauma, or burns through cutaneous wound healing processes [116]. Generally, it involves four main stages, namely coagulation, inflammation, granulation, and remodeling, finally resulting in wound closure (Figure 4) [116,117]. However, wound areas larger than 2 cm^2, wound duration longer that 2 months, and increased wound depth which results in tendon, ligament, or bone exposure are the three main factors which delay or stop the healing process. In this context, research is focused on developing wound dressings not only for protection purposes, but also to promote healing and regeneration processes [118,119]. Therefore, the application of healing-promoting approaches has shown to trigger, accelerate, and enhance wound healing, re-epithelialization, and collagen formation, which subsequently results in reduced scar formation and complications [120,121].

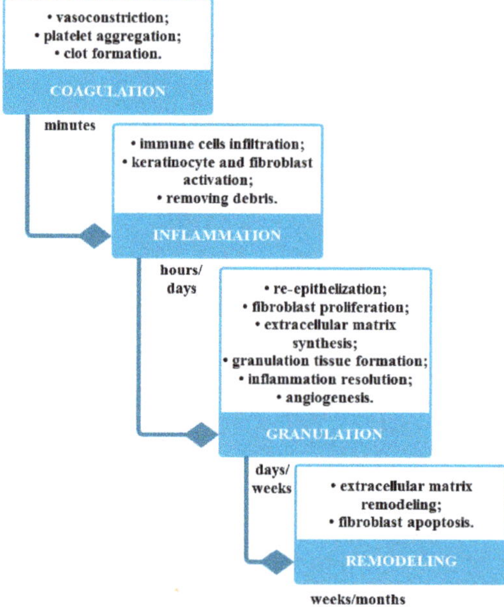

Figure 4. The four main stages involved in the cutaneous wound healing processes and the implicated mechanisms.

Based on their interaction with the wound site, wound dressings can be divided into three main groups, namely inert or passive, interactive, and bioactive. As inert or passive dressings are ordinary materials designed only for covering and providing protection to the wound against pathogen contamination from the external environment [122], they are not suitable for regenerative applications in skin cancer. On the contrary, interactive wound dressings are capable of altering the wound microenvironment, by interacting with the surface and promoting healing processes [123]. Moreover, interactive dressings support all the stages involved in the healing process, such as debris removal, granulation tissue formation, and re-epithelialization, while also decreasing exudate formation and preventing bacterial colonization. Among them, hyaluronic acid, collagen, and alginate-based dressings are the most commonly and extensively investigated products [25,118]. These natural polymers are highly advantageous owing to their cell proliferation and growth, tissue regeneration, non-toxicity, minimal inflammatory and immunological response induction characteristics. Additionally, their chemical structures provide unique physicochemical properties which are fundamental for skin regeneration processes [124].

Hyaluronic acid is a glycosaminoglycan and one of the most important extracellular matrix components ubiquitously found in the connective tissues of all living organisms. Structurally, it is a linear polysaccharide consisting of N-acetyl-d-glucosamine and glucuronic acid. Hyaluronic acid interacts with receptors found on the surface of cells, thus promoting wound repair processes [125–127]. In the wound healing process, high molecular weight hyaluronic acid is produced by platelets which further stimulates fibrinogen deposition and, as a major component of the edema fluid, it promotes neutrophils recruitment for debris removal and tumor necrosis factor-α and interleukin-1β and -8 release. Subsequently, as it is fragmented to low molecular weight hyaluronic acid which will bind to CD44 receptors, leucocytes and monocytes are recruited. Moreover, it will interact with toll-like receptor-2 and -4 present on lymphocytes and macrophages. Finally, hyaluronic acid guides fibroblast invasion and proliferation, which is fundamental for collagen deposition, and differentiation into myofibroblasts for wound contraction [128]. Therefore, hyaluronic acid-based hydrogels have been widely investigated for wound healing purposes, with the most common comprising glycidyl methacrylate-, thiol-, and DNA-functionalized hyaluronic acid [127].

Collagen, another extracellular matrix component, is the most commonly found protein in the body produced by fibroblasts and involved in cellular and molecular cascades involved in wound healing and debridement and tissue regeneration [127]. Collagen dressings are capable of counterbalancing the elevated levels of matrix metalloproteinases which are usually released at the wound site and proteolytically deteriorate native intact and partially degraded fragments of collagen molecules. Currently, collagen is coupled with other natural and synthetic polymers, e.g., salmon milt DNA, anionic polysaccharides, minocycline based hydrogels, α-tocopherulate, and alginic acid, for developing novel dually functional collagen-based wound dressings that combine both wound healing and exudate absorbing properties [127,129,130].

Alginate is natural anionic polymer extracted from brown seaweed widely applied in the biomedical field owing to its biocompatibility, non-toxicity, and low cost [127,131]. Structurally, alginate is linear branchless polysaccharide consisting of various $(1{\rightarrow}4')$-linked β-D-mannuronic acid and α-L-guluronic acid subunit contents [127]. It is widely used as a wound dressing biomaterial as it fulfills the requirements regarding exudate absorption and tissue regeneration. Specifically, owing to its hydrophilic nature, alginate is able to absorb high amounts of wound exudate, while also maintaining the required moisture and exhibiting a hemostatic effect [132,133]. Additionally, it enhances cell migration, increases angiogenesis, promotes collagen type I production, and suppresses pro-inflammatory cytokine concentrations for skin regeneration and prevents bacterial contamination within the wound site [127,132]. Alginate can be easily crosslinked with other organic or inorganic materials, such as calcium, sodium, collagen, and gelatin [127,133], which is reflected by the large number of commercially available alginate-based dressings [132].

However, although they are highly efficient, interactive wound dressings are unable to solve deeper lesions or effectively prevent microbial infections. Therefore, bioactive wound dressings are more promising as they promote healing processes by the gradual release of the biocompounds encapsulated within [25,134–136]. Moreover, for skin cancer applications, the regenerative properties of wound dressings are not sufficient as they require the presence of anti-cancer agents for preventing cancer recurrence. Therefore, wound dressings consisting of biopolymers as the regenerative component and natural anti-cancer agents as the cancer recurrence-preventing component could represent ideal candidates to use for regenerative applications in skin cancer.

In this context, Shukla et al. [137] developed hydrogels consisting of gellan gum and chitosan crosslinked with poly(ethylene glycol) loaded with apigenin. Their results on rat wound models showed a 96.11% release of the bioactive compound within 24 h and significant antioxidant activity. While the unique properties of the hydrogels in terms of biocompatibility, biodegradability, moisture, and antioxidant efficiency are considerably promising for wound healing and regeneration, the release of the bioactive compound should be gradual in order to ensure an optimal concentration at the wound site for longer periods.

Moreover, quercetin-impregnated chitosan and fibrin scaffolds were developed by Karthick et al. and Vedakumari et al., which exhibited ideal mechanical strength for a wound dressing material, in vitro non-toxicity and bactericidal effects against *Escherichia coli* and *Staphylococcus aureus* strains, and accelerated wound healing after topical administration on albino rats [138]. George et al. investigated the therapeutic effects of a chitosan hydrogel crosslinked with dialdehyde cellulose containing phyto-derived quercetin extracted from onion peel waste and zinc oxide nanoparticles. The incorporation of nanoparticles increased drug loading within the hydrogel and inhibited *Staphylococcus aureus* and *Trichophyton rubrum* strains growth. In vitro tests showed good biocompatibility on normal L929 murine fibroblast cells and anti-cancer properties against A431 human skin carcinoma cell lines [139]. Furthermore, Jangde et al. [140] incorporated quercetin into a multiphase hydrogel consisting of Carbopol® (the trade name for carbomers, high molecular weight cross-linked poly(acrylic acid) polymers) and varying gelatin ratio. The most suitable hydrogel was obtained at a ratio of gelatin to Carbopol® of 6 to 4, which exhibited accelerated wound healing and reduced wound closure time albino rat models [141]. Ajmal et al. [142] designed a wound dressing based on poly(ε-caprolactone) nanofibers loaded with ciprofloxacin hydrochloride and quercetin in order to ensure both anti-bacterial and antioxidant properties. Results showed high entrapment efficiency of more than 92% for both biocompounds and a prolonged in vitro release for 7 days. Thus, the hydrogel was able to accelerate wound healing on full thickness wound models in rats by improving collagen deposition and re-epithelialization within 16 days and prevent reactive oxygen species production.

Silymarin was also evaluated as a bioactive compound for wound dressing application. In this regard, Tsai et al. [143] developed a bacterial cellulose nanofiber film onto which silymarin-loaded zein nanoparticles were adsorbed. Their findings proved enhanced antioxidant and anti-bacterial activities for silymarin-containing films, which protected salmon muscle against lipid oxidation and deterioration.

Additionally, Kim et al. developed crosslinked interpenetrating polymer networks hydrogels consisting of poly(N-isopropylacrylamide) and hyaluronic acid for the transdermal delivery of luteolin. Texture and rheometry analyses proved that the 3% crosslinker-containing hydrogel has the most adhesive and stable network, which was further applied for drug release evaluation. Results showed no cytotoxicity and inhibited keratinocyte hyperproliferation in psoriasis, which could further be applied in skin cancer applications in order to prevent cancer cell proliferation [144].

A novel fibrous material consisting of poly(ε-caprolactone) and chitosan containing ferulic acid was prepared by Yakub et al. through electrospinning or electrospinning combined with dip-coating. The composition and design of the matrix influenced ferulic acid incorporation and results showed higher anti-bacterial activity when compared to the application of ferulic acid-containing poly(ε-caprolactone) and chitosan-coated poly(ε-caprolactone) fiber mats against *Staphylococcus aureus* strains. Additionally, the incorporation of ferulic acid within the polymer matrix increased the anti-cancer character against HeLa tumor cells while maintaining its antioxidant activity [145]. Likewise, Poornima et al. investigated the release of quercetin and resveratrol from similar polymer matrices for their anti-inflammatory and pro-angiogenic activities, respectively. The in vitro studies showed a sustained release for both biocompounds of up to 48% and 55%, respectively, for 120 h. Furthermore, in vivo studies on rats resulted in complete wound healing in 15 days when treated with the biomaterials and 20 days for the control [146].

Moreover, caffeic acid has also been widely studied for its potential in regenerative wound dressing applications. In this context, Oh et al. [147] compared the effects of poly (ε-caprolactone), poly (ε-caprolactone) and chitosan, and poly (ε-caprolactone) and chitosan-caffeic acid conjugate fibrous mats fabricated by electrospinning. The chitosan-caffeic acid conjugate-based fibrous mat exhibited significantly increased tensile properties and higher initial cell attachment, cell proliferation, and anti-microbial effects, which proves its potential for wound healing and regeneration. Subsequently, Ignatova et al. designed and fabricated fibrous materials based on poly(3-hydroxybutyrate), quaternized chitosan, κ-carrageenan, and caffeic acid through electrospinning or electrospinning in conjunction with dip-coating and polyelectrolyte complex formation. Results showed that caffeic acid release is influenced by fiber composition. In this regard, caffeic acid-containing mats obtained through polyelectrolyte complex formation exhibited anti-bacterial character against *Staphylococcus aureus* and *Escherichia coli* strains and an enhanced antioxidant activity [148]. Similar results were obtained by the same research group for poly(3-hydroxybutyrate) and polyvinylpyrrolidone poly (3-hydroxybutyrate) containing caffeic acid phenethyl ester [149] and poly (ethylene glycol)-based fibrous materials containing caffeic acid [150].

Jaganathan et al. [151] prepared materials based on *Artocarpus heterophyllus*-derived lignin and chitosan with varying wt% concentrations for biomedical purposes. Lignin significantly improved mechanical stability of the material and biocompatibility towards NIH 3t3 cells, proving the suitability of the biocomposites for regenerative wound dressing applications. Moreover, Zhang et al. [152] introduced lignin into a polyvinyl alcohol and chitosan composite hydrogel for enhancing its mechanical strength and accelerate wound healing processes on murine wound models. Zmejkoski et al. [153] fabricated dressing materials based on bacterial cellulose and coniferyl alcohol composite hydrogel. Their findings showed a sustained release of the bioactive compound, with the highest release within the first hour and a slower release within the following 72 h, and inhibitory and/or bactericidal effects.

Resveratrol is another antioxidant biocompound extensively investigated for wound healing and regeneration. Therefore, Berce et al. [154] synthesized a polymeric sponge consisting of chitosan and sodium hyaluronate for the controlled release of resveratrol. In vitro and in vivo studies confirmed its potential to stimulate tissue regeneration by enhancing granulation formation which facilitates wound healing, while also achieving bacteriostatic effects. Similarly, Hussain et al. developed hyaluronic acid-functionalized chitosan nanoparticles for an efficient topical administration of resveratrol and curcumin. The optimized formulation was characterized by an entrapment efficiency of approximately 90% for both biocompounds and non-Fickian diffusion and sustained release mechanism for in vivo studies [155]. Furthermore, Gokce et al. [156] synthesized resveratrol-loaded hyaluronic acid and dipalmitoylphosphatidylcholine microparticles embedded within a dermal matrix consisting of collagen-laminin. The release of resveratrol was sustained, reaching 70% after 6 h. Application on full-thickness excision diabetic rat models resulted in improved collagen fibers as the addition of resveratrol delayed dermal matrix degradation by collagenases and prevented inflammation due to antioxidant properties. Lakshmanan et al. [157] developed electrospun poly(ε-caprolactone)-based scaffolds containing resveratrol. In vivo experiments full-thickness ischemic mice wound models showed a considerably faster wound closure and re-epithelialization than the collagen-treated and negative control groups. Additionally, the anti-apoptotic and regenerative potential was demonstrated through the activation of thioredoxin-1 and heme oxygenase-1 mediated vascular endothelial growth factor signaling and the expression of bcl-2 in wound edges after treatment.

Therefore, it can be concluded that the combination of biocompatible polymers which are known for both their regenerative and anti-microbial properties with natural anti-cancer agents is a promising strategy for the development of wound dressings that could be applied in skin cancer before or after surgical interventions. In this manner, these dressings could promote tumor regression and prevent cancer recurrence, without causing adverse effects as in the case of conventional topical field treatments.

5. Conclusions and Future Perspectives

For skin cancer treatment, radical tumor excision remains the most effective approach among the available strategies. However, surgical interventions can be disfiguring, requiring additional skin grafts to cover the defects. In this context, post-surgery management should involve the application of wound dressings for promoting skin regeneration and preventing tumor recurrence and microbial infections, which still represents a considerable clinical challenge. While topical field administration is widely applied in skin cancer management, current options are still limited due to the various related side effects and the prolonged treatment periods. In this regard, plant-derived compounds have proved to exhibit DNA protection, antioxidant, anti-inflammatory, chemopreventive, and chemotherapeutic activities and could represent a promising alternative. Therefore, wound dressings consisting of biopolymers as the regenerative component and natural anti-cancer agents as the cancer recurrence-preventing component could represent ideal candidates to use for regenerative applications in skin cancer. As most studies focus on the healing potential, the number of publications on this subject is still relatively limited and further research is fundamental for developing regenerative wound dressings for skin cancer. In this context, research studies that not only assess the antioxidant potential of these bioactive dressings but also the anti-cancer properties should be performed.

Author Contributions: Review writing and revision: T.I.P., C.C., M.R., A.M.G. All authors have read and agreed to the published version of the manuscript.

Funding: This paper was published with the financial support of the University Politehnica of Bucharest, Romania.

Conflicts of Interest: The authors declare no conflict of interest.

References

1. *WHO Report on Cancer: Setting Priorities, Investing Wisely and Providing Care for All*; World Health Organization: Geneva, Switzerland, 2020.
2. Manduku, V.; Akhavan, M.; Asiki, G.; Brand, N.R.; Cira, M.K.; Gura, Z.; Kadengye, D.T.; Karagu, A.; Livinski, A.A.; Meme, H.; et al. Moving towards an evidence-informed cancer control strategy: A scoping review of oncology research in Kenya. *J. Cancer Policy* **2020**, *24*, 100219. [CrossRef]
3. Sakharoff, M. Chapter 23—Buteyko Breathing Technique and Ketogenic Diet as Potential Hormetins in Nonpharmacological Metabolic Approaches to Health and Longevity. In *The Science of Hormesis in Health and Longevity*; Rattan, S.I.S., Kyriazis, M., Eds.; Academic Press: Cambridge, MA, USA, 2019; pp. 257–274. [CrossRef]
4. Laliani, G.; Ghasemian Sorboni, S.; Lari, R.; Yaghoubi, A.; Soleimanpour, S.; Khazaei, M.; Hasanian, S.M.; Avan, A. Bacteria and cancer: Different sides of the same coin. *Life Sci.* **2020**, *246*, 117398. [CrossRef] [PubMed]
5. Fernandes, R.T.S.; França, E.L.; Triches, D.L.G.F.; Fujimori, M.; Machi, P.G.F.; Massmman, P.F.; Tozetti, I.A.; Honorio-França, A.C. Nanodoses of melatonin induces apoptosis on human breast cancer cells co-cultured with colostrum cells. *Biointerface Res. Appl. Chem.* **2019**, *9*, 4416–4423. [CrossRef]
6. Mohammadabadi, R.; Morsali, A.; Heravi, M.M.; Beyramabadi, S.A. Application of quantum chemical calculations in modeling of the supramolecular nanomedicine constructed from host-guest complexes of cucurbit [7] uril with gemcitabine anticancer drug. *Biointerface Res. Appl. Chem.* **2018**, *8*, 3282–3288.
7. Tajan, M.; Vousden, K.H. Dietary Approaches to Cancer Therapy. *Cancer Cell* **2020**, *37*, 767–785. [CrossRef]
8. O'Malley, J.; Kumar, R.; Inigo, J.; Yadava, N.; Chandra, D. Mitochondrial Stress Response and Cancer. *Trends Cancer* **2020**. [CrossRef]
9. Ransohoff, K.J.; Epstein, E.H.; Tang, J.Y. Chapter 100—Vitamin D and Skin Cancer. In *Vitamin D*, 4th ed.; Feldman, D., Ed.; Academic Press: Cambridge, MA, USA, 2018; pp. 863–874. [CrossRef]
10. Wysong, A.; Higgins, S.; Blalock, T.W.; Ricci, D.; Nichols, R.; Smith, F.L.; Kossintseva, I. Defining skin cancer local recurrence. *J. Am. Acad. Dermatol.* **2019**, *81*, 581–599. [CrossRef]
11. Dorrell, D.N.; Strowd, L.C. Skin Cancer Detection Technology. *Dermatol. Clin.* **2019**, *37*, 527–536. [CrossRef]
12. Sajadimajd, S.; Bahramsoltani, R.; Iranpanah, A.; Kumar Patra, J.; Das, G.; Gouda, S.; Rahimi, R.; Rezaeiamiri, E.; Cao, H.; Giampieri, F.; et al. Advances on Natural Polyphenols as Anticancer Agents for Skin Cancer. *Pharmacol. Res.* **2020**, *151*, 104584. [CrossRef]
13. Ho, C.; Argáez, C. CADTH Rapid Response Reports. In *Mohs Surgery for the Treatment of Skin Cancer: A Review of Guidelines*; Canadian Agency for Drugs and Technologies in Health: Ottawa, ON, Canada, 2019.
14. Badash, I.; Shauly, O.; Lui, C.G.; Gould, D.J.; Patel, K.M. Nonmelanoma Facial Skin Cancer: A Review of Diagnostic Strategies, Surgical Treatment, and Reconstructive Techniques. *Clin. Med. Insights Earnose Throat* **2019**, *12*, 1179550619865278. [CrossRef]
15. Marks, D.H.; Arron, S.T.; Mansh, M. Skin Cancer and Skin Cancer Risk Factors in Sexual and Gender Minorities. *Dermatol. Clin.* **2020**, *38*, 209–218. [CrossRef] [PubMed]
16. Gonzalez-Fierro, A.; Dueñas-González, A. Drug repurposing for cancer therapy, easier said than done. *Semin. Cancer Biol.* **2019**. [CrossRef] [PubMed]

17. Abbas, Z.; Rehman, S. An Overview of Cancer Treatment Modalities. *Neoplasm* **2018**, *139*. [CrossRef]

18. Umesha, K.B.; Ningaiah, S.; Lingegowda, N.S.; Basavanna, V.; Doddamani, S. A new approach for the synthesis of 8-((1,3-diphenyl-4,5-dihydro-1H-pyrazole-5 yl)methoxy)quinoline: A novel lead for breast cancer chemotherapy. *Biointerface Res. Appl. Chem.* **2018**, *8*, 3744–3750.

19. Cullen, J.K.; Simmons, J.L.; Parsons, P.G.; Boyle, G.M. Topical treatments for skin cancer. *Adv. Drug Deliv. Rev.* **2019**. [CrossRef]

20. Orthaber, K.; Pristovnik, M.; Skok, K.; Perić, B.; Maver, U. Skin Cancer and Its Treatment: Novel Treatment Approaches with Emphasis on Nanotechnology. *J. Nanomater.* **2017**, *2017*, 2606271. [CrossRef]

21. Todorova, K.; Mandinova, A. Novel approaches for managing aged skin and nonmelanoma skin cancer. *Adv. Drug Deliv. Rev.* **2020**. [CrossRef]

22. Pereira, R.F.; Sousa, A.; Barrias, C.C.; Bayat, A.; Granja, P.L.; Bártolo, P.J. Advances in bioprinted cell-laden hydrogels for skin tissue engineering. *Biomanuf. Rev.* **2017**, *2*, 1. [CrossRef]

23. Stoica, A.E.; Chircov, C.; Grumezescu, A.M. Nanomaterials for Wound Dressings: An Up-to-Date Overview. *Molecules* **2020**, *25*, 2699. [CrossRef]

24. Paduraru, A.; Ghitulica, C.; Trusca, R.; Surdu, V.A.; Neacsu, I.A.; Holban, A.M.; Birca, A.C.; Iordache, F.; Vasile, B.S. Antimicrobial Wound Dressings as Potential Materials for Skin Tissue Regeneration. *Materials* **2019**, *12*, 1859. [CrossRef]

25. Nardini, M.; Perteghella, S.; Mastracci, L.; Grillo, F.; Marrubini, G.; Bari, E.; Formica, M.; Gentili, C.; Cancedda, R.; Torre, M.L.; et al. Growth Factors Delivery System for Skin Regeneration: An Advanced Wound Dressing. *Pharmaceutics* **2020**, *12*, 120. [CrossRef] [PubMed]

26. Kadakia, K.C.; Matusz-Fisher, A.G.; Kim, E.S. Chemoprevention Trials. In *Encyclopedia of Cancer*, 3rd ed.; Boffetta, P., Hainaut, P., Eds.; Academic Press: Oxford, UK, 2019; pp. 352–366. [CrossRef]

27. Salehi, R.; Rasoolzadeh, R. Investigation of capecitabine and 5-fluorouracil anticancer drugs structural properties and their interactions with single-walled carbon nanotube: Insights from computational methods. *Biointerface Res. Appl. Chem.* **2018**, *8*, 3075–3083.

28. Norrlid, H.; Norlin, J.M.; Holmstrup, H.; Malmberg, I.; Sartorius, K.; Thormann, H.; Jemec, G.B.E.; Ragnarson Tennvall, G. Patient-reported outcomes in topical field treatment of actinic keratosis in Swedish and Danish patients. *J. Dermatol. Treat.* **2018**, *29*, 68–73. [CrossRef] [PubMed]

29. Chaplin, S. Topical agents for preventing and treating actinic keratosis. *Prescriber* **2016**, *27*, 32–40. [CrossRef]

30. Zhang, J.Y. Chapter 15—Animal Models of Skin Disorders. In *Animal Models for the Study of Human Disease*, 2nd ed.; Conn, P.M., Ed.; Academic Press: Cambridge, MA, USA, 2017; pp. 357–375. [CrossRef]

31. Čeović, R.; Petković, M.; Mokos, Z.B.; Kostović, K. Nonsurgical treatment of nonmelanoma skin cancer in the mature patient. *Clin. Dermatol.* **2018**, *36*, 177–187. [CrossRef] [PubMed]

32. Coyle, M.J.; Takwale, A. 54—Nonsurgical Management of Non-Melanoma Skin Cancer. In *Maxillofacial Surgery*, 3rd ed.; Brennan, P.A., Schliephake, H., Ghali, G.E., Cascarini, L., Eds.; Churchill Livingstone: London, UK, 2017; pp. 761–764. [CrossRef]

33. Aronson, J.K. Imiquimod. In *Meyler's Side Effects of Drugs*, 6th ed.; Elsevier: Oxford, UK, 2016; p. 27. [CrossRef]

34. Mathews, C.A.; Walker, J.L. 2—Preinvasive Disease of the Vagina and Vulva and Related Disorders. In *Clinical Gynecologic Oncology*, 9th ed.; DiSaia, P.J., Creasman, W.T., Mannel, R.S., McMeekin, D.S., Mutch, D.G., Eds.; Elsevier: Amsterdam, The Netherlands, 2018; pp. 20–37.e23. [CrossRef]

35. Voiculescu, V.M.; Lisievici, C.V.; Lupu, M.; Vajaitu, C.; Draghici, C.C.; Popa, A.V.; Solomon, I.; Sebe, T.I.; Constantin, M.M.; Caruntu, C. Mediators of Inflammation in Topical Therapy of Skin Cancers. *Mediat. Inflamm.* **2019**, *2019*, 8369690. [CrossRef]

36. Pescina, S.; Garrastazu, G.; del Favero, E.; Rondelli, V.; Cantù, L.; Padula, C.; Santi, P.; Nicoli, S. Microemulsions based on TPGS and isostearic acid for imiquimod formulation and skin delivery. *Eur. J. Pharm. Sci.* **2018**, *125*, 223–231. [CrossRef]

37. Wouters, T.; Hendriks, N.; Koeneman, M.; Kruse, A.-J.; van de Sande, A.; van Beekhuizen, H.J.; Gerestein, K.G.; Bekkers, R.L.M.; Piek, J.M.J. Systemic adverse events in imiquimod use for cervical intraepithelial neoplasia – A case series. *Case Rep. Women's Health* **2019**, *21*, e00105. [CrossRef]

38. Al-Mayahy, M.H.; Sabri, A.H.; Rutland, C.S.; Holmes, A.; McKenna, J.; Marlow, M.; Scurr, D.J. Insight into imiquimod skin permeation and increased delivery using microneedle pre-treatment. *Eur. J. Pharm. Biopharm.* **2019**, *139*, 33–43. [CrossRef]

39. McMillin, G.A.; Wadelius, M.; Pratt, V.M. 11—Pharmacogenetics. In *Principles and Applications of Molecular Diagnostics*; Rifai, N., Horvath, A.R., Wittwer, C.T., Eds.; Elsevier: Amsterdam, The Netherlands, 2018; pp. 295–327. [CrossRef]

40. Murshed, H. Chapter 11—Radiation and Combined Modality Therapy. In *Fundamentals of Radiation Oncology*, 3rd ed.; Murshed, H., Ed.; Academic Press: Cambridge, MA, USA, 2019; pp. 191–199. [CrossRef]

41. Kronfol, M.M.; McClay, J.L. Chapter 14—Epigenetic biomarkers in personalized medicine. In *Prognostic Epigenetics*; Sharma, S., Ed.; Academic Press: Cambridge, MA, USA, 2019; Volume 15, pp. 375–395.

42. Ledderhof, N.J.; Caminiti, M.F.; Bradley, G.; Lam, D.K. Topical 5-fluorouracil is a novel targeted therapy for the keratocystic odontogenic tumor. *J. Oral Maxillofac. Surg.* **2017**, *75*, 514–524. [CrossRef]

43. Lally, A.; Jenkins, S.N.; Zwald, F. Chapter 34—Non-Malignant and Malignant Skin Lesions in Kidney Transplant Patients. In *Kidney Transplantation–Principles and Practice*, 7th ed.; Morris, P.J., Knechtle, S.J., Eds.; Content Repository Only: Philadelphia, PA, USA, 2014; pp. 550–568. [CrossRef]

44. Werbel, T.; Cohen, P.R. Topical Application of 5-Fluorouracil Associated with Distant Seborrheic Dermatitis-like Eruption: Case Report and Review of Seborrheic Dermatitis Cutaneous Reactions after Systemic or Topical Treatment with 5-Fluorouracil. *Derm. Ther.* **2018**, *8*, 495–501. [CrossRef] [PubMed]

45. Hanke, C.W.; Albrecht, L.; Skov, T.; Larsson, T.; Østerdal, M.L.; Spelman, L. Efficacy and safety of ingenol mebutate gel in field treatment of actinic keratosis on full face, balding scalp, or approximately 250 cm2 on the chest: A phase 3 randomized controlled trial. *J. Am. Acad. Dermatol.* **2020**, *82*, 642–650. [CrossRef] [PubMed]

46. Athanasakis, K.; Boubouchairopoulou, N.; Tarantilis, F.; Tsiantou, V.; Kontodimas, S.; Kyriopoulos, J. Cost-effectiveness of Ingenol Mebutate Gel for the Treatment of Actinic Keratosis in Greece. *Clin. Ther.* **2017**, *39*, 993–1002. [CrossRef] [PubMed]

47. Erlendsson, A.M.; Karmisholt, K.E.; Haak, C.S.; Stender, I.-M.; Haedersdal, M. Topical corticosteroid has no influence on inflammation or efficacy after ingenol mebutate treatment of grade I to III actinic keratoses (AK): A randomized clinical trial. *J. Am. Acad. Dermatol.* **2016**, *74*, 709–715. [CrossRef] [PubMed]

48. Black, A.T. Chapter 13—Dermatological Drugs, Topical Agents, and Cosmetics. In *Side Effects of Drugs Annual*; Ray, S.D., Ed.; Elsevier: Amsterdam, The Netherlands, 2016; Volume 38, pp. 129–141.

49. Peake, B.M.; Braund, R.; Tong, A.Y.C.; Tremblay, L.A. 5—Impact of pharmaceuticals on the environment. In *The Life-Cycle of Pharmaceuticals in the Environment*; Peake, B.M., Braund, R., Tong, A.Y.C., Tremblay, L.A., Eds.; Woodhead Publishing: Sawston, Cambridge, UK, 2016; pp. 109–152. [CrossRef]

50. Singer, K.; Dettmer, K.; Unger, P.; Schönhammer, G.; Renner, K.; Peter, K.; Siska, P.J.; Berneburg, M.; Herr, W.; Oefner, P.J.; et al. Topical Diclofenac Reprograms Metabolism and Immune Cell Infiltration in Actinic Keratosis. *Front. Oncol.* **2019**, *9*. [CrossRef] [PubMed]

51. Thomas, G.J.; Herranz, P.; Cruz, S.B.; Parodi, A. Treatment of actinic keratosis through inhibition of cyclooxygenase-2: Potential mechanism of action of diclofenac sodium 3% in hyaluronic acid 2.5%. *Dermatol. Ther.* **2019**, *32*, e12800. [CrossRef]

52. Nunes, A.R.; Vieira, Í.G.P.; Queiroz, D.B.; Leal, A.L.A.B.; Maia Morais, S.; Muniz, D.F.; Calixto-Junior, J.T.; Coutinho, H.D.M. Use of Flavonoids and Cinnamates, the Main Photoprotectors with Natural Origin. *Adv. Pharmacol. Sci.* **2018**, *2018*, 5341487. [CrossRef]

53. Chowdhury, W.; Arbee, S.; Debnath, S.; Bin Zahur, S.; Akter, S. The Role of Antioxidant Molecules in Prevention and Potent. *Oxid Med. Cell Longev.* **2014**. [CrossRef]

54. Piotrowski, I.; Kulcenty, K.; Suchorska, W. Interplay between inflammation and cancer. *Rep. Pract. Oncol. Radiother.* **2020**, *25*, 422–427. [CrossRef]

55. Teleanu, R.I.; Chircov, C.; Grumezescu, A.M.; Volceanov, A.; Teleanu, D.M. Antioxidant Therapies for Neuroprotection—A Review. *J. Clin. Med.* **2019**, *8*, 1659. [CrossRef]

56. George, V.C.; Vijesh, V.V.; Amararathna, D.I.M.; Lakshmi, C.A.; Anbarasu, K.; Kumar, D.R.N.; Ethiraj, K.R.; Kumar, R.A.; Rupasinghe, H.P.V. Mechanism of Action of Flavonoids in Prevention of Inflammation-Associated Skin Cancer. *Curr. Med. Chem.* **2016**, *23*, 3697–3716. [CrossRef] [PubMed]

57. de Oliveira Júnior, R.G.; Ferraz, C.A.A.; e Silva, M.G.; de Lavor, É.M.; Rolim, L.A.; de Lima, J.T.; Fleury, A.; Picot, L.; Quintans, J.D.S.S.; Júnior, L.J.Q. Flavonoids: Promising Natural Products for Treatment of Skin Cancer (Melanoma). *Nat. Prod. Cancer Drug Discov.* **2017**, *1*, 161.

58. Duman, A.; Mogulkoc, R.; Baltaci, A.K.; Sivrikaya, A. The effect of 3′,4′-dihydroxyflavonol on plasma oxidant and antioxidant systems in testis ischemia-reperfusion injury in rats. *Biointerface Res. Appl. Chem.* **2018**, *8*, 3441–3445.

59. Amawi, H.; Ashby, C.R.; Tiwari, A.K. Cancer chemoprevention through dietary flavonoids: What's limiting? *Chin. J. Cancer* **2017**, *36*, 50. [CrossRef] [PubMed]

60. Furman-Toczek, D.; Zagórska-Dziok, M.; Dudra-Jastrzębska, M.; Kruszewski, M.; Kapka-Skrzypczak, L. A review of selected natural phytochemicals in preventing and treating malignant skin neoplasms. *J. Pre-Clin. Clin. Res.* **2016**, *10*, 127–130. [CrossRef]

61. Kopustinskiene, D.M.; Jakstas, V.; Savickas, A.; Bernatoniene, J. Flavonoids as Anticancer Agents. *Nutrients* **2020**, *12*, 457. [CrossRef]

62. Madunic, J.; Vrhovac Madunic, I.; Gajski, G.; Popić, J.; Garaj-Vrhovac, V. Apigenin: A dietary flavonoid with diverse anticancer properties. *Cancer Lett.* **2018**, *413*, 11–22. [CrossRef]

63. Sung, B.; Chung, H.Y.; Kim, N.D. Role of Apigenin in Cancer Prevention via the Induction of Apoptosis and Autophagy. *J. Cancer Prev.* **2016**, *21*, 216–226. [CrossRef]

64. Mirzoeva, S.; Tong, X.; Bridgeman, B.B.; Plebanek, M.P.; Volpert, O.V. Apigenin Inhibits UVB-Induced Skin Carcinogenesis: The Role of Thrombospondin-1 as an Anti-Inflammatory Factor. *Neoplasia* **2018**, *20*, 930–942. [CrossRef]

65. Zhao, G.; Han, X.; Cheng, W.; Ni, J.; Zhang, Y.; Lin, J.; Song, Z. Apigenin inhibits proliferation and invasion, and induces apoptosis and cell cycle arrest in human melanoma cells. *Oncol. Rep.* **2017**, *37*, 2277–2285. [CrossRef]

66. Vafadar, A.; Shabaninejad, Z.; Movahedpour, A.; Fallahi, F.; Taghavipour, M.; Ghasemi, Y.; Akbari, M.; Shafiee, A.; Hajighadimi, S.; Moradizarmehri, S.; et al. Quercetin and cancer: New insights into its therapeutic effects on ovarian cancer cells. *Cell Biosci.* **2020**, *10*, 32. [CrossRef]

67. Kashyap, D.; Garg, V.K.; Tuli, H.S.; Yerer, M.B.; Sak, K.; Sharma, A.K.; Kumar, M.; Aggarwal, V.; Sandhu, S.S. Fisetin and Quercetin: Promising Flavonoids with Chemopreventive Potential. *Biomolecules* **2019**, *9*, 174. [CrossRef] [PubMed]

68. Ajit, D.; Simonyi, A.; Li, R.; Chen, Z.; Hannink, M.; Fritsche, K.L.; Mossine, V.V.; Smith, R.E.; Dobbs, T.K.; Luo, R.; et al. Phytochemicals and botanical extracts regulate NF-κB and Nrf2/ARE reporter activities in DI TNC1 astrocytes. *Neurochem. Int.* **2016**, *97*, 49–56. [CrossRef] [PubMed]

69. Harris, Z.; Donovan, M.G.; Branco, G.M.; Limesand, K.H.; Burd, R. Quercetin as an Emerging Anti-Melanoma Agent: A Four-Focus Area Therapeutic Development Strategy. *Front. Nutr.* **2016**, *3*, 48. [CrossRef] [PubMed]

70. Schadich, E.; Hlaváč, J.; Volná, T.; Varanasi, L.; Hajdúch, M.; Džubák, P. Effects of Ginger Phenylpropanoids and Quercetin on Nrf2-ARE Pathway in Human BJ Fibroblasts and HaCaT Keratinocytes. *Biomed. Res. Int.* **2016**, *2016*, 2173275. [CrossRef] [PubMed]

71. Chaiprasongsuk, A.; Onkoksoong, T.; Pluemsamran, T.; Limsaengurai, S.; Panich, U. Photoprotection by dietary phenolics against melanogenesis induced by UVA through Nrf2-dependent antioxidant responses. *Redox Biol.* **2016**, *8*, 79–90. [CrossRef] [PubMed]

72. Wei, L.; Wu, P.; Zhou, X.-Q.; Jiang, W.-D.; Liu, Y.; Kuang, S.-Y.; Tang, L.; Feng, L. Dietary silymarin supplementation enhanced growth performance and improved intestinal apical junctional complex on juvenile grass carp (Ctenopharyngodon idella). *Aquaculture* **2020**, *525*, 735311. [CrossRef]

73. Dorjay, K.; Arif, T.; Adil, M. Silymarin: An interesting modality in dermatological therapeutics. *Indian J. Dermatol. Venereol. Leprol.* **2018**, *84*, 238.

74. Ng, C.Y.; Yen, H.; Hsiao, H.-Y.; Su, S.-C. Phytochemicals in Skin Cancer Prevention and Treatment: An Updated Review. *Int. J. Mol. Sci.* **2018**, *19*, 941. [CrossRef]

75. Buddhan, R.; Manoharan, S. Diosmin reduces cell viability of A431 skin cancer cells through apoptotic induction. *J. Cancer Res. Ther.* **2017**, *13*, 471–476. [CrossRef]

76. Choi, J.; Lee, D.H.; Park, S.Y.; Seol, J.W. Diosmetin inhibits tumor development and block tumor angiogenesis in skin cancer. *Biomed. Pharmacother. Biomed. Pharmacother.* **2019**, *117*, 109091. [CrossRef] [PubMed]

77. Boisnic, S.; Branchet, M.-C.; Gouhier-Kodas, C.; Verriere, F.; Jabbour, V. Anti-inflammatory and antiradical effects of a 2% diosmin cream in a human skin organ culture as model. *J. Cosmet. Dermatol.* **2018**, *17*, 848–854. [CrossRef] [PubMed]

78. Hostetler, G.L.; Ralston, R.A.; Schwartz, S.J. Flavones: Food sources, bioavailability, metabolism, and bioactivity. *Adv. Nutr.* **2017**, *8*, 423–435. [CrossRef] [PubMed]

79. Srivastava, A.; Srivastava, P.; Pandey, A.; Khanna, V.K.; Pant, A.B. Chapter 24—Phytomedicine: A Potential Alternative Medicine in Controlling Neurological Disorders. In *New Look to Phytomedicine*; Ahmad Khan, M.S., Ahmad, I., Chattopadhyay, D., Eds.; Academic Press: Cambridge, MA, USA, 2019; pp. 625–655. [CrossRef]

80. Calderón-Oliver, M.; Ponce-Alquicira, E. Chapter 7—Fruits: A Source of Polyphenols and Health Benefits. In *Natural and Artificial Flavoring Agents and Food Dyes*; Grumezescu, A.M., Holban, A.M., Eds.; Academic Press: Cambridge, MA, USA, 2018; pp. 189–228. [CrossRef]

81. Tuli, H.S.; Tuorkey, M.J.; Thakral, F.; Sak, K.; Kumar, M.; Sharma, A.K.; Sharma, U.; Jain, A.; Aggarwal, V.; Bishayee, A. Molecular Mechanisms of Action of Genistein in Cancer: Recent Advances. *Front. Pharmacol.* **2019**, *10*. [CrossRef] [PubMed]

82. Irrera, N.; Pizzino, G.; D'Anna, R.; Vaccaro, M.; Arcoraci, V.; Squadrito, F.; Altavilla, D.; Bitto, A. Dietary Management of Skin Health: The Role of Genistein. *Nutrients* **2017**, *9*, 622. [CrossRef]

83. Moolakkadath, T.; Aqil, M.; Ahad, A.; Imam, S.S.; Praveen, A.; Sultana, Y.; Mujeeb, M.; Iqbal, Z. Fisetin loaded binary ethosomes for management of skin cancer by dermal application on UV exposed mice. *Int. J. Pharm.* **2019**, *560*, 78–91. [CrossRef]

84. Pal, H.C.; Oca, M.K.M.d.; Sharma, P.; Pearlman, R.L.; Afaq, F. Abstract 5107: Fisetin treatment reduces tumor growth and metastasis by modulating PI3K/AKT and MEK/ERK pathways in a $BRAF^{V600E}/PTEN^{NULL}$ mouse model of melanoma. *Cancer Res.* **2017**, *77*, 5107. [CrossRef]

85. Yao, X.; Jiang, W.; Yu, D.; Yan, Z. Luteolin inhibits proliferation and induces apoptosis of human melanoma cells in vivo and in vitro by suppressing MMP-2 and MMP-9 through the PI3K/AKT pathway. *Food Funct.* **2019**, *10*, 703–712. [CrossRef]

86. Omar, S.H. Chapter 4—Biophenols: Impacts and Prospects in Anti-Alzheimer Drug Discovery. In *Discovery and Development of Neuroprotective Agents from Natural Products*; Brahmachari, G., Ed.; Elsevier: Amsterdam, The Netherlands, 2018; pp. 103–148. [CrossRef]

87. Chen, J.; Yang, J.; Ma, L.; Li, J.; Shahzad, N.; Kim, C.K. Structure-antioxidant activity relationship of methoxy, phenolic hydroxyl, and carboxylic acid groups of phenolic acids. *Sci. Rep.* **2020**, *10*, 2611. [CrossRef]

88. Kumar, N.; Goel, N. Phenolic acids: Natural versatile molecules with promising therapeutic applications. *Biotechnol. Rep.* **2019**, *24*, e00370. [CrossRef]

89. Bare, Y.; Krisnamurti, G.C.; Elizabeth, A.; Rachmad, Y.T.; Sari, D.R.T.; Gabrella Lorenza, M.R.W. The potential role of caffeic acid in coffee as cyclooxygenase-2 (COX-2) inhibitor: In silico study. *Biointerface Res. Appl. Chem.* **2019**, *9*, 4424–4427. [CrossRef]

90. Ha, S.J.; Lee, J.; Park, J.; Kim, Y.H.; Lee, N.H.; Kim, Y.E.; Song, K.M.; Chang, P.S.; Jeong, C.H.; Jung, S.K. Syringic acid prevents skin carcinogenesis via regulation of NoX and EGFR signaling. *Biochem. Pharmacol.* **2018**, *154*, 435–445. [CrossRef] [PubMed]

91. Abotaleb, M.; Liskova, A.; Kubatka, P.; Büsselberg, D. Therapeutic Potential of Plant Phenolic Acids in the Treatment of Cancer. *Biomolecules* **2020**, *10*, 221. [CrossRef]

92. Peres, D.D.A.; Sarruf, F.D.; de Oliveira, C.A.; Velasco, M.V.R.; Baby, A.R. Ferulic acid photoprotective properties in association with UV filters: Multifunctional sunscreen with improved SPF and UVA-PF. *J. Photochem. Photobiol. B Biol.* **2018**, *185*, 46–49. [CrossRef] [PubMed]

93. Balupillai, A.; Nagarajan, R.P.; Ramasamy, K.; Govindasamy, K.; Muthusamy, G. Caffeic acid prevents UVB radiation induced photocarcinogenesis through regulation of PTEN signaling in human dermal fibroblasts and mouse skin. *Toxicol. Appl. Pharmacol.* **2018**, *352*, 87–96. [CrossRef]

94. Zeng, N.; Hongbo, T.; Xu, Y.; Wu, M.; Wu, Y. Anticancer activity of caffeic acid n-butyl ester against A431 skin carcinoma cell line occurs via induction of apoptosis and inhibition of the mTOR/PI3K/AKT signaling pathway. *Mol. Med. Rep.* **2018**, *17*, 5652–5657. [CrossRef]

95. Pelinson, L.P.; Assmann, C.E.; Palma, T.V.; da Cruz, I.B.M.; Pillat, M.M.; Mânica, A.; Stefanello, N.; Weis, G.C.C.; de Oliveira Alves, A.; de Andrade, C.M.; et al. Antiproliferative and apoptotic effects of caffeic acid on SK-Mel-28 human melanoma cancer cells. *Mol. Biol. Rep.* **2019**, *46*, 2085–2092. [CrossRef]

96. Srinivasulu, C.; Ramgopal, M.; Ramanjaneyulu, G.; Anuradha, C.M.; Suresh Kumar, C. Syringic acid (SA)—A Review of Its Occurrence, Biosynthesis, Pharmacological and Industrial Importance. *Biomed. Pharmacother.* **2018**, *108*, 547–557. [CrossRef]

97. Zduńska, K.; Dana, A.; Kolodziejczak, A.; Rotsztejn, H. Antioxidant Properties of Ferulic Acid and Its Possible Application. *Ski. Pharmacol. Physiol.* **2018**, *31*, 332–336. [CrossRef]

98. Espíndola, K.M.M.; Ferreira, R.G.; Narvaez, L.E.M.; Silva Rosario, A.C.R.; da Silva, A.H.M.; Silva, A.G.B.; Vieira, A.P.O.; Monteiro, M.C. Chemical and Pharmacological Aspects of Caffeic Acid and Its Activity in Hepatocarcinoma. *Front. Oncol.* **2019**, *9*. [CrossRef]

99. Lee, S.C.; Tran, T.M.T.; Choi, J.W.; Won, K. Lignin for white natural sunscreens. *Int. J. Biol. Macromol.* **2019**, *122*, 549–554. [CrossRef] [PubMed]

100. Supanchaiyamat, N.; Jetsrisuparb, K.; Knijnenburg, J.T.N.; Tsang, D.C.W.; Hunt, A.J. Lignin materials for adsorption: Current trend, perspectives and opportunities. *Bioresour. Technol.* **2019**, *272*, 570–581. [CrossRef] [PubMed]

101. Bajwa, D.S.; Pourhashem, G.; Ullah, A.H.; Bajwa, S.G. A concise review of current lignin production, applications, products and their environmental impact. *Ind. Crop. Prod.* **2019**, *139*, 111526. [CrossRef]

102. Wu, W.; Liu, T.; Deng, X.; Sun, Q.; Cao, X.; Feng, Y.; Wang, B.; Roy, V.A.L.; Li, R.K.Y. Ecofriendly UV-protective films based on poly(propylene carbonate) biocomposites filled with TiO2 decorated lignin. *Int. J. Biol. Macromol.* **2019**, *126*, 1030–1036. [CrossRef]

103. Nagapan, T.S.; Ghazali, A.R.; Basri, D.F.; Lim, W.N. Photoprotective Effect of Stilbenes And Its Derivatives against Ultraviolet Radiation-Induced Skin Disorders. *Biomed. Pharmacol. J.* **2018**, *11*, 1199–1208. [CrossRef]

104. Padmanabhan, P.; Correa-Betanzo, J.; Paliyath, G. Berries and Related Fruits. In *Encyclopedia of Food and Health*; Caballero, B., Finglas, P.M., Toldrá, F., Eds.; Academic Press: Cambridge, MA, USA; Oxford, UK, 2016; pp. 364–371. [CrossRef]

105. Martinez, K.B.; Mackert, J.D.; McIntosh, M.K. Chapter 18—Polyphenols and Intestinal Health. In *Nutrition and Functional Foods for Healthy Aging*; Watson, R.R., Ed.; Academic Press: Cambridge, MA, USA, 2017; pp. 191–210. [CrossRef]

106. Fernández-Quintela, A.; González, M.; Aguirre, L.; Milton-Laskibar, I.; Léniz, A.; Portillo, M.P. Chapter 17—Resveratrol and Protection in Hepatic Steatosis: Antioxidant Effects. In *The Liver*; Patel, V.B., Rajendram, R., Preedy, V.R., Eds.; Academic Press: Boston, MA, USA, 2018; pp. 199–209. [CrossRef]

107. Lee, Y.; Lee, J.-Y. Chapter 8—Protective Actions of Polyphenols in the Development of Nonalcoholic Fatty Liver Disease. In *Dietary Interventions in Liver Disease*; Watson, R.R., Preedy, V.R., Eds.; Academic Press: Cambridge, MA, USA, 2019; pp. 91–99. [CrossRef]

108. Iqbal, J.; Abbasi, B.A.; Ahmad, R.; Batool, R.; Mahmood, T.; Ali, B.; Khalil, A.T.; Kanwal, S.; Afzal Shah, S.; Alam, M.M.; et al. Potential phytochemicals in the fight against skin cancer: Current landscape and future perspectives. *Biomed. Pharmacother.* **2019**, *109*, 1381–1393. [CrossRef]

109. Xiao, Q.; Zhu, W.; Feng, W.; Lee, S.S.; Leung, A.W.; Shen, J.; Gao, L.; Xu, C. A Review of Resveratrol as a Potent Chemoprotective and Synergistic Agent in Cancer Chemotherapy. *Front. Pharmacol.* **2019**, *9*. [CrossRef]

110. Kim, S.M.; Kim, S.Z. Biological activities of resveratrol against cancer. *J. Phys. Chem. Biophys.* **2018**. [CrossRef]

111. Chen, R.-J.; Kuo, H.-C.; Cheng, L.-H.; Lee, Y.-H.; Chang, W.-T.; Wang, B., Jr.; Wang, Y.-J.; Cheng, H.-C. Apoptotic and Nonapoptotic Activities of Pterostilbene against Cancer. *Int. J. Mol. Sci.* **2018**, *19*, 287. [CrossRef]

112. Chen, R.-J.; Lee, Y.-H.; Yeh, Y.-L.; Wu, W.-S.; Ho, C.-T.; Li, C.-Y.; Wang, B., Jr.; Wang, Y.-J. Autophagy-inducing effect of pterostilbene: A prospective therapeutic/preventive option for skin diseases. *J. Food Drug Anal.* **2017**, *25*, 125–133. [CrossRef]

113. Kershaw, J.; Kim, K.-H. The Therapeutic Potential of Piceatannol, a Natural Stilbene, in Metabolic Diseases: A Review. *J. Med. Food* **2017**, *20*, 427–438. [CrossRef] [PubMed]

114. Du, M.; Zhang, Z.; Gao, T. Piceatannol induced apoptosis through up-regulation of microRNA-181a in melanoma cells. *Biol. Res.* **2017**, *50*, 36. [CrossRef] [PubMed]

115. Lopez, A.T.; Carvajal, R.D.; Geskin, L. Secondary prevention strategies for nonmelanoma skin cancer. *Oncology* **2018**, *32*, 195–200. [PubMed]

116. Shedoeva, A.; Leavesley, D.; Upton, Z.; Fan, C. Wound Healing and the Use of Medicinal Plants. *Evid. Based Complement. Altern. Med.* **2019**, *2019*, 2684108. [CrossRef]

117. Fazil, M.; Nikhat, S. Topical medicines for wound healing: A systematic review of Unani literature with recent advances. *J. Ethnopharmacol.* **2020**, *257*, 112878. [CrossRef]

118. Weller, C.D.; Team, V.; Sussman, G. First-Line Interactive Wound Dressing Update: A Comprehensive Review of the Evidence. *Front. Pharmacol.* **2020**, *11*, 155. [CrossRef]

119. Saghazadeh, S.; Rinoldi, C.; Schot, M.; Kashaf, S.S.; Sharifi, F.; Jalilian, E.; Nuutila, K.; Giatsidis, G.; Mostafalu, P.; Derakhshandeh, H.; et al. Drug delivery systems and materials for wound healing applications. *Adv. Drug Deliv. Rev.* **2018**, *127*, 138–166. [CrossRef]

120. Shpichka, A.; Butnaru, D.; Bezrukov, E.A.; Sukhanov, R.B.; Atala, A.; Burdukovskii, V.; Zhang, Y.; Timashev, P. Skin tissue regeneration for burn injury. *Stem Cell Res. Ther.* **2019**, *10*, 94. [CrossRef]

121. Simões, D.; Miguel, S.P.; Ribeiro, M.P.; Coutinho, P.; Mendonça, A.G.; Correia, I.J. Recent advances on antimicrobial wound dressing: A review. *Eur. J. Pharm. Biopharm.* **2018**, *127*, 130–141. [CrossRef]

122. Andreu, V.; Mendoza, G.; Arruebo, M.; Irusta, S. Smart Dressings Based on Nanostructured Fibers Containing Natural Origin Antimicrobial, Anti-Inflammatory, and Regenerative Compounds. *Materials* **2015**, *8*, 5154–5193. [CrossRef]

123. Weller, C.; Team, V. Interactive dressings and their role in moist wound management. In *Advanced Textiles for Wound Care*; Elsevier: Amsterdam, The Netherlands, 2019; pp. 105–134.

124. Rahimi, M.; Noruzi, E.B.; Sheykhsaran, E.; Ebadi, B.; Kariminezhad, Z.; Molaparast, M.; Mehrabani, M.G.; Mehramouz, B.; Yousefi, M.; Ahmadi, R.; et al. Carbohydrate polymer-based silver nanocomposites: Recent progress in the antimicrobial wound dressings. *Carbohydr. Polym.* **2020**, *231*, 115696. [CrossRef] [PubMed]

125. Naseri-Nosar, M.; Ziora, Z.M. Wound dressings from naturally-occurring polymers: A review on homopolysaccharide-based composites. *Carbohydr. Polym.* **2018**, *189*, 379–398. [CrossRef] [PubMed]

126. Chircov, C.; Grumezescu, A.M.; Bejenaru, L.E. Hyaluronic acid-based scaffolds for tissue engineering. *Rom. J. Morphol. Embryol. Rev. Roum. Morphol. Embryol.* **2018**, *59*, 71–76.

127. Homaeigohar, S.; Boccaccini, A.R. Antibacterial biohybrid nanofibers for wound dressings. *Acta Biomater.* **2020**, *107*, 25–49. [CrossRef] [PubMed]

128. Graça, M.F.P.; Miguel, S.P.; Cabral, C.S.D.; Correia, I.J. Hyaluronic acid—Based wound dressings: A review. *Carbohydr. Polym.* **2020**, *241*, 116364. [CrossRef]

129. Fleck, C.A.; Simman, R. Modern Collagen Wound Dressings: Function and Purpose. *J. Am. Coll. Certif. Wound Spec.* **2010**, *2*, 50–54. [CrossRef]

130. Axibal, E.; Brown, M. Surgical Dressings and Novel Skin Substitutes. *Dermatol. Clin.* **2019**, *37*, 349–366. [CrossRef]

131. Ezzat, H.A.; Hegazy, M.A.; Nada, N.A.; Ibrahim, M.A. Effect of nano metal oxides on the electronic properties of cellulose, chitosan and sodium alginate. *Biointerface Res. Appl. Chem.* **2019**, *9*, 4143–4149. [CrossRef]

132. Koehler, J.; Brandl, F.P.; Goepferich, A.M. Hydrogel wound dressings for bioactive treatment of acute and chronic wounds. *Eur. Polym. J.* **2018**, *100*, 1–11. [CrossRef]

133. Varaprasad, K.; Jayaramudu, T.; Kanikireddy, V.; Toro, C.; Sadiku, E.R. Alginate-based composite materials for wound dressing application:A mini review. *Carbohydr. Polym.* **2020**, *236*, 116025. [CrossRef] [PubMed]

134. Schoukens, G. 5—Bioactive dressings to promote wound healing. In *Advanced Textiles for Wound Care*, 2nd ed.; Rajendran, S., Ed.; Woodhead Publishing: Sawston, Cambridge, UK, 2019; pp. 135–167. [CrossRef]

135. Mandla, S.; Davenport Huyer, L.; Radisic, M. Review: Multimodal bioactive material approaches for wound healing. *APL Bioeng.* **2018**, *2*, 021503. [CrossRef] [PubMed]

136. Felgueiras, H.P.; Tavares, T.D.; Amorim, M.T.P. Biodegradable, spun nanocomposite polymeric fibrous dressings loaded with bioactive biomolecules for an effective wound healing: A review. *IOP Conf. Ser. Mater. Sci. Eng.* **2019**, *634*, 012033. [CrossRef]

137. Shukla, R.; Kashaw, S.; Jain, A.; Lodhi, S. Fabrication of Apigenin loaded Gellan Gum–Chitosan Hydrogels (GGCH–HGs) for effective diabetic wound healing. *Int. J. Biol. Macromol.* **2016**, *91*. [CrossRef]

138. Vedakumari, W.S.; Ayaz, N.; Karthick, A.S.; Senthil, R.; Sastry, T.P. Quercetin impregnated chitosan–fibrin composite scaffolds as potential wound dressing materials—Fabrication, characterization and in vivo analysis. *Eur. J. Pharm. Sci.* **2017**, *97*, 106–112. [CrossRef] [PubMed]

139. George, D.; Maheswari, P.U.; Begum, K.M.M.S. Synergic formulation of onion peel quercetin loaded chitosan-cellulose hydrogel with green zinc oxide nanoparticles towards controlled release, biocompatibility, antimicrobial and anticancer activity. *Int. J. Biol. Macromol.* **2019**, *132*, 784–794. [CrossRef]

140. Mastropietro, D.; Park, K.; Omidian, H. 4.23 Polymers in Oral Drug Delivery. In *Comprehensive Biomaterials II*; Ducheyne, P., Ed.; Elsevier: Oxford, UK, 2017; pp. 430–444. [CrossRef]

141. Jangde, R.; Srivastava, S.; Singh, M.R.; Singh, D. In vitro and In vivo characterization of quercetin loaded multiphase hydrogel for wound healing application. *Int. J. Biol. Macromol.* **2018**, *115*, 1211–1217. [CrossRef]

142. Ajmal, G.; Bonde, G.V.; Thokala, S.; Mittal, P.; Khan, G.; Singh, J.; Pandey, V.K.; Mishra, B. Ciprofloxacin HCl and quercetin functionalized electrospun nanofiber membrane: Fabrication and its evaluation in full thickness wound healing. *Artif. Cells Nanomed. Biotechnol.* **2019**, *47*, 228–240. [CrossRef]

143. Tsai, Y.-H.; Yang, Y.-N.; Ho, Y.-C.; Tsai, M.-L.; Mi, F.-L. Drug release and antioxidant/antibacterial activities of silymarin-zein nanoparticle/bacterial cellulose nanofiber composite films. *Carbohydr. Polym.* **2018**, *180*, 286–296. [CrossRef]

144. Kim, A.R.; Lee, S.L.; Park, S.N. Properties and in vitro drug release of pH- and temperature-sensitive double cross-linked interpenetrating polymer network hydrogels based on hyaluronic acid/poly (N-isopropylacrylamide) for transdermal delivery of luteolin. *Int. J. Biol. Macromol.* **2018**, *118*, 731–740. [CrossRef]

145. Yakub, G.; Ignatova, M.; Manolova, N.; Rashkov, I.; Toshkova, R.; Georgieva, A.; Markova, N. Chitosan/ferulic acid-coated poly(ε-caprolactone) electrospun materials with antioxidant, antibacterial and antitumor properties. *Int. J. Biol. Macromol.* **2018**, *107*, 689–702. [CrossRef]

146. Poornima, B.; Korrapati, P.S. Fabrication of chitosan-polycaprolactone composite nanofibrous scaffold for simultaneous delivery of ferulic acid and resveratrol. *Carbohydr. Polym.* **2017**, *157*, 1741–1749. [CrossRef] [PubMed]

147. Oh, G.-W.; Ko, S.-C.; Je, J.-Y.; Kim, Y.-M.; Oh, J.; Jung, W.-K. Fabrication, characterization and determination of biological activities of poly(ε-caprolactone)/chitosan-caffeic acid composite fibrous mat for wound dressing application. *Int. J. Biol. Macromol.* **2016**, *93*, 1549–1558. [CrossRef]

148. Ignatova, M.; Manolova, N.; Rashkov, I.; Markova, N. Quaternized chitosan/κ-carrageenan/caffeic acid–coated poly(3-hydroxybutyrate) fibrous materials: Preparation, antibacterial and antioxidant activity. *Int. J. Pharm.* **2016**, *513*, 528–537. [CrossRef] [PubMed]

149. Ignatova, M.; Manolova, N.; Rashkov, I.; Markova, N. Antibacterial and antioxidant electrospun materials from poly(3-hydroxybutyrate) and polyvinylpyrrolidone containing caffeic acid phenethyl ester—"in" and "on" strategies for enhanced solubility. *Int. J. Pharm.* **2018**, *545*, 342–356. [CrossRef]

150. Ignatova, M.G.; Manolova, N.E.; Rashkov, I.B.; Markova, N.D.; Toshkova, R.A.; Georgieva, A.K.; Nikolova, E.B. Poly(3-hydroxybutyrate)/caffeic acid electrospun fibrous materials coated with polyelectrolyte complex and their antibacterial activity and in vitro antitumor effect against HeLa cells. *Mater. Sci. Eng. C* **2016**, *65*, 379–392. [CrossRef]

151. Jaganathan, G.; Manivannan, K.; Lakshmanan, S.; Sithique, M.A. Fabrication and characterization of Artocarpus heterophyllus waste derived lignin added chitosan biocomposites for wound dressing application. *Sustain. Chem. Pharm.* **2018**, *10*, 27–32. [CrossRef]

152. Zhang, Y.; Jiang, M.; Zhang, Y.; Cao, Q.; Wang, X.; Han, Y.; Sun, G.; Li, Y.; Zhou, J. Novel lignin–chitosan–PVA composite hydrogel for wound dressing. *Mater. Sci. Eng. C* **2019**, *104*, 110002. [CrossRef]

153. Zmejkoski, D.; Spasojević, D.; Orlovska, I.; Kozyrovska, N.; Soković, M.; Glamočlija, J.; Dmitrović, S.; Matović, B.; Tasić, N.; Maksimović, V.; et al. Bacterial cellulose-lignin composite hydrogel as a promising agent in chronic wound healing. *Int. J. Biol. Macromol.* **2018**, *118*, 494–503. [CrossRef]

154. Berce, C.; Muresan, M.-S.; Soritau, O.; Petrushev, B.; Tefas, L.; Rigo, I.; Ungureanu, G.; Catoi, C.; Irimie, A.; Tomuleasa, C. Cutaneous wound healing using polymeric surgical dressings based on chitosan, sodium hyaluronate and resveratrol. A preclinical experimental study. *Colloids Surf. B Biointerfaces* **2018**, *163*, 155–166. [CrossRef]

155. Hussain, Z.; Pandey, M.; Choudhury, H.; Ying, P.C.; Xian, T.M.; Kaur, T.; Jia, G.W.; Gorain, B. Hyaluronic acid functionalized nanoparticles for simultaneous delivery of curcumin and resveratrol for management of chronic diabetic wounds: Fabrication, characterization, stability and in vitro release kinetics. *J. Drug Deliv. Sci. Technol.* **2020**, *57*, 101747. [CrossRef]

156. Gokce, E.H.; Tuncay Tanrıverdi, S.; Eroglu, I.; Tsapis, N.; Gokce, G.; Tekmen, I.; Fattal, E.; Ozer, O. Wound healing effects of collagen-laminin dermal matrix impregnated with resveratrol loaded hyaluronic acid-DPPC microparticles in diabetic rats. *Eur. J. Pharm. Biopharm.* **2017**, *119*, 17–27. [CrossRef] [PubMed]

157. Lakshmanan, R.; Campbell, J.; Ukani, G.; O'Reilly Beringhs, A.; Selvaraju, V.; Thirunavukkarasu, M.; Lu, X.; Palesty, J.A.; Maulik, N. Evaluation of dermal tissue regeneration using resveratrol loaded fibrous matrix in a preclinical mouse model of full-thickness ischemic wound. *Int. J. Pharm.* **2019**, *558*, 177–186. [CrossRef] [PubMed]

MDPI

St. Alban-Anlage 66

4052 Basel

Switzerland

Tel. +41 61 683 77 34

Fax +41 61 302 89 18

www.mdpi.com

Cancers Editorial Office

E-mail: cancers@mdpi.com

www.mdpi.com/journal/cancers

www.ingramcontent.com/pod-product-compliance
Lightning Source LLC
LaVergne TN
LVHW070117170726
843515LV00007B/1706